Library of
Davidson College

A Company of Scientists

A Company of Scientists

Botany, Patronage, and Community at the Seventeenth-Century Parisian Royal Academy of Sciences

Alice Stroup

UNIVERSITY OF CALIFORNIA PRESS
Berkeley · Los Angeles · Oxford

University of California Press
Berkeley and Los Angeles, California

University of California Press
Oxford, England

Copyright © 1990 by The Regents of the University of California

Library of Congress Cataloging-in-Publication Data

Stroup, Alice.
    A company of scientists : botany, patronage, and community at the Seventeenth-century Parisian Royal Academy of Sciences / Alice Stroup.
    p. cm.
    Includes bibliographical references.
    1. Académie des sciences (France)—History. 2. Botany—Research—France—Paris—History. I. Title.
Q46.S76 1990
506.044—dc20   89-39532
                                                                                     CIP

Printed in the United States of America

1 2 3 4 5 6 7 8 9

The paper used in this publication meets the minimum requirements of American National Standard for Information Sciences—Permanence of Paper for Printed Library Materials, ANSI Z39.48-1984 ∞

For Timothy Stroup

Contents

List of Illustrations   xi

Acknowledgments   xiii

PART I. THE SETTING

1. Portrait of an Institution   3
   Understanding the Institution   4
   Public Image   5
   Behind the Public Image   8

PART II. THE INSTITUTION AND ITS PATRONAGE

2. Members and Protectors   13
   Composition of the Early Academy   15
   Students of Plants   17
   The Protectors and Their Spokesmen   23
3. Models for a Company of Scientists   27
   Sociocultural Contexts   27
   Bureaucratic Models   29
   Conclusion   33
4. The Material Benefits of Membership:
   Pensions and Quarters   34

The Functions of Pensions	35
Bibliothèque du Roi and Jardin Royal	38
Observatoire	43
Conclusion	45

5. **Research Subventions and Ministerial Control** — 46

Colbert: The Generous Foundation	47
Louvois: Declining Interest and Support	51
Pontchartrain: A Penurious Revival	56
Conclusion	60

PART III. BOTANICAL RESEARCH AT THE ACADEMY

6. **The Natural History of Plants: Rival Conceptions** — 65

Changing Ways of Thinking about Plants	65
Proposals for a Natural History of Plants	70
Research for the Natural History	79
Editorial Rivalry	83
Conclusion	87

7. **Justifying the Chemical Analysis of Plants** — 89

The Controversy over Distillation	90
The Method of Distillation	93
Why Distillation?	95
The Goals of Chemical Analysis	98
Publicity and Discretion	100
Conclusion	101

8. **Ministerial Intervention and an Unexpected Outcome** — 103

The Lost Second Installment	103
Ministerial Intervention	107
A New Editor	113
Conclusion	115

9. **Analogical Reasoning: The Model** — 117

The Nature and Functions of Analogical Reasoning	118
The Circulation of the Blood	125
Analogical Reasoning in the Harveian Model	128
Conclusion	130

10. **Analogical Reasoning: The Theory** — 131

The Circulation of the Sap	131

Contents ix

      Pretheoretic Plausibility   133
      Pushing the Analogy to Its Limits   134
      Solving the Problem of Crucial Dissimilarities   137
      Explaining the Rise of Sap   139
      Conclusion   142

11. Chemical and Mechanical Explanation of Physiological Processes   145
      Generation and Reproduction   145
      Germination, Maturation, and the Role of External Factors   149
      How Plants Grow and How They Are Nourished   151
      Conclusion   153

12. The New Instruments and Botany   155
      Early Botanical Microscopy at the Academy   157
      Plants and the Air Pump   159
      Conclusion   165

PART IV. THE ACADEMY AND THE LARGER COMMUNITY

13. Medical Motivations and Social Responsibility   169
      Medical Interests   171
      Ergotism, Illness of the Poor   174
      Conclusion   179

14. Scientific Paris at the End of the Century   180
      The Scientific Community   180
      Modest Public Interest in Science   182
      Scientific Goods and Services in Paris, 1660–1700   185
      Conclusion   198

15. Academicians and the Larger Scientific Community   199
      Private Contacts between Academicians and Other Savants   200
      Institutional Regulation of Contacts   204
      The Character and Benefits of Contacts   209
      Conclusion   215

PART V. THE EFFECTS OF PATRONAGE

16. The Academy as an Instrument of the Crown   221

Abbreviations Used in the Appendix, Notes, and Bibliography — 227

Appendix: The Record of Expenditure, 1666–1699 — 229

Notes — 285

Bibliography — 339
   Primary Sources: Manuscripts, Drawings, and Paintings — 339
   Primary Sources: Printed and Engraved — 342
   Secondary Sources — 352

Index — 371

Illustrations

PLATES

1. The Academy and Its Protectors 6
2. The Chemical Laboratory of the Academy 40
3. A Dissection and Microscopic Observations 42
4. Ergotted Rye 177
5. Paris circa 1700 188

FIGURES

1. Herba mimosa frutescens/Sensitive en arbrisseau 68
2. Lychnis umbellifera montana helvetica/Lychnis à umbelles 72
3. Cymbalaria 76
4. Melo vulgaris/Melon 77
5. Sanicula, sive Diapensia/Sanicle 78
6. Gentianella alpina verna magno flore 81
7. Mandragora mas/Mandragore 84
8. Mandragora flore sub caeruleo purpurascente, C.B./Mandragore 85
9. Buglossum creticum bullatum, flore vario odorato/Buglosse de Candie à feuilles bosselées 86
10. Lactuca canadensis, altissima, latifolia, flore leucophaeo/Laitue de Canada, à large feuille, à fleur gris de perle 105
11. Nasturtium aquaticum supinum/Cresson d'eau 106
12. Aubépine (plante chinoise) 112

xi

Acknowledgments

I became interested in the Academy of Sciences in a graduate seminar with the late C. Doris Hellman at The City University of New York in 1969, where I examined the work of Edme Mariotte. Over the years my ideas about how to treat the Academy have expanded. During the 1970s I concentrated on the Academy's botanical research. In 1976 I began to investigate the Academy's finances and in 1981 found proof that the Academy was funded during the 1690s. At the same time I explored the Academy's relations with the rest of the scholarly community. My aim has been to understand the Academy in the most inclusive context possible, while making its botanical studies the core of my research.

During these twenty years I have accumulated numerous debts of gratitude for intellectual and moral encouragement, constructive criticism, and financial support. It is a pleasure to express here my sincere thanks to the many individuals and institutions that have helped me.

My research in Oxford and Paris as a graduate student was supported by scholarships and grants for research and travel from the Regents of the State of New York; the Graduate Center of The City University of New York; and the Taylor Institution, the Committee for Graduate Studies, and the University Chest of Oxford University. A summer stipend from the National Endowment for the Humanities supported my research in Parisian archives during the summer of 1981, and the National Science Foundation provided a semester's leave of absence in 1982 for the drafting of chapters 2 through 5. The American Philosophical Society (1981, 1982), the American Council of Learned Societies (1983), and the National Science Foundation (1984) funded research on patronage of the Academy. A Professional Development Grant from Bard College supported me during the drafting of chapter 14. Bard College also supplied funds for research, for the illustrations and tables, and for proofreading.

Over the years I have tested various hypotheses about the Academy at scholarly meetings, and I am grateful to my audiences and commentators for their astute comments. Parts of the argument about the Academy as a company were presented at the annual meeting of the History of Science Society (Toronto, October 1980); about the Academy's chemical research, at the semicentennial meeting of the History of Science Society (Norwalk, October 1974); about cooperation, at Harvard University (May 1978); about analogical reasoning, at the Joint Atlantic Seminar in the History of Biology (Princeton, April 1981); about the pensions of academicians, at the annual meeting of the American Historical Association in Los Angeles and in New York at the Metropolitan Seminar for the History of Technology (December 1981); and about subtle forms of power in the Academy at the Northeastern Anthropological Association Meeting (Hartford, March 1984). Portions of chapters 1 and 16 were presented at the Folger Library's conference on the ascendancy of culture during the reign of Louis XIV (March 1985). I discussed my earliest work on Mariotte at the Metropolitan New York History of Science Society (New York, February 1969) and at Oxford University (Hilary Term, 1970).

Several persons advised me as I studied the Academy. A. C. Crombie supervised my doctoral dissertation at Oxford University on the Academy's botanical research; his advice and example have supported me at all stages of the project. Michael S. Mahoney and Simone Balayé generously shared ideas and research findings on the finances of the Academy and the Bibliothèque du roi, respectively. J. D. North and Everett Mendelsohn posed questions that led me to investigate new areas. The late C. Doris Hellman encouraged my original interest in Mariotte and the Academy. For criticizing early drafts of individual chapters, I am grateful to Gerard L'E. Turner, F. L. Holmes, and Allen G. Debus. Roger Hahn, Bert Hansen, John W. Olmsted, Karen Reeds, Jacques Roger, Richard S. Westfall, the late Henry Guerlac, and the anonymous referees for the National Science Foundation and the University of California Press offered criticisms that stimulated second thoughts and revisions. Friends and colleagues in the United States, England, France, and Finland gave moral support and learned advice. Special thanks are due to the late Elizabeth C. Patterson and to colleagues at Bard College, particularly John C. Fout; their encouragement was most welcome under trying circumstances.

The staffs of libraries and archives were always helpful. I have worked at the Pharmaceutical Society of Great Britain, the British Library (then the British Museum), and University College (London); the Bodleian Library and Museum of the History of Science (Oxford); the Archives de l'Académie des Sciences (Institut de France), Archives Nationales, Bibliothèque du Muséum National d'Histoire Naturelle, and Bibliothèque Nationale (Paris); the Hunt Institute for Botanical Documentation (Pittsburgh); the New York Public Library (and its Wertheim Study) and the New York Botanical Garden; and the Widener, Houghton, Arnold Arboretum, and Gray Herbarium Libraries of Harvard University. Åbo Akademis Bibliotek extended many courtesies during the year and several summers I lived in Åbo [Turku], Finland. Jane Hryshko at the Bard College Library and Margareta Blumenthal at Åbo Akademis Bibliotek ordered essential materials from other libraries. Mme Gauja, M. Pierre Berthon, Mme Claudine Pouret, Mlle Geneviève Darrieus, and their staff unfailingly welcomed me in my constant use of Archives of the Académie des Sciences over many years. I am grateful to the Permanent Secretaries of the Académie des Sciences of the Institut de France for permission to quote from manuscripts in the Archives of the Académie des Sciences.

It is also a pleasure to thank Kathleen E. Duffin, who found biographical information about many of the Academy's acquaintances; Armelle de Crépy, who collected information about Parisian neighborhoods; Tabetha Ewing, Kristina Mickelson, Laura Muller, and Willie G. Pannell, Jr., who helped prepare the final text; the Bard students who read the typescript aloud, backwards, to help me proofread; and the typesetters, who were accurate and vigilant. At the University of California Press, Diana Feinberg, Susan Gallick, Bettyann Kevles, Elizabeth Knoll, and Shirley Warren have provided tactful and efficacious advice about pruning the text and have encouraged and prodded me in the difficult last stages of preparing the book for the public.

The greatest thanks of all go to my husband, Timothy Stroup, whose moral and intellectual support has made my work possible.

Responsibility for errors of judgment or fact is of course my own. I hope that my work will provoke additional inquiries into the early Academy of Sciences, which remains a fertile terrain for exploration.

Annandale-on-Hudson, New York   A. S.
May 1989

PART I
The Setting

CHAPTER I

Portrait of an Institution

This is a book about ideas, an institution, and an intellectual community and about the individuals who participated in the development of all three.

The Parisian Académie royale des sciences was established by Louis XIV on the advice of Jean Baptiste Colbert, his minister of finance, in December 1666. An absolute monarch who saw everything as a potential instrument of statecraft, Louis displayed by this act his support for scholarship. His munificence was also a calculated maneuver for glorifying his reign. He expected the Academy to enhance his regal reputation while providing concrete benefits for commerce, industry, medicine, and warfare. Members of the Academy sought practical gains, but they wanted these to be grounded in correct theories. They welcomed royal financing of their scientific investigations, were honored by their official status, and relied on royal subventions to augment their personal incomes. The Academy thus embodied both royal and scholarly expectations.

The Academy of Sciences has survived to the present day and has played a significant role in the life and thought of the last three centuries, making numerous theoretical and practical contributions to science. Its early history is important, for during the first three decades of its existence the crown secured the Academy's financial base, members learned to balance the conflicting demands of state and of scholarship, and the institution established a corporate identity.

The Academy was founded, furthermore, when science was in transition. Theories were challenged, novel apparatus was devised, and perplexing new phenomena were observed. Scientific language was inadequate, and the logic of scientific explanation was itself a topic of discussion. Science was scarcely regarded as a profession in its own right: the word "scientist" had not even been coined. Instead scientific savants called themselves "geometers," "astronomers," "chemists," "botanists," and, especially, "natural philosophers." As such, they thought of themselves as practitioners of useful skills or as philosophers of nature. Few supported themselves, however, through their scientific activities. Thus the creation of a scientific institution by a king who paid savants for doing scientific research was a departure from tradition. It made academicians the envy of their contemporaries and affected the conduct of scientific research.

UNDERSTANDING THE INSTITUTION

In founding the Academy, the crown created an organization that would foster learning. But like any institution, the Academy evolved a distinctive character that reflected its origins, composition, and accomplishments. It developed written and unwritten procedures, enjoyed acclaim and suffered opprobrium, grew and declined. Its members contributed in different ways to the work—some were more diligent, others more imaginative, and a handful assumed leadership while the rest were made to follow—and a few were highly rewarded or esteemed. The institution became associated with particular sites and molded them to its own purposes, but the sites also affected its work and procedures. It interacted with persons and groups outside it—patrons, savants, aspirants, suppliers, or other organizations—and those exchanges affected it. It existed for purposes—scholarly, political, and utilitarian, for example—that may have changed over time but offer a standard for judging how well it worked.

Institutions usually preserve plentiful evidence, and the Academy was no exception. Some of this was private and intended for internal purposes, such as minutes, financial records, and correspondence. There was also public evidence, including publications and monuments. Members are part of the evidence. Their biographies reveal patterns of recruitment and criteria for advancement. Their interests helped qualify them for membership and then shaped the institution itself.

In analyzing an institution, the historian seeks access to the minds of its members and sponsors. Thus its program is paramount. Goals, methods, and actual accomplishments may be traditional or innovative, can evoke respect or skepticism among contemporaries, and may reveal the values of members or their sponsors. It is not enough to isolate the work of the few acknowledged "stars," whose work was acclaimed in their lifetimes and since. Nor can the historian concentrate only on the successful projects of the organization. The "dead ends" of history—whether they be failed careers or ideas that fell stillborn into the world—may reveal more about the spirit of a given period than either the luminaries or the ideas that seem to be precursors of modern thought. An institution is the creature of individuals and of society, of deliberate acts as well as accidents, of reasonable assumptions and untenable prejudices. The historian must discover such elements and try to understand how, singly and together, the institution, its members, and its work developed as they did.

PUBLIC IMAGE

The most intelligible introduction to the Académie royale des sciences is through published portraits. Sébastien Le Clerc developed the Academy's public image in several contemporary engravings. The best known of these dates from 1671 and shows what seems to be a visit of Louis XIV and Colbert to the Academy (plate 1). Others depict academicians examining objects in microscopes, discussing, performing a dissection, working in the chemical laboratory, and carrying out other scientific tasks. These engravings were intended both for nonscientific audiences—the king-patron, the recipients of presentation copies of engravings commissioned by the crown, and collectors—and for more knowledgeable readers of the Academy's books, which the prints illustrated. Each engraving presents a self-contained portrait, each is a deliberate public image of a royal establishment, and all allude to traits that are essential to understanding the Academy.

Le Clerc's formal portrait of the king, Colbert, and the Academy is both factual and fantastic. The artist at once portrays and misrepresents the Academy in ways that have stimulated and perplexed historians.[1] On the factual level, Le Clerc represents the Academy's experimentalist credo, its accomplishments, and its members. A lavish display of objects suggests the Academy's interests. Skeletons of dissected animals adorn the walls. Scientific instruments are everywhere. Maps, laboratory apparatus, plants, and models of machines reveal that the Academy's research program was broadly defined. The engraver portrayed academicians accurately, differentiating subjects by their garb, so that the viewer can distinguish clerical and lay academicians, identify members of the royal family, and grasp the social status of each person.[2]

It is not surprising that a portrait of the Academy by Le Clerc should reveal such an eye for detail and sensitivity to nuance. Le Clerc himself was

Plate 1. The Academy and Its Protectors. In the front are, from left to right, the prince de Condé, the duc d'Orléans, Louis XIV, and Colbert. Cassini, gesturing with his left hand, is between Condé and the king's brother, while Perrault stands between Louis and Colbert. (Watson, "Early Days"; drawn by Goyton, engraved by Le Clerc; photograph courtesy of Bibliothèque Nationale, Paris.)

not only a skilled draftsman and engraver but also an engineer who studied mathematics, natural philosophy, and cosmography. Furthermore, the Academy sponsored some of the best scientific illustration of the century, and Le Clerc helped make its anatomical illustrations acknowledged masterpieces of the time.

Le Clerc also portrayed the Academy in ways that are misleading, although even his misrepresentations convey truths about the institution. For example, he stressed the Academy's experimentalist bias at the expense of its fairly cautious theorizing. He did not delineate the entire scope of the Academy's research but rather created the impression of catholic interests and emphasized subjects, especially those with practical applications, on which Colbert spent the largest sums of money. The Academy did not limit its scientific inquiries, but its patrons preferred some fields over others, and they made their preferences evident in material ways that Le Clerc captured.

By grouping academicians and showing them in conversation, Le Clerc conveyed a collaborative spirit at the Academy, where members worked together. What Le Clerc did not show was that personal rivalries and professional disagreements enervated academicians, many of whom found it more productive to work individually than in teams.

The physical setting of Le Clerc's portrait is also misleading. The view outside the room where academicians are assembled shows the Observatory. In fact, no such prospect was possible from any of the Academy's three principal locations—the King's Library, the King's Garden, and the Observatory—none of which could be seen from any other. By including the Observatory in the portrait, Le Clerc sacrificed accurate topography but conveyed royal munificence, for the Observatory was constructed entirely with royal funds.

Finally, the central event did not occur as Le Clerc suggests, for the king did not visit the Academy until ten years *after* Le Clerc depicted the supposed occurrence. Even then, in December 1681, Louis was a reluctant visitor; although Colbert had long tried to persuade him to see his creation at first hand, the king continually dragged his heels. No scientific amateur, he lacked the knowledge to ask the kinds of well-informed and detailed questions that James II of England, for example, would pose on a visit to the Observatory in 1690. Louis was a patron of science not out of intellectual interest but out of self-interest predicated on the practical advantages that would accrue from his intervention. His appearance in the engraving does not correspond to the facts but is meant to convey, through artistic license, a royal seal of approval.[3]

Le Clerc's portrait of the Academy contributes to a program of royal propaganda. He makes it obvious that the king was the Academy's generous patron, entitled to share whatever acclaim its work received. The resulting public image of the Academy is that of a splendidly equipped royal foundation, dedicated to the experimental ideal and to the cooperative pursuit of broad interests. He makes it obvious that the Academy has a dual function: to make scientific discoveries and to honor the king.

BEHIND THE PUBLIC IMAGE

We began with a picture. Now let us turn to the descriptive and theoretical writings of savants, the terse entries of royal expenditure, the secrets preserved in private documents. Where illustration ends, the word begins.

This book makes two kinds of inquiry into the Academy. One is internalist or scientific, the study of ideas. Seventeenth-century savants observed and participated in what has since been called the scientific revolution. It was a time when theories about the world, and ways of seeing and analyzing it, changed fundamentally. The scientific revolution is normally defined with reference to the physical sciences, but the biological sciences were also transformed, partly because savants sought models for the life sciences from biology rather than from physics, mechanics, technology, or mathematics. Academicians were in the vanguard of this trend with their research on the anatomy and physiology of plants, and even their natural history of plants incorporated innovative elements.

The other line of inquiry is traditionally called externalist; it addresses connections between ideas and society. Externalist histories of science usually look for a causal relationship between scientific ideas and underlying social, economic, political, or religious structures. The causal implications of such research have made it controversial, for historians of science can be as reluctant as modern scientists to concede that scientific ideas have any but an intellectual genesis. Yet as contemporary examples show, science is not value free, and neither admission to a scientific career nor status within the scientific community is solely dependent on the quality of one's mind or work. In the seventeenth century, social background was an important determinant of an individual's access to the scientific community.[4]

But externalist inquiries may transcend the correlation of class or religion with career patterns or styles of scientific thought. The relationships among academicians or between them and their ministerial protectors influenced ideas and the functioning of the Academy. Academicians' contacts with individuals—friends, patrons, savants, or craftsmen—outside the Academy form a distinctive pattern, characteristic of interactions between an intellectual elite and those excluded from the group. The Academy as a corporate entity was under various obligations, and individual members felt personal responsibilities to society; such duties affected the way they selected problems for study. Inquiry along these lines helps clarify the varied strands of thought that molded the Academy.

Botanical research, the internalist focus of this inquiry, serves as a barometer of both scientific thought and the competing interests that affected science in the Academy. Academic studies of plants took their inspiration from discoveries of new flora, from Bacon and Descartes, from the great botanical and zoological compendia of the previous two centuries, from Harvey, and from chemical theories of the composition of the world. If its inspiration was internalist, botanical research was vulnerable to externalist influences. Academicians alternately collaborated and feuded. Ministerial intervention sometimes encouraged and sometimes sabotaged research. Research was generously supported for a while and then funds were withheld. Academicians tried to balance pure and applied science and to define their responsibilities to the wider community. Official protectors encouraged whatever seemed most useful to the king or the kingdom. As both academicians and protectors sought to reconcile their quest for pure knowledge with their desire to improve the conditions of life, they changed the emphasis of their research. Ideas about the purpose of the Academy changed between its foundation in 1666 and its reorganization in 1699, and these changes affected the conduct of research. In the process academicians developed new theories and new methods for studying plants. By the end of the seventeenth century, they and their ministerial partners had redefined the organization and goals of the Academy, and botanical research was instrumental to that redefinition. The interplay of intellectual and external influences in the Academy thus forces the historian to consider both in order to illuminate the ideas and individuals which composed the institution.

Scientific ideas are the product of many influences, not all of which are intellectual. They are created by individuals with friends and enemies, mentors and patrons, acquaintances and unseen strangers, any of whom may induce the savant to ask particular questions or use special methods. Members of an institution like the Academy were a hothouse variety of scientist, for they lived and worked together; they were related by common goals, common patrons, and common sites of scholarly activity. They affected one another by collaborating, feuding, and advising. They also affected others, who were not admitted to their ranks; they were envied and challenged, and they were called on to referee disputes or to solve technical problems.

The Academy became a resource affecting the entire kingdom. It repaid its debt to the crown by producing new ideas and information, by developing a pattern of scholarly conduct that helped professionalize natural philosophy, by carrying out specific commissions for the crown, and by advising magistrates and royal ministers. Ideas, institutional dynamics, finances, governmental pressure, socioeconomic trends, and demand for science all played a part in the story of that development. This book attempts to unite some of those elements into a coherent whole and thereby to enrich our understanding of both the parts and the ensemble.

PART II

The Institution and Its Patronage

CHAPTER 2

Members and Protectors

The essential facts about the early Academy of Sciences are straightforward. It was founded in 1666 by Louis XIV at the behest of Jean Baptiste Colbert, minister of finance and of the navy, who championed the Academy until his death in 1683. Michel François Le Tellier, marquis de Louvois, minister of war, was responsible for the Academy from 1683 until his death in 1691. A less enthusiastic protector than Colbert had been, Louvois presided over a decline in the institution that academicians lamented. In 1691 Louis Phélypeaux de Pontchartrain succeeded Louvois as ministerial protector of the Academy. Although weak finances thwarted Pontchartrain's initial efforts to revive the institution, he sponsored a formal recognition and reorganization of the Academy in January 1699 that set the institution on a new footing. The thirty-three years from founding until reorganization provide the chronological focus of this book.

The Academy was organized hierarchically. The king was its patron and head. The ministerial protector was in charge of its funding, housing, and recruitment. Academicians were responsible for research and writing, and they also assessed new technology and advised the crown on technical matters. Academicians bore unequal honors and responsibilities. Two celebrities, the Dutch mathematician Christiaan Huygens and the Italian (later naturalized) astronomer Jean Dominique Cassini, had the highest pay and the greatest influence. Lower in the hierarchy were the regular working members—natural philosophers, anatomists, botanists, chemists, geometers, astronomers, mechanicians, and permanent secretaries—

with moderate pay. Their work ranged from the empirical to the theoretical, and they directed projects and administered the Academy. At the bottom were the student members, badly paid or not paid at all, who had ill-defined responsibilities and no formal path of advancement within the Academy. Some were mere assistants while others were independent researchers whose papers were highly regarded by fellow members. More ambiguous were the associate, honorary, or corresponding memberships available to savants who did not live in Paris. These posts were distinguished but lacked remuneration. Finally, academicians hired assistants who were not members of the Academy; these included the surgeon's or apothecary's apprentices who worked at dissections or in the laboratory, as well as the mathematicians who helped survey and map France. Rank in the Academy and access to the Academy's protectors were related.

From 1666 through 1696, the Academy established three formal administrative positions for its members. The first, created in 1666, was that of permanent secretary.[1] Its original occupant was Jean Baptiste Du Hamel, relieved by Jean Gallois for two years starting in 1668 and replaced in 1697 by Bernard Le Bovier de Fontenelle. The second post, that of president—a member who served as intermediary between academicians and ministerial protector—developed at first informally. Outsiders thought either Carcavi or Huygens was president in the early years. But Colbert had close ties to several academicians, and from the 1670s Du Hamel wrote annual reports for him.[2] Louvois used specific intermediaries, first the undependable abbé de Lannion[3] and later Henri Bessé de La Chapelle. They kept him up to date with the Academy's activities and informed academicians of his wishes. Pontchartrain formalized the arrangement when in 1691 he appointed Jean Paul Bignon with the title of president of the Academy. The third position to evolve was that of treasurer. It was an onerous post, requiring the occupant to pay the Academy's expenses out of his own pocket and then to request reimbursement from the crown. The Couplet family bore this responsibility, with Claude Antoine Couplet at first taking it on informally and finally in 1696 receiving the empty title that was surely insufficient recompense for his troubles.[4]

The Academy's purposes and activities were complex. As an academy of sciences, it studied mathematics, the nature of the world, and the principles of machines. As a royal institution, its work was to bring honor to the king and benefits to the kingdom. As a participant in the larger scientific community, it sought opportunities to exchange ideas and information.

Its activities were both regular and varied. The Academy met at the Bibliothèque du roi twice a week, on Wednesdays and Saturdays, except for a six- to eight-week vacation in the fall. When the Academy was healthy, meetings lasted four or five hours, but when morale was bad, members could barely fill two hours.[5] Between meetings, academicians conducted research or tried to solve learned problems. They performed these tasks in the Bibliothèque du roi, in the Observatory, in the Jardin royal, at home, elsewhere in Paris and France, and abroad. The meetings were given over to reports, demonstrations, and discussions. Saturdays were reserved for natural philosophy, Wednesdays for mathematical sciences, and academicians were expected to attend and contribute to both. Minutes were kept for all of the seventeenth-century meetings except from 1670 through 1674.

The early Academy was small, as Le Clerc's formal portrait suggests. Sixty-two members were appointed before 1699, and there were never more than thirty-four, or fewer than nineteen, members in any given year. Since participants were fewer than those eligible to attend, the working Academy—numbering from one dozen to two dozen academicians—was intimate. Its size made the Academy susceptible to the influence of a few members, and new appointments affected research and morale. Thus, recruitment was crucial to the health of the Academy, yet the methods of and criteria for selecting academicians remain little known. From even the most minimal surviving biographical details of academicians—such as their regional origins, age at entry, education, and responsibilities outside the Academy—it is clear, however, that the careers of academicians reflect the usual patterns of education and advancement that prevailed in seventeenth-century France.

COMPOSITION OF THE EARLY ACADEMY

The origins and careers of academicians exemplify several general trends in early modern France, indeed Europe.[6] These include regional disparities in literacy, the drift of provincial talent to large capital cities, the rise of the liberal professions in esteem and economic status, and government demand for the services of an educated elite.

Most academicians came from the north of France, where literacy was higher than elsewhere in France.[7] Of the forty-two French academicians whose birthplaces are known, 70 percent were born in northern France, including sixteen who were born in or near Paris.[8] Six came from Normandy, principally from cities—Rouen, Caen, Dieppe—but also from small towns in stock-breeding regions. Two came from the Maine, two from Anjou, and one each from Brittany and Burgundy. The twelve academicians from the south came mainly from the Lyonnais, Languedoc, and Provence. Three were born in Lyon, and two nearby; Avignon, Toulon, and Aix-en-Provence each contributed an academician; and the rest came from small towns in Lower Languedoc, the Lower Auvergne, the Rouergue, and Lower Navarre.

Academicians tended to come from cities. Twenty-seven French academicians, or 64 percent of those whose birthplace is known, were born in cities, about half of them in Paris. The careers of many reflect the drain of provincial talent to the capital.[9] From 30 to 50 percent of each minister's appointments came to Paris from the provinces to establish themselves. Paris was a magnet not only for financiers, lawyers, and courtiers but also for ambitious intellectuals. Like London, it became a center of conspicuous consumption, a place where high culture was appreciated by the wealthy.[10] Once established, the Academy itself attracted savants to Paris, but by the eighteenth century Parisians dominated the working Academy.[11]

The education of academicians varied according to their social origins and the careers for which their families destined them. The marquis de l'Hospital kept his love of geometry a secret from other nobles of the sword, and the orphaned Bourdelin had to teach himself Latin and Greek. But the majority of academicians, like the other educated elite of the period, enjoyed a taste for letters and were trained in the classics. Such circles agreed that learning suited the magistracy and that the sciences were a kind of erudition, along with poetry, music, and letters. Older academicians, steeped in these traditions, could be torn by conflicting intellectual values. Thus they revered certain ancient accomplishments, but sought to dispel ancient misconceptions about nature. Claude Perrault, for example, translated Vitruvius and employed classical principles in his own architecture but used his dissections to disprove ancient claims about the salamander and the pelican. As a rule, academicians with a classical education were the theoreticians of the Academy and commanded higher pensions among the regulars.[12]

Within the social hierarchy of the realm, most academicians came from the upper half of the third estate and represented the liberal professions. Many served municipal or princely governments. Eighteen, or about 29 percent, were physicians, surgeons, or apothecaries. Twenty, or roughly a third, taught—as professors of mathematics at the University of Paris, as lecturers at the Jardin royal, as mathematics instructors to the youths of the Grande écurie, as teachers of hydrography in the port cities of Marseilles and Rochefort, and as tutors to members of the royal family. The royal treasury paid the stipends for many of these positions.

More than half the academicians had ties to government. At least thirty-eight served a royal, regional, or municipal government in some capacity. In addition to the teaching posts already mentioned at the Jardin royal and Collège royal, academicians held positions in the Bibliothèque du roi, served as royal almoner or as inspector of royal buildings, provided medical services to the French and Spanish courts, or were royal engineers. The families of a few boasted upwardly mobile councilors of state or members of *parlement* or the *grand conseil* and were among the wealthy and powerful bureaucratic elite. Some academicians participated in diplomatic missions. In these respects, the Academy was representative of natural philosophers throughout Europe in the late seventeenth century. Some of its members were known to the king or his ministers in an official capacity before they entered the Academy. Membership in the Academy also led to additional appointments. Thus, academicians were part of the power structure of seventeenth-century Paris.

Their regional origins, education, social rank, and access to persons of influence helped academicians discover their scientific aptitude and opened the doors of the Academy to them. But other savants with similar backgrounds did not become members of the Academy, so that contemporaries speculated about the criteria for admission. It was said that Paracelsians, Jesuits (under Colbert), and the regular clergy (under Pontchartrain) were excluded.[13] The earliest appointments clearly favored older men of stature, such as La Chambre, once the favorite of Séguier and Richelieu, at seventy years of age; Gilles Personne de Roberval, at sixty-four; Bernard Frenicle de Bessy at sixty-one; and Samuel Cottereau Duclos at sixty-eight.[14] Some of these savants served the Academy only briefly and sporadically, and by the late 1670s many of the original members were no longer active. Thereafter academicians tended to be younger, and the practice of nepotism meant that certain families established scholarly dynasties. As new members became noticeably younger, older academicians worried about a decline in the institution, concluding that membership was no longer a reward for achievement but an opportunity for developing talent.[15] Savants coveted membership in the Academy and tried to gain the attention of its protectors, but many were disappointed.

STUDENTS OF PLANTS

Academicians as a group reflected broader trends. But it was as individuals that they made their mark on the Academy and the scholarly world, and it is as individuals that they will become familiar in the present study.

Of the sixty-two academicians appointed before 1699, twenty-two, or 35 percent, contributed to botanical studies, and it will be helpful to focus attention on them at a more personal level. This group included four botanists (Nicolas and Jean Marchant, Denis Dodart, and Joseph Pitton de Tournefort), two natural philosophers (Edme Mariotte and Claude Perrault), five of the Academy's seven chemists (Claude Bourdelin, Jacques Borelly, Moyse Charas, Samuel Cottereau Duclos, and Guillaume Homberg), a mineralogist (Morin de Toulon), and two anatomists (Joseph Guichard Du Verney and Daniel Tauvry). Six academicians whose principal work was in the mathematical sciences (Jean Dominique Cassini, Jean Gallois, Christiaan Huygens, Philippe and Gabriel Philippe de La Hire, and Sédileau) also discussed plants. Finally, both permanent secretaries (Jean Baptiste Du Hamel and Bernard Le Bovier de Fontenelle) wrote extensively about botany in their histories of the early Academy.[16]

The principal designer of botanical research was Claude Perrault (1613–1688). He championed the term "la botanique," and his January 1667 proposal influenced botanical studies at the Academy for the rest of the century. He and his brothers advised Colbert, but the academician was no favorite of Louvois, who had Perrault's house razed to clear ground for a new library. A physician who practiced medicine only for family, friends, and the poor, and an architect who designed several royal structures including the controversial Observatory, Perrault directed the Academy's acclaimed *Histoire des animaux* and interpreted comparative anatomy mechanistically. The Perrault family, which counted Christiaan Huygens among its friends, was representative of the French upper middle class that supplied lawyers, scholars, and bureaucrats during the reign of Louis XIV.[17]

The second major botanical theorist was the Burgundian prior Edme Mariotte (c. 1620–1684), whose debate with Perrault about the circulation of sap exposed their different methods. While Mariotte was both theorist and experimenter, Perrault mostly speculated in the abstract about plants and borrowed Mariotte's data. Mariotte was a polymath who studied hydrostatics and air pressure, developed a theory of colors, and invented surveying instruments. His work was indebted to Boyle, some of whose writings he translated for the Academy, and to a network of scholarly correspondents from Aberdeen to Warsaw.[18]

The Academy's first chemical theorist was Samuel Cottereau Duclos (1598–1685). He designed and directed the Academy's laboratory in the Bibliothèque du roi. In it he studied mineral waters, analyzed the chemical constituents of plants and animals, and developed the alchemical ideas that he abjured, along with his Protestant religion, in the last days of his life. One of Colbert's elder statesmen of science, he was disliked by Louvois, who did not pay his pension at the end. Before becoming an academician, Duclos had run his own laboratory in Paris; among his pupils was Nicaise Le Febvre, whose popular chemical textbook owed much to Duclos's methods.[19]

Denis Dodart (1634–1717) directed the Academy's natural history of plants from the early 1670s until the 1690s. Much of Dodart's other work in the Academy — on diseases of the poor, nutrition, and the effects of fasting — was stimulated by his medical, social, and religious concerns. Dodart owed his place in the Academy, won before he was forty, to his connections with the Perrault family. He earned it by reviving the institution during its early slump. Known at the end of his life to the duke of Saint-Simon as a "very learned and quite saintly man," Dodart was a committed Jansenist who used his medical consultations to the king to defend his coreligionists. His friends included Jean Racine, Antoine Arnauld, Pierre Nicole, the duc de Roannez, and others associated with Port Royal.[20]

Perrault, Mariotte, Duclos, and Dodart dominated theoretical research on plants. But they depended on Claude Bourdelin to analyze plants in the laboratory and on Nicolas and Jean Marchant to cultivate and describe them.

Claude Bourdelin (1621–1699) was responsible for nearly all of the Academy's chemical analyses. He refined chemical techniques, especially for analyzing oils, and kept detailed records of his experiments and expenses. The Academy ignored his sole programmatic paper, however, partly because his ideas were too narrowly medical. Born near Lyon and orphaned at an early age, Bourdelin became an influential Paris apothecary and ensured good positions for his sons, in whose educations he enlisted Du Hamel and La Hire. He counted Racine among his friends.[21]

The Marchants, father and son, cultivated rare plants for the Academy's natural history. Together with Dodart they also composed descriptions of plants, their cultivation, and uses. As Dodart's role in the project grew, that of the Marchants shrank. Nicolas Marchant (?–1678) had served Gaston, duke of Orléans, and with Perrault encouraged the Academy to model its history of plants after work begun under the duke. Jean Marchant (?–1738) continued his father's work but never brought it to fruition; perhaps that is why Fontenelle wrote no eulogy for him.[22]

By 1689 Perrault, Mariotte, Duclos, and Nicolas Marchant were dead, and Bourdelin could not keep up his previous pace. In 1691 Pontchartrain appointed a chemist and a botanist to revitalize the Academy's botanical research. The chemist was Guillaume Homberg (?–1715), who had nearly been admitted to the Academy by Colbert. He became an influential member, enlivening meetings with his varied papers and initial optimism about Bourdelin's analyses of plants. Homberg was also interested in mining, astronomy, scientific instruments and machines, history, and languages, including Hebrew. He learned by touring the continent, so that he could meet scholars and trade in scientific novelties. Like many contemporaries, he pursued his scientific interests against the wishes of his family. Happily, the Academy provided him a new family, for in 1708 he married Dodart's daughter. An entrepreneur and risk-taker, Homberg's biography suggests his courage and strong will.[23]

The botanist Pontchartrain appointed was Joseph Pitton de Tournefort (?–1708), the most renowned of all the early Academy's researchers in this field and the first academician to travel abroad for botanical research. Tournefort's brief career was distinguished. Having arrived in Paris from Aix via Montpellier with Guy Crescent Fagon's support, he obtained appointments at the Academy, the Jardin royal, and the Collège royal. He published several influential books and developed the principal botanical taxonomy before Linnaeus. His interest in chemistry took him to Nicolas Lémery's courses and to Bourdelin's laboratory. Tournefort also collected shells, seeds, and fruits. He willed eight thousand dried plants to the king for the Academy's use and left his botanical books to Bignon, whose personal physician he had been.[24]

These academicians collaborated with one another in studying plants, but others made individual contributions. The latter were often more active in the mathematical section of the Academy. Their botanical contributions were episodic and peripheral and developed as a result of reading, observation, and conversation.

Jacques Borelly (?–1689) was interested in the chemical composition of soils and in plant nutrition, and he favored analysis by solvents. Overshadowed by Duclos and director of the laboratory for only a few years, Borelly never came into his own as a chemist at the Academy. Outside the Academy, Borelly attended Montmor's and Bourdelot's scientific meetings and explained chemical vocabulary to Antoine Furetière for the latter's dictionary. Borelly also published articles about astronomy and telescope lenses of his own manufacture. Cassini and Huygens had a low opinion of his lenses, but Louvois raised his pension and moved him into Duclos's apartment in the Bibliothèque du roi after the older chemist's death.[25]

Seventy-three when he joined the Academy, Moyse Charas (1619–1698) worked on poisons, antidotes, opium, and vipers. But he was more an honored guest than a working academician. Charas had enjoyed a distinguished medical career in England, Holland, and Spain; he lectured on chemistry and published popular works on chemical techniques and pharmaceutics. Like Homberg, he was a Protestant whose appointment to the Academy followed his conversion to Catholicism.[26]

Joseph Guichard Du Verney (1648–1730) was the only academician before Tournefort to discuss Malpighi's ideas about plant physiology. Known for his treatise on the ear, he was the first to teach osteology and the diseases of bones at the Jardin royal, where his lecture-demonstrations were very popular. The son of a provincial doctor, he trained at Avignon and built his career in Paris. There he attended Lémery's course on chemistry and participated in Bourdelot's and Denis's scientific meetings. By dissecting the brain in these private societies, Du Verney earned his reputation as a promising young anatomist. Du Verney later became the dauphin's tutor in natural philosophy and entertained the court with his dissections. He willed a large collection of anatomical preparations to the Academy.[27]

Daniel Tauvry (1669–1701) analyzed resins and gums, plant products that were thought to come from sap. An anatomist who came from the provinces to build a career in Paris, he did not long survive his success, for he died less than three years after his appointment to the Academy. Although he attended meetings regularly and shared Du Verney's skepticism about Jean Méry's views on the circulation of the blood in the fetus, Tauvry contributed few papers to the Academy.[28]

Morin (?–1707), about whom little is known except that he came from Toulon, was appointed with the title of botanist but was more interested in mineralogy and porcelain. He was often absent from meetings of the Academy, but contributed a paper on a plant found in Provence.[29]

The mathematician Christiaan Huygens (1629–1695), the most highly regarded of all academicians, influenced botany by observing plants with his new scientific instruments. Huygens came from an influential and wealthy family in Holland. Best known for his work on clocks, theoretical mathematics, and light, his very presence dignified the Academy during its early years.[30]

The astronomer Philippe de La Hire (1640–1718) studied the rise of sap and the origins of petrified wood. A Parisian by birth and the son of the painter Laurent de La Hyre, La Hire taught mathematics at the Collège royal and was a member of the Académie royale d'architecture. For the Academy of Sciences, he worked on the extension of the meridian and the map of the kingdom, surveyed for the waterworks at Versailles, edited the works of deceased colleagues for publication, and kept Huygens informed about the Academy after 1681.[31]

Gabriel Philippe de La Hire (1677–1719), son of Philippe, wrote a paper on how vines grip walls. He entered the Academy as a student astronomer at the age of seventeen and also followed in his father's footsteps by becoming professor royal of architecture.[32]

Sédileau (?–1693), whose first name and biography are unknown, studied orange trees and their diseases. A mathematician influenced by Ignace Gaston Pardies, Sédileau translated and annotated Frontinus's treatise on aqueducts, wrote meteorological and astronomical papers, designed several instruments, and fashioned the terrestrial map on the floor of the western tower of the Observatory.[33]

Even the astronomer Jean Dominique Cassini (1625–1712) contributed to the Academy's work on plants, if only by discussing the medical uses of plant products during the late 1680s, when other botanical research was largely eclipsed. Cassini was also interested in insects and blood transfusion and had visited the Accademia del Cimento. Like Huygens, he was a force to be reckoned with in the Academy: he built a formidable team of astronomers, began mapping the kingdom and the world, and in his seventy-sixth year traveled to the borders of France to extend the meridian. He kept Louis XIV interested in the Academy by stressing the practical applications of astronomical observations.[34]

Jean Gallois (1632–1707) contributed to botany indirectly. He publicized it by mentioning the Jesuits' observations of flora and fauna in a summary of their reports; more important, he encouraged academicians' research and argued for additional engravings of plants. A classicist and a geometer known for his elegant style, he wrote papers on an air gun and on geometry. This Paris-born abbot was member of the Académie française and professor of Greek at the Collège royal; he participated in Bourdelot's conferences, served as Colbert's librarian, and edited the *Journal des sçavans*.[35]

The two permanent secretaries—Jean Baptiste Du Hamel (1623–1706) and Bernard Le Bovier de Fontenelle (1657–1757)—influenced the Academy's research principally through their effect on corporate morale. They wrote scholarly and popular treatises—Du Hamel in Latin, Fontenelle in French—reviewed manuscripts for publication, wrote the Academy's history, and maintained the minutes. Both were Normans.

A lawyer's son who joined the Congregation of the Oratory, Du Hamel was royal almoner and held church posts before becoming an academician. On a diplomatic mission to England, he met Fellows of the Royal Society and bought a microscope for the Academy, even though he had relinquished his academician's pension.[36]

Educated by the Jesuits and intended at first for a legal career, Fontenelle pursued a literary career. He became a member of the Académie française and the Académie des inscriptions and a popularizer of Cartesianism and the sciences. His history of the seventeenth-century Academy was both more and less than a translation of Du Hamel's Latin account, and he began the custom of issuing annual reports of the Academy's accomplishments and eulogizing academicians after their deaths.[37]

In summary, these academicians made different contributions to plant study. Their work ranged from abstract theory to rigorous experiment and observation, from suggestion to dedicated personal labor, from the traditional to the innovative, from the technical to the general. Although they represent different generations, they had much in common. All but two were French by birth, but only five were born in Paris. Most came from the upper ranks of the third estate, although two (Tournefort and Cassini) claimed to be gentlemen; the fathers of at least two (Perrault and Du Hamel) were lawyers. Several joined the Academy before the age of forty. Two (Jean Marchant and G. P. de La Hire) were the sons of academicians who had also contributed to the Academy's botanical research. Some had close ties to the Jansenists (Dodart, Homberg, and Perrault). Three were Catholic clergy (Mariotte, Du Hamel, and Gallois), three (Charas, Duclos, and Homberg) converted from Protestantism, and the paternal grandmother of one (Tournefort) came from a Jewish family. Two (Tournefort and Homberg) studied the sciences despite their parents' wishes, two (Cassini and Homberg) became naturalized subjects of Louis XIV. Bourdelin, La Hire, and Tournefort were orphaned or lost one parent before they were twenty. Most were polymaths and many were physicians. These academicians traveled, especially in France, England, and Italy, but also in Holland, Spain, eastern Europe, and Sweden. Many enjoyed other royal appointments or ties to government, as adviser to ministers, physician to members of the royal family, professor at the Collège royal, demonstrator at the Jardin royal, or royal almoner.

THE PROTECTORS AND THEIR SPOKESMEN

Responsibility for the Academy's successes and failures must be shared between the researchers and the protectors. The ministers and their spokesmen influenced research through appointments, financing, and interference, both subtle and open. The king also affected the Academy, albeit in ways that are now often obscure.

As spokesmen, La Chapelle and Bignon kept the ministers informed about the Academy, submitted the *estats* requesting payment of pensions, and defended academicians' requests for additional financial support for research. They also conveyed the wishes of the protector to the Academy. Doing the job well divided their loyalties. Henri Bessé de La Chapelle (?–1694) was assistant (*commis*) to Louvois, serving as controller general of royal buildings, inspector of fine arts, and overseer of the Academies of Sciences and Inscriptions. A member of both academies, he was a geometer in the former. Little is known of his life. Because Louvois's relations with the Academy were uneasy—he reduced its budget and size and mistrusted its members who were partisans of Colbert—the responsibilities of his spokesman were difficult. In January 1686, for example, La Chapelle criticized botanists and chemists on Louvois's behalf, unintentionally provoking a decline in the Academy's study of plants.[38]

Fortunately for the abbé Jean Paul Bignon (1662–1743), his uncle Pontchartrain was sympathetic to improving the Academy. A staunch advocate of the Academy, Bignon sought to regularize its procedures and increase its funding; he tried to ensure that it was treated fairly by comparison with the other academies. He recommended savants for membership, bolstering botanical research by selecting Tournefort and Homberg in 1691. Armed with reports on its personnel, projects, and expectations, he argued the Academy's case to Pontchartrain, citing precedent. The weak condition of the royal treasury forced Bignon to justify every request for funds and to appoint academicians without pensions. He pleaded for payment of what was owed academicians but was more successful in obtaining permission for them to publish. He proposed a merit system which Pontchartrain vetoed in favor of the seniority system.[39]

Bignon came from "a distinguished family of magistrates and royal librarians that stood at the very center of the *robin* society" in Paris. Like Du Hamel a member of the Congregation of the Oratory, Bignon's sinecures made him a wealthy man. In the eighteenth century, he directed the book trade, edited the *Journal des sçavans*, and became royal librarian. Thus, Bignon controlled much of French intellectual life from 1691 until his death.[40]

The three ministerial protectors of the Academy—Colbert, Louvois, and Pontchartrain—influenced research both deliberately and accidentally. Their appointments, financial support, persuasion, and encouragement to publish all affected institutional health. They arranged for special benefits or assistance to the Academy through bureaucratic and diplomatic channels and could arbitrate disputes among academicians.[41] Each valued botanical studies differently.

Jean Baptiste Colbert (1619–1683) appointed fifteen of the Academy's students of plants and supported the natural history of plants generously until the early 1680s. He was interested in natural philosophy. Thus he invited academicians to his estate for learned conversations and visited the Academy well before the king did. But he was not above exploiting its members for familial advantage, for he had Du Hamel write the book that won Colbert's son admission to the Académie française.[42]

François Michel Le Tellier, marquis de Louvois (1639–1691) did not share Colbert's enthusiasm for the sciences or the Academy, and his awkward management demoralized the institution. Nevertheless, he favored the biological sciences, especially their medical applications and comparative anatomy. He appointed no new botanists or chemists, however, and canceled publications on anatomy and botany. His personal interest in certain subjects injured academic research by threatening its independence.[43]

Louis Phélypeaux de Pontchartrain (1643–1727) and his family valued learning both for its own sake and for its benefits to the state. As protector of the Academy, he was serious about rejuvenating it as inexpensively as possible. He appointed six botanists and chemists and ordered them to investigate the natural history of plants. He also subsidized publication of Tournefort's treatises.[44]

Behind the scenes was the king. Advocate of personal rule, good at figures, interested in details, eager to catch out his ministers in an oversight, and fond of telling them what they already knew, Louis XIV (1638–1715) spent much of his day closeted with royal officials. The king's dislike of unfamiliar faces and his personal attention to the affairs of state gave certain ministers, including the three protectors of the Academy, great power.

Some of the Academy's projects must have pleased this monarch who loved gardens, rare plants, and exotic animals, who was vainglorious and hungry for tax revenue, and who sought to expand his kingdom. Academicians studied rare plants and dissected animals from the royal menagerie. They mapped both the tax district around Paris and the entire kingdom, and they studied military technology. Their elegantly illustrated books smoothed diplomacy when Louis presented them as marks of royal favor. When the king visited the Academy in 1681, academicians demonstrated some experiments and apparatus and gave him a list of manuscripts ready for publication. But none of the botanical texts on the list was published. Indeed, academicians never got all the support they wanted, for the king's favor, ministerial interest, and the health of the royal treasury fluctuated.

For Louis the Academy was a potential source of honor and invention. Thus, he told Cassini that he wanted the Academy to make France as illustrious in the realm of letters as it was in warfare, and when Cassini explained how astronomy could reform geography and navigation, the monarch was attentive. Louis was curious to witness the spectacular or the curious, including the comet of 1664, a large burning mirror in 1669, and the dissection of an elephant in 1683. But Colbert persuaded him only with difficulty to visit the Academy in 1681 and the Observatory in 1682, and when rain interrupted the second visit the king never returned. There was little need for him to visit the Academy's headquarters, however, for the Academy came to court whenever Louis granted Cassini an audience, when Dodart attended Louis as his physician, or when Du Verney and other academicians tutored members of the royal family. Louis ruled his kingdom personally, taking an interest in details and serving as his own prime minister and superintendent of finances, but his impersonal sponsorship of the Academy suggests that it was relatively unimportant to him. Although Louis approved of the Academy, had certain expectations of it, and was fascinated by some of its more arcane activities, he delegated the responsibility for running it almost entirely to the ministerial protectors.[45]

Like the kingdom of which it was part, the Academy had its place in a hierarchy of power and privilege. At the apex stood the king, with ministerial protectors mediating between him and the academicians. But the distinction between academicians and protectors went beyond relationships of power and responsibility and, as will be seen in the next chapter, permeated perceptions and expectations of the Academy.

CHAPTER 3

Models for a Company of Scientists

The Académie royale des sciences was founded by partners from two spheres: the commonwealth of letters and the royal government of France. These partners had different expectations of the Academy, because of their allegiances to distinct worlds. All agreed on the principal goals of the new institution, but members of the Academy stressed scholarly and professional ideals, while the ministers underscored benefits to king and kingdom. Although their attitudes differed, each had experience in the other's sphere, and all held many presuppositions in common. Ministers of state respected literature, science, and the arts; savants served the royal government. The areas of agreement were sufficiently large to establish the new institution, but not large enough to prevent conflict. The inevitable struggles for control influenced the vigor and structure of the early Academy of Sciences.

SOCIOCULTURAL CONTEXTS

Natural philosophy fit awkwardly with many seventeenth-century values. Inconsistencies at the court, for example, highlight the ambiguous position of the sciences. The king preferred pious and hierarchical books but supported the new, experimental science practiced at the Jardin royal and the Academy. Some courtiers disliked science on the grounds that its practitioners were coarse and pedantic, its language technical and ignoble. Others maintained private laboratories or observatories and sponsored scientific meetings.[1]

Being natural philosophers made some academicians social and cultural misfits. Some pursued their careers against the wishes of their families. If their research was dirty, smelly, or dangerous, it opened them to scorn as "sooty empirics," while a fascination with geometry or Cartesianism might seem arcane or subversive. Scientific ideas could be at odds with the classicist intellectual values espoused within the social milieus of many academicians. When the new natural philosophy challenged tradition, savants had to weigh conservative or elitist social tendencies against innovative scholarly impulses. In contrast, the Academy's ministerial protectors were fairly typical of their circles, where amateur interest in science was prominent. In general academicians challenged, ministerial protectors enacted, the values of their different social orders.[2]

The Academy was therefore a creature of varied cultural spheres. Its members and protectors inhabited a world of conflicting ideals, at once pious and critical, unconventional and hierarchical. These partners brought very different expectations to the founding of the institution.

Language is a clue to attitudes about the Academy and to its antecedents. Two words described the institution: "l'Académie" and "la Compagnie." The official name was "l'Académie royale des sciences," and it was affirmed by the *règlement*, or regulations, of 1699. But academicians usually spoke of "la Compagnie." The *Journal des sçavans* referred to "the Company that meets in the King's Library," and the Academy recorded its decisions in the minutes as rulings of "the Company."[3]

In late seventeenth-century France, an academy might be a craft organization; a place where aristocratic youths learned how to ride; or, in an ironic or abusive sense, a public place where illegal games were played. But in learned circles, an academy was a group of scholars who met regularly, and it stemmed from a heritage that included Plato and Bacon.

The word "compagnie" had a broader usage. Thus, in seventeenth-century France, there were trading companies, companies for administering justice and for directing hospitals, military companies, and the Company of Jesus. Above all, the word "compagnie" suggests the corporate bodies characteristic of French towns. These companies—of notaries, lawyers, professors at the Collège royal, Parisian secular clergy with doctorates, or the king's legal officers, for example—had their "own statutes, accumulated privileges, leaders, rules of assembly, financial structure and corporate mentality" and were foci of urban professional prestige.[4]

The Academy of Sciences combined features of the professional corporations with those of such scholarly antecedents as the sixteenth-century French academies, the Baconian House of Solomon, contemporary private

Parisian academies, and the Florentine and London scientific societies.[5] By imitating the urban corporations, the Academy professionalized the scholarly institution.

Ideals of behavior and scholarship espoused by the scholarly world—universities, libraries, publishers, and circles of like-minded savants—also shaped the Academy. Some ideals came from the classics. Modesty, for example, was a recurrent theme in the eulogies of academicians, as a personal trait becoming to scholars.[6] Academicians were animated by other expectations—that they publish and serve the public, that they obtain patronage and conduct research collectively—articulated by Descartes, Bacon, and others.[7] Both precept and practice recommended collective research. Research teams, especially where a master-disciple relationship existed, could boast impressive accomplishments, such as the Maurist editions of religious texts.[8]

Seventeenth-century savants valued intellectual discourse. They also anticipated useful applications of scientific knowledge, encouraged scholars to educate the public, and hoped for state support. The Academy incorporated these values. Academicians described their earliest projects in Baconian language, and they planned and researched collectively. Their correspondence was far-flung and they sponsored worldwide expeditions. They emulated the experimentalism of the Accademia del Cimento and the Royal Society, and they enjoyed substantial financial support from the French crown. Finally, they emphasized the Baconian and Cartesian favorites, natural history and geometry, respectively. The Academy embodied prevailing scholarly and professional ideals.

BUREAUCRATIC MODELS

Colbert took advice from scholars and professionals when he founded the Academy, but he also heeded goals and prototypes more familiar from those spheres—the economy, the navy, and the royal buildings—for which he was responsible as Louis's minister. For him the Academy was more than a shelter for learning. It was also an instrument of reform and propaganda, susceptible to his usual bureaucratic practices.[9]

Even if Colbert had read Bacon's *New Atlantis* or was familiar with Descartes's proposal for greater public support of the sciences, what interested him was not Baconianism or Cartesianism per se but rather his own "generous plan for a universal reform" with respect to "those matters pertaining to the maintenance and tranquility of the State."[10] Colbert's every official act was intended to promote the glory of the king and increase the wealth of his realm.[11] The tradition of royal patronage, the crown's increased control over publishing, Colbert's respect for learning, and his multiple responsibilities as royal minister empowered him to influence French intellectual and cultural life. Here as in other spheres, his economic goals and expertise colored Colbert's official acts.

Colbert's economic policies have been labeled mercantilist by later writers. As an economic philosophy, mercantilism aimed to increase the wealth of one's own country; because resources were limited, this was necessarily at the expense of foreign lands. Thus mercantilism was based on and inflamed proto-nationalistic feelings. As practiced by Colbert, it included the establishment of companies for overseas trade, regulation and improvement of manufacturing, and importation of foreign workers, "taking care that the government got its money's worth for any aid granted."[12] French policies consolidated the powers of the state at the expense of local initiative.[13] Unlike savants who valued Baconian or Cartesian precepts and solved scientific problems, Colbert was motivated principally by patriotic and propagandistic goals in an economic context. The Academy owed a debt, therefore, not only to a scholarly tradition but also to Colbert's official program.

Colbert linked the economy of France to a broader plan to reform justice and develop the arts and sciences. He sought to overcome "ignorance" in "the sciences," where he believed the "abuses" were more significant even "than those of justice and finance."[14] Seeking a tool for reform, Colbert was enthusiastic about the academy as a type of institution. This is clear from his letters patent of 1676, which proclaimed the purposes of his academies in the following terms:

> Because the splendor and happiness of a State consist not only in maintaining the glory of arms abroad, but also in displaying at home an abundance of wealth and in causing the arts and sciences to flourish, we have been persuaded for many years to establish several academies for both letters and sciences.[15]

During the 1660s and 1670s, Colbert founded or took under his protection many academies. When he contemplated one for the sciences, an adviser suggested that he "ask other persons from the various academies to give a model of their own" and to assist in planning.[16] The idea was part of a pattern for reforming and organizing cultural life in the kingdom to benefit the king.

Before establishing the Academy of Sciences, Colbert had experimented with traditional patronage in the form of *pensions et gratifications* sent to scholars all over Europe. Such grants seemed at first to be "the best way of putting men of letters and artists in the service of the grandeur of the king."[17] Chapelain advised Colbert to reward Italians as well as the Dutch, so that having increased "the glory of the King in these northern countries" he might achieve "the same result for the southern provinces, that is, Florence and Pisa."[18] The gifts were repaid in the coin of scholars, for Hevelius dedicated his first book on comets to Colbert and both his *Cométographie* and *Machine Céleste* to Louis XIV, as expressions of gratitude.[19]

Colbert's advisers, however, soon argued that more formal, public, and systematic support for the arts and sciences would enhance the reputation of everyone involved. Anticipating that an academy of sciences would increase French "renown in the world," one gloated: "what glory to the King and what honor for Mgr. Colbert." Colbert would "enhance it above all the others and give it advantages that will make clear the hand by which it is sustained." He wanted the French Academy to surpass its rivals. It was to be "the most learned and most celebrated in the world," and the king would be applauded for its accomplishments. As the Observatory was built, it too found a place in a list of projects that would increase "grandeur and magnificence" in the kingdom. An account of the Academy's activities was solicited for inclusion in the official history of the reign, and the first histories of the Academy announced that the institution hoped to honor its king. Savants were sensitive to these competitive motives: an Avignonese wishing to flatter Huygens wrote tactlessly that by acquiring the Dutch mathematician for France, the king had outdone his conquest of Holland. The patriotism which "infused . . . and colored" mercantilism was at work in the establishment of the Academy. Even academicians and their associates were aware of their role in the competition among states for intellectual primacy.[20]

Manipulating an Academy for propagandistic ends was less cumbersome than corresponding with a dozen or more individual recipients of awards, as Chapelain did, to ensure that they repaid largesse with homage. After the Academy began to flourish, Colbert diminished the program of pensions for independent scholars, finally neglecting traditional patronage in favor of the more controlled venture. By the mid-1670s, when the crown stopped pensioning foreign scientists, patronage had adopted new habits.

Having shifted patronage from individuals to institutions, Colbert drew on his bureaucratic experience for ideas about how to run the Academy. To stimulate the stagnant French economy, he offered monetary incentives to the directors of companies; he also imported skilled workers and subsidized manufacture of luxury products.[21] To stimulate French science, therefore, he raided faltering private societies for their best members, to whom he paid pensions; he imported highly regarded savants from Holland, Bologna, and Denmark; and he subsidized research and publication.

Like the manufacturing and trading companies, the Academy was intended to be useful. Although this is sometimes said to have been a blemish imposed by Louvois,[22] Colbert and his advisers had designed the Academy with its utility to the kingdom in mind. They chose as academicians men whose aggregate skills would "make the royal academy as noble as it is useful." It was Colbert who ordered the Academy to examine the drinking water at Versailles and who encouraged La Hire to dissect the fish along the coasts of Brittany and Normandy "because this work will be very useful." Colbert wanted results, in the form of a map of the tax district around Paris, a method of determining longitude, or a natural history of plants. His expectations were nourished by mercantilist presuppositions and by the propaganda of early modern scientists, technologues, and amateurs, for whom utility was an article of faith.[23]

The Academy enjoyed privileges similar to those of the manufacturing and trading companies. The honor of a royal visit to the Academy was arranged in 1681, just as had been done for the Gobelins in 1667. Colbert exempted academicians from taxes in regions where they were making observations for the great map of the kingdom, and he used the power of his office and his extensive contacts to obtain what the Company needed. Like certain master craftsmen, many academicians were housed in royal buildings.[24]

As he did with economic ventures, Colbert supervised the Academy's activities. Cassini spoke to Francis Vernon of this control. In 1670, hoping that his ephemerides would soon be published, Cassini admitted that "It depends upon the orders and determinations of Monsr Colbert upon whom all the motions of the Royal Academie are to bee calculated; For the measure of their times are sett by him." Moreover, he attributed the generosity of pensions and paucity of members to the king's wish "not only to have a Titular butt an effectuall influence upon his royall Academie."[25] This Louis accomplished through the ministerial protector, who reviewed annual reports and proposals, appointed academicians, and regulated finances and publication. These methods Colbert and his successors adapted from the bureaucratic world that was their principal concern.

The Academy of Sciences was of course very different from trading or manufacturing companies. Membership in the Academy depended on connections and scientific talent, not wealth. The Academy was not intended to monopolize, but to reform. It had no religious, legal, or political powers, although it controlled members' rights to publish. It was not intended to sell a commodity or to make a profit, and its publications circulated as gifts rather than through purchase.

Nevertheless, the Academy shared some traits with the manufacturing and trading companies. Like the overseas trading companies, it exploited the natural resources of the colonies and sponsored expeditions that retrieved materials from foreign lands in order to enrich knowledge in the mother country. Like manufacturing companies, the Academy produced luxury and practical goods: hypotheses, data, scholarly publications, and useful inventions. Both Academy and companies aggrandized France and the king at the expense of foreign rivals. Both protected participants' claims or rights, enjoyed royal financial support, were closely supervised by royal ministers, owed their existence to ministerial initiative, and were meant to benefit the kingdom. As with the economy so in the learned world, the crown championed those activities of the third estate which seemed advantageous to king and kingdom.[26]

The Academy was indebted to the economic policies of Colbert, whose companies influenced his academies. The French statist economic tradition supplied both the justification and the procedures for sponsoring the Academy of Sciences.

CONCLUSION

The Academy was a company of scientists who shared with its ministerial founder certain practical and theoretical goals. But different prototypes inspired members and protectors. The professional model emphasized exclusivity, power, and prestige for its members. The scholarly model called for experimentation, observation of nature, cooperation among savants, communication of scientific knowledge, and patronage. The bureaucratic model stressed government control, benefits to king and kingdom, and reform of knowledge and practice. These exemplars encouraged the activities that characterized the seventeenth-century Academy: self-aggrandizement, research and debate, dissemination of information, and assistance to the crown.[27] Yet the Academy was not constricted by its antecedents. Rather, as will be seen, fluctuations in the research budget, changing ministerial policies, the actual progress of research, and personal relationships also contributed to the institution's character.

CHAPTER 4

The Material Benefits of Membership: Pensions and Quarters

Statism and autocratic pride led the French crown to fund the Academy of Sciences. Colbert was not the first minister of finance nor Louis the first ruler to ally natural philosophy and the state. What was new was the extent of support. Existing rival societies challenged the French to outdo them: in Florence the princes provided facilities and participated in research, and it was rumored that the Royal Society received material aid from Charles II. Colbert and Louis would not forgo a single weapon in the contest for supremacy.[1]

Louis funded the seventeenth-century Academy of Sciences at a level similar to the annual income of the wealthiest monastery in France.[2] His support functioned in two ways: personally, by providing rewards for individuals, and institutionally, by guaranteeing subventions for research and publication. At the personal level, many members of the Academy received pensions or were lodged in royal buildings. At the institutional level, the crown supported the Academy's program by constructing an Observatory, furnishing a chemical laboratory, buying equipment, hiring research assistants, paying for expeditions in France and abroad, and printing academicians' books and articles. The surviving financial records reveal the extent and importance of royal patronage. They also shed light on the internal operations of the Academy and suggest how patronage influenced the institution's work and morale.

This and the following chapter analyze royal funding of the early Academy in three categories: the pensions paid to academicians, the physical plant and equipment of the institution, and the program of research and publication. The financial record, however, is incomplete and sometimes combines expenditure for the Academy with that for other royal institutions. Moreover, the Academy's budget does not wholly reflect its research program, for certain activities incurred few costs. Nevertheless, an examination of its finances is valuable because royal funding clarifies the value of the institution to its members, the effects of its quarters, the changing fortunes of the Academy, and the influence of patronage on research.

THE FUNCTIONS OF PENSIONS

The pensions paid to academicians mark a significant break with tradition. They resemble only superficially the *pensions* and *gratifications* that the crown paid to savants and artists as individuals and that it could summarily halt. Academicians received their pensions because they were members of an Academy. Once appointed, most remained members for life and were entitled to annual pensions so long as they worked. The association of a pension with membership in an academy, and the continuity of entitlement, broke with earlier practices. The effect was to separate the pension from the arbitrary will of a prince or the commissioning of specific works. Instead pensions were connected with research done in the Academy, which grew to have traditions and prerogatives of its own. This helped establish scientific research and writing as professions; it also built the corporate identity of the Academy.[3]

Pensions did not guarantee financial independence. The Academy was rarely the sole source of income for its members, and for some it provided no income whatsoever. Membership in the Academy fell into four categories vis à vis pensions. There were highly paid celebrities, competitively paid regulars, modestly paid students, and unpaid honorary, associate, or student members. Only the two celebrities—Huygens and Cassini—received pensions generous enough to provide a comfortable living. Both were foreigners who worked in the mathematical sciences and received allowances for moving to Paris; they enjoyed higher social and economic status than all but their noble colleagues in the Academy. Their large pensions—of 6,000 to 9,000 livres a year—brought them status both inside and outside the Academy.[4]

Except for the celebrities, academicians were pensioned at levels similar to those prevailing in the other royal academies, in the Collège royal, and among other members of the liberal professions who received *gages* and *pensions* from the crown. Unlike the celebrities, all but two of the regulars and students were French. Regulars received from 300 to 2,000 livres, and students, when they were pensioned at all, from 300 to 1,000 livres (table 1). Those who earned 1,500 to 2,000 livres a year would have found that their pensions, at least until the mid-1680s, provided sufficient leisure to devote themselves to research, unless they had families. But many regulars and students found their pensions inadequate and depended on other income.[5] The position of student members was ambivalent, for there was no established path of advancement within the Academy. They did best to use their membership as a first step to a good career outside it.[6] Nor was there any established mechanism whereby an academician could increase his pension, and although some academicians got "raises," the pensions of others were cut.

Finally, some academicians received no pensions at all. For the most part, they were foreign (Leibniz, Tschirnhaus, and Guglielmini), nonresident (Fantet de Lagny and Chazelles), or noble (L'Hospital); others were pensioned by the crown in another capacity (La Chapelle, Thévenot, and Bignon). But even active members who attended meetings and received no other stipends from the crown might not be pensioned for several years (Le Febvre and Varignon), and under Pontchartrain students were no longer entitled to stipends. Furthermore, when academicians took leaves of absence, whether for reasons of health (Huygens) or to assume different responsibilities temporarily (Du Hamel), they lost their pensions.

Pensions reflected a hierarchy within the Academy. The higher an academician's pension, the more likely he was to command the esteem of the Academy's protectors,[7] to wield power within the institution, to present theoretical papers, or to direct the research of others. The best-paid academicians tended to have better access to the king and ministers and to have more elevated social status.

The size and value of pensions fell during the seventeenth century. Colbert was the most generous, establishing exalted levels for Huygens and Cassini and paying a higher average pension to other academicians—about 1,400 livres—than did either of his successors. Louvois and Pontchartrain paid an average of about 1,000 livres to academicians other than Cassini. Louvois reduced expenditure on pensions in three ways: he did not replace all deceased or excluded members; he pensioned a smaller proportion of the Academy; and he paid smaller stipends, offering amounts in the range formerly reserved for student members. Even though Pontchartrain raised some low pensions and paid formerly unpensioned academicians, his stipends continued to be modest.[8]

It was no accident that pensions declined from the early 1680s. This was deliberate policy, resulting from a faltering economy, falling tax revenues, and increasing military expenditure. Moreover, from the crown's point of view, once the Academy was established it was not necessary to maintain pensions at a very high level, so long as academicians continued to work in a manner that enhanced the reputation of the king.

Economic hardships exacerbated the decline of pensions. The value of the livre began to fall in the late 1680s.[9] Worse, for several years during the 1690s the crown failed to pay academicians their pensions at all, offering finally to make good the debt in the form of annuities.[10] Lodging became a more significant benefit as pensions declined in value.

Academicians and their protectors disagreed about the function of pensions. Colbert used them to recognize "merit and reputation," while Louvois and Pontchartrain transformed them into incentives or modest supplements to income. But academicians believed that pensions should provide "the peace of mind and leisure" required for their work, and by the 1690s most found their pensions too small.[11]

The celebrities and regular members formed the core of the Academy, which remained small throughout the century.[12] Thus, pensions were important to the morale of individual members and of the entire Academy, since the pensioned members were also the working members. It was principally they who used the Academy's facilities for research, collaborated in team projects, and shared ideas at meetings. Fluctuations in the size of the Academy, and especially in the proportion of working members, affected institutional vigor. If there were too few members, they could not complete ambitious projects or surprise one another with new ideas. Quarrels or the loss of a member due to travel, illness, or death were felt keenly in this small society whose members lived and worked together. Academicians tried to produce science as an ensemble, and blows to the equilibrium of the company had the greater impact because the society was intimate.[13]

The right to appoint and pension members of the Academy was an effective means of controlling the institution. Ministers paid higher stipends to savants they particularly valued, appointed more scholars in favored disciplines, and determined how large the working Academy would be. By treating academicians generously and assuring relative stability in the size of the Academy, Colbert placed the new institution on a firm footing. His successors economized by reducing and delaying pensions at a time when inflation was further diminishing their monetary value. Louvois also allowed the number of members to decline, until by 1690 academicians were worried about the small size of the Academy. When Pontchartrain tried to renew the Academy, he did so by adding new members, not by paying them more.

BIBLIOTHÈQUE DU ROI AND JARDIN ROYAL

Beyond the personal benefit of salary for academicians, one of the chief material supports was the guarantee of a place to meet and work. Fixed establishments where savants could meet, experiment, make observations, store equipment and notebooks, and display natural history specimens— these provided the essential institutional nucleus. Because academicians lived in its quarters, the Academy was more than simply a scientific institution. It was a society in microcosm. More than any contemporary learned institution it touched most aspects of its members' lives, and after their deaths it even dissected many of them.[14]

The early Academy depended on three principal locations for its work in Paris: the Bibliothèque du roi, the Jardin royal, and the Observatoire. In the first two of these sites, the Academy shared working space, personnel, and expenses with other royal institutions.[15] The Library and the Garden already existed when the Academy was founded, and both held resources useful to academicians. The crown allied both with the fledgling scientific society, as for example when it integrated the Academy's work into the ambitious engraved history of the reign that was run from the Library.[16]

The close ties of the Academy with the Jardin royal were usually amicable. Several academicians, including Méry, Boulduc, Charas, Du Verney, and Tournefort, taught at the Garden. Du Verney also lived and maintained an anatomy room there, where he dissected many of the animals sent to the Academy. Academicians studied the Garden's plants and stored curious objects there, like a petrified tree or a coconut. The Academy actually controlled certain parts of the Jardin royal, including Jean Marchant's *petit jardin*, a "logement," and a room for its skeletons. But the Marchants' ambiguous status at the Garden caused jurisdictional conflicts, and the two institutions redefined their ties during the 1690s when Fagon suppressed Marchant's permit for the *petit jardin* but offered to regularize the Academy's position at the Jardin royal.[17]

The Academy's earliest home was the Bibliothèque du roi. It was located on the rue Vivienne just north of the Palais royal, in a neighborhood dominated by the wealthy and powerful nobility of the robe. Surrounded by the *hôtels* of Colbert, Louvois, and Pontchartrain, the Bibliothèque du roi and the Academy housed in it could be closely supervised by their ministerial protectors.[18] Although the Library was never meant to be the Academy's permanent home, the two institutions had more in common than quarters transiently shared. At first the Academy's very identity was tied up with its lodgings, for it was known as "the Company that meets in the King's Library." At least one member of the Academy thought it was housed in the Library so that the scientists could refer to the books there,[19] and one academician—first Carcavi and later Thévenot—was also *commis à la garde de la bibliothèque* and looked after small expenses for both institutions (table 12c–d). As a result, many instruments purchased for the Academy were said to be for the Bibliothèque or Cabinet du roi, and until the Academy moved to the Louvre in 1699, its wood, candles, paper, and pens were supplied out of the Library's funds.[20]

The houses belonged to the Colbert family, but the crown paid to ready them for their scholarly tenants. Painting, interior fixtures, and a sundial cost more than 9,000 livres in 1666 and 1667. Then there were yearly payments to maintain and repair the cesspool, pump, well, cabinets, and bookshelves. Rent, refurbishment, and maintenance of the buildings and grounds cost the royal treasury substantial sums (table 12a–e, h–i).

Academicians controlled the garden and several rooms at the Library. The former served for botany, meteorological experiments, and astronomical observations, although Cassini complained that the city air impeded his work.[21] Several rooms were set aside for meetings, collections of scientific apparatus and specimens, the laboratory, and apartments. The Academy met on a lower floor in the smallest room, called "la salle de l'Academie" and recognized by its green door, which housed books on natural philosophy. There were also a chemical laboratory, a space for dissections, and living quarters for several savants.[22]

The laboratory, in use day and night, became a focus of the Academy's activity.[23] The chemists experimented and also prepared medicines there for colleagues, and visitors came to share arcane lore. Constructed according to Duclos's requirements, it was fitted out with furnaces, specially designed cabinets and tables, apparatus and glassware, and chemical reagents (plate 2).[24] The crown spent about 1,000 livres a year (table 7) to build, supply, and modernize it.[25]

At first academicians also conducted dissections and vivisections in the Library, fitting a table with straps to restrain live subjects and using surgical instruments made by the cutlers André and André Guillaume Gérard.[26] The menagerie at Versailles assured a plentiful, if unpredictable, supply of exotic animals. When one died, a messenger notified the anatomists and carters delivered the carcass to the Bibliothèque du roi. After August 1686,

Plate 2. The Chemical Laboratory of the Academy. Among the academicians seated or standing around the table should be Bourdelin, Duclos, N. Marchant, and possibly Dodart, Du Hamel, and Gallois. The older man on the right who reaches toward the table with a vial in his right hand resembles portraits of Duclos. In the shelves behind academicians and visitors is apparatus, including on the top left an alembic and below it two aludels and a bell jar, and on the right, a mortar and pestle. One assistant works at the window with a balance, another at the furnace on which is an alembic at the far right. (Watson, "Early Days"; Eklund, *Incompleat Chymist*; engraved by Le Clerc; photograph courtesy of Bibliothèque Nationale, Paris.)

however, it was usually sent on to an academician's own quarters—Du Verney's house at the Jardin royal or Méry's rooms at the Invalides—a sign of the declining role of the Library in the Academy's work.[27]

Although Le Clerc portrayed a bright and tidy area for the Academy's dissections and vivisections in the Library (plate 3), Martin Lister, describing the dissection room in Du Verney's house, imparts more of the atmosphere in which academicians must have worked:

> a private Anatomy Room is to one not accustomed to this kind of manufacture, very irksome if not frightful: Here a Basket of Dissecting Instruments, as Knives, Saws, &c. and there a Form with a Thigh and Leg flayed, and the Muscles parted asunder: On another Form an Arm served after the same manner: Here a Trey full of Bits of Flesh, for the more minute discovery of the Veins and Nerves; and every where such discouraging Objects. So, as if Reason and the Good of Mankind, did not put Men upon this Study, it could not be endured: for Instinct and Nature most certainly abhors the Employment.[28]

Not surprisingly, the anatomists consumed every year dozens of pints of *eau de vie* in their dissections, requiring alcohol, as Bourdelin explained, "not only to apply to the viscera because of the stench, but also for drinking and washing their hands."[29] Unhappily for Jean Pecquet, who regarded the drink as a universal remedy, it became an *eau de mort*.[30]

Incongruous deliveries and offensive odors must have disturbed other users of the Library, and it is not surprising that Cassini banned such activities from the Observatory. Nor were these the only disruptions to scholarly calm. As home to Huygens, Carcavi, Clément, Duclos (then Borelly), and the laboratory assistant, the Library was the scene of daily life, including cooking, parties, quarrels, sickness, and death.

Huygens enjoyed a large apartment, "very noble, and well for Air, upon the Garden"; it included a kitchen, a cellar, and four other rooms. He lived and slept and kept his books in two rooms immediately below the library; his instruments and machines were two floors down. Letters to his family reveal his way of life there. Huygens lived well, decorating the dining room with soft gilded leathers that looked like brocades, keeping a carriage and horses, and hiring a coachman and a cook. He also entertained frequently; a party that lasted until one in the morning included the Perrault brothers and some women who sang and played the harpsichord.[31]

Living as neighbors in the Library brought academicians together socially. During the 1660s, Huygens, Carcavi, and the astronomer Auzout dined, gambled, and conversed.[32] But when Huygens quarreled in 1671 with Carcavi and Carcavi's son because they borrowed his barouche without permission, this diminished camaraderie and threatened to injure the

Plate 3. A Dissection and Microscopic Observations. Huygens uses a lens at the window; Gayant dissects while Pecquet, Perrault, and perhaps Mariotte (holding spectacles on his nose) discuss his work. Colbert observes from the right. (Watson, "Early Days"; engraved by Le Clerc; photograph courtesy of Bibliothèque Nationale, Paris.)

Academy.³³ Work and private life overlapped: Huygens abruptly interrupted his letters home to attend dissections, and his colleagues tried to protect him from scholarly pressures when he was ill. Clément heard Duclos's startling deathbed renunciation of a lifetime's Protestantism and alchemical research, serving less as proxy for the dying man's family than as witness to a political act by an academician whose views were no longer tolerated.³⁴ Royal patronage affected even the academicians' manner of living and dying.

OBSERVATOIRE

The Observatory was as important to the Academy as the Bibliothèque du roi. With its underground caverns, astronomical and meteorological apparatus, and collections of data from scientific expeditions, the Observatory resembled the House of Solomon. Yet it had serious disadvantages. Situated in the countryside at the southern edge of Paris, walled, and monitored by liveried veterans, it was physically remote from other Parisian intellectual centers. Designed by Perrault—who lacked architectural models or astronomical experience and had effected the expulsion of the one academician qualified to advise him—it was unsuitable for Cassini's work and had to be adapted. Originally intended to house all the activities and members of the Academy, it became the preserve of astronomers and mathematicians. Otherwise, only Huygens and Mariotte used it regularly, and even some astronomers preferred to work elsewhere.³⁵

Despite its shortcomings, the Observatory became the second major hub of the Academy. Academicians lived and worked in its cold and drafty quarters: Cassini's children were born there, and nearby Saint Jacques du Haut Pas became the parish church of the mathematicians and astronomers.³⁶ Work on the Observatory began in 1667, and so great was the need for these facilities that academicians began working there in 1668. In 1671 the impatient Cassini moved in, although the Observatory was scarcely ready for habitation or regular observations. Thuret began maintaining its clocks in 1672, and in 1673 Jean Patigny was preparing astronomical engravings there.³⁷ By 1687, when the building was completed and the Marly tower in place, the Observatory had cost more than 720,000 livres (table 2). Altogether its construction and maintenance accounted for 34 percent of the Academy's budget during the seventeenth century (table 17).³⁸

The building and its grounds were more than an astronomical observatory. Like their mythical forebears in Solomon's House, academicians exploited the site for quite varied investigations. Underground tunnels, grottoes, pools, and pits — the remnants of disused quarries — were fortified with walls of cut stone and a statue of the Virgin and then used for experiments and meteorological observations. Behind the building stretched a terrace on which academicians mounted telescope masts, including the Marly tower. The aqueduct of Arcueil, which ran beside the walls, supplied subjects for study. From the staircase, Mariotte and La Hire experimented in 1683 with falling bodies, reproducing their tests at a nearby well. Nearly every day Cassini observed the direction of the wind from nearby windmills and recorded temperature, barometric pressure, and the state of the sky. With Mariotte he compared air pressure in the underground tunnels and on top of the building. From 1683 La Hire examined the declination of the magnetic needle at the Observatory. Sédileau and Cusset studied rainfall with apparatus made by Villette and Hubin that had been installed on the platform-roof of the Observatory (table 3b).[39] The walls and floors of the building recorded data: Cassini made one room a giant sundial and transformed the floor of another into an immense universal map.[40]

Architectural showpiece and symbol of royal patronage, the Observatory attracted many visitors despite its isolation. Germain Brice praised it in his guide book, Blondel included it among Parisian architectural monuments, and Martin Lister and John Locke mentioned it in their journals. A remarkable central staircase connecting the Observatory with its subterranean galleries was admired by such visitors as the duchess of Luxembourg and the prince and princess of Bournonville. The *salle des machines* on the second floor displayed models of machines and military engines, maps, and instruments, many formerly kept in the cramped quarters at the Library. Visitors also inspected the Academy's maps of the moon or of the night sky over Paris and studied mathematics and mechanics, parting with an appreciation of the Academy's observations and practical functions.[41]

The apparatus at the Observatory represented a considerable investment by the crown. Research equipment for the astronomers included pendulum clocks made by Thuret, telescopes supplied by Le Bas, lenses purchased from Borelly, Hartsoeker, and Divini and Campani, a quadrant bought from Picard's estate, an azimuthal circle made by Migon, and various mathematical instruments by Lagny, Le Guern, Gosselin, and Sevin.[42] There were also instruments that recorded knowledge. These included a "talking ephemerides," designed by Roemer and made by Thuret, that demonstrated the motions of the planets according to the Copernican system; a machine that demonstrated the causes of eclipses;

torship; not until the eighteenth century could he revive the astronomical projects that Louvois had halted.

Expenditure on the Academy falls into three research categories and three chronological periods. The research categories are, first, the mathematical sciences, especially astronomy; second, natural philosophy, including anatomy, botany, chemistry, and mineralogy; and third, practical projects, such as mapping, mechanics, and surveying. The financial record scarcely reflects work in theoretical mathematics, however, and yields incomplete information in all research categories. Despite its usefulness, it is only a partial guide to the Academy's program.

The chronological periods correspond to the protectorships of Colbert, Louvois, and Pontchartrain. Colbert determined expenditure from 1666 until his death in September 1683. Louvois became responsible for the Academy the month Colbert died and began to authorize spending on its behalf that year, but he allowed payments for the Academy to fall into arrears. When Pontchartrain took charge in 1691, he demonstrated his good faith by immediately paying overdue items. Colbert and Louvois shared responsibility for expenditure during fiscal year 1683, therefore, while Louvois and Pontchartrain shared it during fiscal year 1691. Despite some ambiguities and lacunae the financial record permits comparison of the three ministerial protectors of the early Academy, and it helps review the early Academy's substantive accomplishments.

COLBERT: THE GENEROUS FOUNDATION

Overview. From 1666 to 1683, Colbert spent 1,578,787 livres, or an average of 87,700 livres a year, directly on the Academy. Of that amount, 643,708 livres, or 41 percent, was for pensions to academicians and their assistants; 713,704 livres, or 45 percent, for the Observatory; and 221,374 livres, or 14 percent, for research (table 14). In addition, the Academy benefited from 207,349 livres spent jointly on it and other royal buildings and institutions (table 12).

The mathematical sciences were more costly than either natural philosophy or practical projects (table 14). Colbert spent 92,322 livres, or 42 percent of the research budget, on the mathematical sciences. By comparison he spent 56,110 livres for natural philosophy and 51,483 livres for practical projects, which accounted for 25 percent and 23 percent, respectively, of the research budget. The remaining 10 percent, or 22,231 livres, went toward expenses whose precise purpose is unknown (table 11).

The Mathematical Sciences. The mathematical sciences dominated not only the Academy's budget but also its membership. Twenty-two of the thirty-six academicians Colbert appointed, or 61 percent, were astronomers, geometers, or mechanicians. Colbert particularly favored the astronomers, who needed costly apparatus (table 3a, d), expeditions (table 4b, d), and engravings (table 9a–c). Many of the small expenses also paid for their work (table 11b–c).

The silver planisphere that Cassini and Butterfield made exemplifies the Academy's attitude toward theoretical astronomy. On its back was a mechanism that illustrated the Copernican, Tychonic, and Ptolemaic systems. The point was to show how similar the three systems were despite their very different hypotheses. This instrument symbolizes the fictionalist attitude of most academicians toward astronomical theories. A recurring theme in their writings is that it made little practical difference whether one accepted a heliocentric or a geocentric universe. This indifference to the underlying physical implications of competing astronomical theories helps explain why Cassini missed the point of reports from his assistants who had to shorten their pendulums near the equator. The Academy worried about publishing raw data and leaving hypothesizing to others, who would thereby unfairly gain credit to the detriment of the Company. But pragmatism and cosmological agnosticism led to just such a result and also turned the astronomers toward the practical projects that appealed to the crown.

Natural Philosophical Research. Natural philosophy's share of the research budget did not reflect its share of the membership. Fourteen academicians, or 39 percent of Colbert's appointments, were botanists, anatomists, chemists, or natural philosophers. They controlled only 25 percent of the research budget, however, partly because they did not require expeditions or precision instruments. The laboratory (table 7), anatomical research (table 6a), and engravings (table 8) accounted for 88 percent of their funds and supported Perrault's comparative anatomy of animals, Dodart's natural history of plants, Duclos's analyses of French mineral waters, and other inquiries. Small expenses (table 11a, c) also benefited these projects, while the costly burning mirror (table 3c) was peripheral to the Academy's research but interesting to the king.

Practical Projects. The Academy was meant to be useful. Academicians expected their natural histories to improve medicine by correcting and amplifying pharmacopoeia, by clarifying human anatomy through comparative studies, or by identifying the components of mineral waters. Astronomy and mathematics too were no purely theoretical exercises but also the handmaidens of navigation and cartography. Colbert brought

Cassini from Italy and Huygens from Holland partly because they might solve the problem of determining longitude at sea, one of his interests as secretary of the navy and as champion of expanded overseas trade.[1] It is no surprise, then, that Colbert devoted a significant portion of the research budget to utilitarian interests.

There were four practical projects—technological, cartographic, architectural, and hydraulic—each of which the Academy sought to put on a sound theoretical foundation. Each also reflects the interests of the patrons. Colbert, for example, enlisted the Academy to reform industrial, agricultural, and military technology by charging members to study theoretical and applied mechanics. This work took three principal forms: academicians collaborated on a book about the principles of mechanics and their applications, they assessed inventions, and they collected models of machines (table 5).[2] Other projects had more immediate appeal to a vainglorious monarch. Architectural display, for example, glorified the reign. Thus, Colbert appointed to the Academy Perrault and Blondel, rival architects who designed several royal monuments, and Perrault's edition of Vitruvius illustrated classical architectural principles with buildings constructed by the crown (table 10b).[3]

The most important practical projects that Colbert initiated, however, were cartographic and hydraulic. Maps of France were notoriously inaccurate, yet Louis took pride in his kingdom and wanted to know its exact extent and Colbert needed correct maps to assist his economic and fiscal reforms. Colbert, therefore, appointed to the Academy astronomers and practitioners who could address this problem. They worked simultaneously on a world map and a map of the kingdom. For the former, they compiled the coordinates of cities and towns on both sides of the Atlantic, around the Mediterranean, and in the Far East, inking the sites onto the map on the Observatory floor. For the latter, they extended the meridian in France and mapped the Atlantic coast and the *généralité,* or administrative district, of Paris and its environs (tables 4a, 10a). As trial balloons for the map of France, these undertakings had three main advantages: the *généralité de Paris* was a small but central part of the kingdom, the Academy had established the meridian there (table 4b), and academicians could demonstrate vividly that extensive corrections to existing maps were necessary. Although these preliminary efforts were successful, they were also harbingers of the huge expenses that would be necessary to complete the map of the kingdom. Mapping the environs of Paris, for example, cost more than 21,000 livres, or 10 percent of Colbert's total spending on research (table 14), and took ten years to complete. Bringing the larger project to fruition would require further extension of the meridian, triangulation along and then east and west of it, and topographic surveying, all associated with enormous costs. But from the mid-1670s, as Louis's wars deflected funds from the Academy, adequate funding was no longer available. For the remainder of the century academicians were armchair cartographers, correcting their world map with coordinates sent from abroad, teaching others to use the data and methods academicians had developed, and awaiting permission to revive work on the meridian.[4]

The hydraulic project provided direct support to king and court, for its goals were to guarantee the supply and quality of water for Versailles and to design fountains. The Academy surveyed rivers, analyzed the chemical composition of waters, studied hydraulic machines, and identified promising sites for aqueducts; members also developed the principles of decorative fountains for the gardens at Versailles. The work was costly, requiring surveying instruments, travel, and overnight accommodation. Academicians ran up bills of nearly 2,000 livres for incidentals, 3,500 livres for horses and carriages, and 1,335 livres for room and board at an inn in Versailles while they "worked to verify the surveying for the construction of aqueducts in the environs of Versailles" (tables 3b, 4c, 10e).[5]

Colbert relied on the Academy as a research institution and as a source of practical skills. Its members were physicians, surveyors, architects, and engineers, eager to improve those disciplines.[6] Pensions covered both theoretical and practical work, so that academicians were inexpensive consultants, even when the crown paid bonuses. The Academy could consider the work from several different angles, as its efforts on the water supply of Versailles reveal. From the crown's point of view, their interdisciplinary skills, commitment to theory and practice, and dependency on the king made academicians the ideal consultants.

Reimbursement of Academicians. Although the royal treasury underwrote the costs of the Academy's research, academicians and their suppliers usually had to make extensive outlays and then request reimbursement. Only rarely did the crown advance funds, as when Mariotte got 200 livres for experiments related to the water supply of Versailles in 1682 (table 4c), or when Richer, Meurisse, and Deglos prepared for voyages (table 4d). Bourdelin's notebook offers a glimpse into the standard practices. He recorded the details of each purchase and every few months submitted a formal request, prepared by a notary, for repayment. Sometimes the crown paid him and his suppliers directly, but often it paid intermediaries such as Carcavi (or Homberg and Fontenelle in the 1690s).[7]

Academicians must have had some general authorization to purchase for the Academy, but they needed special permission for certain items—for instance, engravings of plants—and requests for reimbursement had to provide details. Certain academicians acted as purchasing agents for the Academy. Thus, Perrault spent 4,000 livres, much of it probably for the natural history of animals (table 11a, 1674), and when Du Hamel went to London he bought 500 livres' worth of books and microscopes for the Academy (table 11a, 1669). But Couplet bore the heaviest responsibility for the institution's finances. He paid for many of the small expenses, purchasing animals for dissections, machines and apparatus for experiments, and seeing to repairs of equipment and physical plant at the Observatory (table 11). Finally, Nicolas Clérambault (table 11a), Carcavi, and Thévenot also bought for the Academy, sometimes mingling the Academy's and the Library's small expenses (table 12c–d).

Summary. Colbert spent about 250,000 livres, or an average of 12,300 livres a year (table 14), on the Academy's research program. Expenditure fluctuated from year to year. It peaked from 1667 to 1672, because of astronomical expeditions, the map of the environs of Paris, and engravings. It plummeted from 1678 until 1683, because of the Dutch wars. During the 1660s and early 1670s, Colbert paid for expenses soon after they were incurred, but by the late 1670s, payments began to fall one or two years behind.

The Academy's research budget reflects Colbert's preferences. Astronomy and practical projects were clear favorites. The Academy was a reservoir of talent on which Louis and Colbert drew for technical expertise, and academicians themselves usually found scientific merit in these practical challenges. Far from being a disinterested and unalloyed supporter of basic research, Colbert demanded both practical and theoretical returns on the king's investment. By tapping the Academy for its technical advice and by reducing expenditure on the Academy during the last years of his ministry, Colbert set precedents for his successors that were more influential than his initial generosity.

LOUVOIS: DECLINING INTEREST AND SUPPORT

Overview. Louvois was ministerial protector of the Academy from September 1683, when he bought the controllership of bâtiments from Colbert's son, until his death in July 1691. During these nine years he reduced the size and budget of the Academy, appointed representatives to convey his wishes, intervened to shape research, and finally lost interest in the Academy as his personal standing with the king deteriorated. Although Louvois has been blamed for stressing utility over theory, he differed from Colbert and Pontchartrain only in degree. The damage Louvois did to the Academy came from his relative lack of interest in its work, from his reduction of financial support, and from his attempts to direct the methods of research, at least as much as from any imposition of utilitarian goals.

Louvois spent 238,354 livres, or an average of 26,484 livres a year, directly on the Academy. Of that amount, 171,833 livres, or 72 percent, was for pensions to academicians and their assistants; 12,335 livres, or 5 percent, for the Observatory; and 54,185 livres, or 23 percent, for research (table 15). In addition, the Academy benefited from 98,837 livres spent jointly on it and other royal buildings and institutions (table 12).

Natural philosophy cost more than the mathematical sciences and practical projects combined (table 15). Louvois spent 29,380 livres, or 54 percent of the research budget, on natural philosophy. In contrast he spent 16,510 livres for the mathematical sciences and 6,462 livres for practical projects, which accounted for 30 percent and 12 percent, respectively, of the research budget. The remaining 3 percent, or 1,832 livres, went toward transport of animals for dissection, repairs to equipment and lodging, and other expenses whose precise purpose is unknown (table 11). The institution's finances reveal a new set of ministerial preferences under Louvois, who redirected the Academy's efforts toward natural philosophy.[8]

Louvois spent less each year on the Academy than had Colbert. Colbert spent an average of 35,762 livres a year on pensions, Louvois only 19,093 livres. Colbert committed 39,650 livres a year to the Observatory, Louvois 1,371 livres. Colbert paid 12,300 livres a year for the Academy's research, while Louvois paid only 6,020 livres a year (table 17). These raw comparisons exaggerate Louvois's economy, however, for he took over an institution with quarters, equipment, and publications, whereas Colbert built the Academy from nothing.

Louvois inherited an intrinsically cheaper institution, but he also deliberately reduced the Academy's budget in several ways. To minimize pensions, the largest single expense once the Observatory was completed, Louvois diminished the number of academicians and the levels of their pensions. In addition, he simply halted cartographic expeditions[9] and research on determining longitude at sea. He also canceled plans to publish the Academy's astronomical and anatomical treatises.[10] Finally, he delayed payments, so that when he died a substantial debt had accumulated for which Pontchartrain became responsible. Academicians resented the cessations, reductions, and delays and blamed Louvois and his wars for damaging the Academy.[11]

The Mathematical Sciences. Although Louvois shifted more of the Academy's financial resources to natural philosophy, he appointed more mathematicians than natural philosophers. Thus sixteen of twenty-seven academicians during his protectorship were active principally in the mathematical sciences. Louvois also spent more on mathematical and astronomical instruments, which cost 13,440 livres, than on any other subcategory of research (table 15). Most of the instruments, however, equipped the Jesuit missionary-scientists whose researches in the Far East the Academy sponsored (table 3e). Louvois also paid for the last of the work on the meridian (table 4b) and for small expenses of the Observatory (table 11b).

Their expeditions canceled, the astronomers worked at the Observatory. They prepared earlier research for publication; studied eclipses, sunspots, the satellites of Jupiter and Saturn, and the parallax of Mars; analyzed reports from the provinces and abroad; and tested objective lenses of great focal lengths. Cassini wrote on the libration of the moon and the history of astronomy, and he made some observations in the north of France during the late 1680s.[12]

Louvois made Jesuit missionaries the Academy's proxies in the Far East. The Academy trained the Jesuits, who used their mathematical and astronomical knowledge as a passport into foreign lands. To repay their debt, the Jesuits sent the Academy data: measurements of latitude and longitude, astronomical observations, and reports about flora and fauna, calendars, alphabets, and numerical systems. Hoping to get accurate calculations of longitude for their world map, academicians had emphasized proper astronomical techniques, especially for observing the satellites of Jupiter. The crown fitted out the Jesuits lavishly with instruments (table 3e): the China mission of 1685 took along "books, mathematical instruments, pendulum clocks and other kinds of clocks," while a second group destined for China carried a large microscope with three lenses, two burning mirrors, one thermometer, one barometer, a mounted telescope with thirteen lenses, two pendulum clocks, and some mathematical instruments. The Siam and India mission was equipped with eighty-four telescope lenses, three burning mirrors, and other instruments as well, to judge from its reports. Nicolas Hartsoeker supervised production of the glassware for all these instruments.[13]

Unfortunately several obstacles impeded the scientific work of the

Jesuits. On the one hand, the voyages were long and unpleasant, the missionaries became ill, and two were imprisoned by the Dutch; much of their time was occupied in learning oriental tongues and in preaching. Some of the Chinese data were destroyed when the Dutch confiscated them and when a French ship was lost.[14] On the other hand, when French Jesuits extended the protection of the French king to other members of their order in the Far East, the foreign Jesuits sent their observations to the Academy as well. Despite problems the Jesuits provided much useful information in the 1680s and 1690s. From it Thomas Gouye, after consultation with academicians, edited two treatises on astronomical and mathematical topics and de Beze prepared a short pamphlet on flora and fauna.[15] These works were published under Louvois and Pontchartrain. Designed initially under Colbert to supplement the Academy's own cartographic voyages, the partnership between the Jesuits and the Academy became a substitute for the Academy's own expeditions.[16] Under Louvois the collaboration became an inexpensive vestige of the Academy's more ambitious projects, the legacy of Colbert's practice of enlisting officials to assist the Academy.

Natural Philosophical Research. Natural philosophy commanded 54 percent of the Academy's research budget, but only 41 percent of the members under Louvois. Although Louvois spent more a year than Colbert on natural philosophy, he wasted one-third of the money on mediocre burning mirrors (tables 3c, 15). Despite that failure, however, Louvois favored natural philosophy in several ways. Planning to publish installments of the *Histoire des animaux* and *Mémoires des plantes* (table 8), he appointed a new anatomist, pensioned the engraver Chastillon for his services to the Academy, and allowed botanists to revise existing plates. He also increased Borelly's pension and supported the *petit jardin* (table 6b), and he personally instructed the Academy as to the conduct of research on plants. However, some good intentions were subject to retrenchment: he abruptly canceled publication of the *Histoire des animaux* in the late 1680s because he was absorbed in the war efforts and lost interest in the Academy.[17]

In addition to dissecting animals from the menagerie and preparing the ill-fated third volume of the natural history of animals, anatomists pursued more focused individual studies. They examined the eye and the ear, the circulation of the blood, digestion and the digestive tract, respiration, and the persistence of nervous reactions in dead animals. They also performed autopsies on several persons, young and old, military and civilian, including the painter Le Brun and their own Mariotte and Perrault.[18]

For the laboratory, which was central to the Academy's natural historical research, the early years under Louvois were a period of crisis, caused by the infirmities of individual chemists. Duclos was disaffected by Dodart's appropriation of the natural history of plants, his health was failing, and as a Protestant he was out of favor with Louvois, who did not pension him after 1684. Bourdelin worked in his own laboratory instead of the Academy's on grounds of age and ill health.[19] As a result, his expenses and reimbursements fell drastically (table 7a), the latter because Louvois refused to pay for Bourdelin's laboratory assistant. Bourdelin regretted that the right to work at home was tempered by an increased financial burden on himself:

> It is noteworthy that I have been allowed to work at home for the Academy. At the same time, M. de La Chapelle has told me twice that I will not be paid for an assistant, even though I need one as much as if I directed the laboratory. But it is necessary to put up with this.[20]

The dwindling finances of chemical research chart the decline of Bourdelin but not necessarily of the laboratory. Borelly stepped into the breach, taking over the Academy's laboratory from Bourdelin and moving into Duclos's apartment, but no record of his expenses is known.

Practical Projects. Although Louvois is reputed to have promoted utilitarian research at the Academy, he spent substantially less on it—both annually and as a percentage of the total he allocated to research—than had Colbert (tables 14, 15). Louvois continued only one of Colbert's projects generously: the survey of rivers for the water supply at Versailles (tables 3b and 4c). This he supervised attentively, promoting both its theoretical and its practical aspects. At his request, the Academy surveyed and planned diversions of rivers, sought the origins of rivers in springs and rainfall, studied hydraulic machinery, and translated Frontinus's treatise on Roman aqueducts.[21] Louvois's determination to water Versailles encouraged the Academy to study hydrology and bore fruit in papers by Sédileau, Varignon, and La Hire during the 1690s.[22]

Louvois compromised two other projects—maps and machines—that Colbert had initiated. The former he effectively gutted by canceling the extension of the meridian. As for the latter, Louvois was ambivalent. During his protectorship, the Academy assessed fewer inventions than it had previously,[23] and the collection of models disappeared as a distinct category of expenses in the buildings account (table 5). Moreover, Louvois refused to mount a public exhibition of the Academy's collection.[24] On the other hand, he pensioned the engineer Dalesme as inventor to the Academy (table 1, ii),[25] which continued to study new military, navigational, manufacturing, and timekeeping devices. Models of some machines were deposited at the Observatory, so that the collection of models grew inexpensively during the 1680s. Above all, Louvois focused the Academy's attention on hydraulic technology, in order to advance his pet project of supplying Versailles with water.[26] In conclusion, Louvois defined the Academy's technical consultancy more narrowly than had Colbert, and he stressed the water supply of Versailles to the near exclusion of other practical projects.

Summary. Louvois's protectorship was anomalous for the Academy. He maintained the institution, appointed a few new members, and added the Marly tower to the Observatory. He funded astronomical observations, research on natural history, and practical projects. His utilitarian demands on the Academy were single-minded, however, and he economized on the pensions of some new members and reduced expenditure in nearly all categories of research. Only for the natural histories did Louvois's average annual expenditure exceed that of Colbert. During his ministry, the imbalance of expenditure on astronomy and natural history shifted, with natural history receiving a larger share of the Academy's financial resources. Yet he undermined his own initiatives by canceling publications on anatomy and astronomy. Finally, Louvois injured the morale of the Academy by intervening ineptly into the research program. The problem was not that he altogether lacked interest in the Academy's work. Rather he was impatient for practical results. Overly close supervision, narrow goals, and reduced funding, followed by indifference in his final years, led to the decline of the Academy in the late 1680s.

PONTCHARTRAIN: A PENURIOUS REVIVAL

Overview. When Pontchartrain assumed control of the Academy in 1691, it was badly demoralized.[27] Members were owed their pensions for 1689 and 1690, treatises actually in press had been suppressed, and the number of members had fallen. More sympathetic to science and technology than Louvois had been, Pontchartrain tried to revive the Academy by immediately appointing energetic and highly qualified savants, approving publications, and resuming pensions and research subventions. The War of the League of Augsburg, however, allowed him to fund the Academy only sporadically and at minimal levels throughout the 1690s.[28] As a result, financing for the Academy declined steeply. Pontchartrain has nonetheless been regarded as a champion of the Academy, because he preserved and revived it during difficult times, made shrewd appointments, and launched its most ambitious publishing program.

From 1691 through 1699, Pontchartrain budgeted 322,849 livres, or an average of 35,872 livres a year, directly on the Academy. Of that amount, 286,017 livres, or 88.6 percent, was for pensions to academicians and their assistants; 2,873 livres, or nearly 1 percent, for the Observatory; and 35,960 livres, or 10.5 percent, for research (table 16). Because the crown converted many pensions into annuities, however, the actual direct cash outlay was less than the amount budgeted (table lc). In addition, the Academy benefited from 80,887 livres spent jointly on it and other royal buildings and institutions (table 12).

Natural philosophy was more costly than the mathematical sciences or practical projects (table 16). Pontchartrain spent at least 10,747 livres, or 31.6 percent of the research budget, on natural philosophy. In contrast he spent 4,677 livres for practical projects and 2,811 livres for the mathematical sciences, which accounted for 13.8 and 8.3 percent, respectively, of the research budget. The remaining 46.3 percent, or 15,725 livres, went toward small expenses, a category that defies elucidation; most of this sum was paid in 1699, suggesting that the crown delayed reimbursing academicians for months or years (table 11). Since small expenses account for so much of Pontchartrain's research subvention, it is dangerous to generalize from the financial records about his preferences. Other indicators suggest that he treated the Academy evenhandedly and channeled academicians' energies into publishing more than into research.[29]

Although Pontchartrain continued to fund the Academy, no mean feat during what was arguably the worst decade of the reign, he did so at markedly reduced levels. He pensioned a smaller proportion of academicians and drastically reduced research subventions, even though he increased membership. He actually spent less in each category than had either Colbert or Louvois. First, while Pontchartrain budgeted an average of 31,780 livres a year for pensions — as compared with 35,762 livres under Colbert and 19,093 livres under Louvois — he actually disbursed less. That was because many academicians had to take their pensions for 1692, 1693, 1694, and 1695 as annuities. Pontchartrain's average annual outlay for pensions was, therefore, smaller than normal.[30] Second, Pontchartrain spent on average only 319 livres a year on the Observatory, in contrast with 39,650 livres under Colbert and 1,371 livres under Louvois. Colbert built the Observatory, Louvois added the Marly tower to it, and Pontchartrain maintained it and paid for the salary and livery of the porter. Third, Pontchartrain spent only 3,773 livres a year for research, by comparison with Colbert who spent an average of 12,299 livres a year and Louvois who spent 6,020 livres (table 17).

Pontchartrain revived the Academy as inexpensively as possible. From 1691 until 1694 he made new appointments and encouraged academicians to publish. In 1699 he sponsored a formal *règlement* and, finally, a new infusion of money, and in the following year he authorized Cassini and his team of astronomers to extend the meridian. Before 1699, Pontchartrain applied the limited funds available to pensions, publication, and maintenance. After 1699, he was able to expand the research program.

The Mathematical Sciences. Like his predecessors, Pontchartrain appointed more mathematicians than natural philosophers. Thus twenty-four of thirty-nine academicians, or 62 percent, were active principally in the mathematical sciences. Like Louvois, however, Pontchartrain apparently spent little on their research, focusing on the maintenance of scientific instruments (table 3d). The only known special purchases are six telescope lenses made by Nicolas Hartsoeker and a pendulum clock "supplied to Sr Couplet the son for the observations he has been ordered to make in Portugal" (table 3a).[31]

The principal new impetus in the mathematical sciences came from infinitesimal calculus, which the mathematicians debated. The astronomers continued to work as they had done under Louvois. They observed eclipses and the satellites of Jupiter, compared eastern and western calendars, catalogued fixed stars, and calculated solar and lunar diameters. Cassini and his son made observations and studied the declination of a magnetic needle during their travels in Italy, France, Holland, and England from 1694 to 1698. Above all, the astronomers awaited permission to extend the meridian. In keeping with Pontchartrain's policy of publishing as much as possible, Cassini's memoirs, the reports of Jesuits in the Far East, and several astronomical articles were printed.[32]

Natural Philosophical Research. Only fourteen of thirty-nine academicians, or 36 percent, were natural philosophers. But Pontchartrain seems to have spent more on their work than on the mathematical sciences. The record of expenditure is far from complete, however, with respect to the laboratory, the *petit jardin*,[33] and engravings (tables 6b, 7, 8, and 12i). Payments for botanical illustrations, for example, reflect neither the forty to sixty-nine plates, which normally cost 90 livres apiece, that Chastillon completed, nor the plates for Tournefort's *Élémens de botanique,* rumored to have cost 12,000 livres. If engravings actually cost 15,000 to 18,000 livres more than the treasury accounts reveal, then Pontchartrain's average annual expenditure on natural philosophy was closer to that of Louvois and Colbert.[34]

New academicians scrutinized previous chemical research. They also studied mineralogy (table 6d) and tried fresh approaches to botany, notably Tournefort's influential classification of plants. Otherwise the Academy's natural philosophy continued under Pontchartrain much as it had during the previous two decades. Anatomists resurrected the third installment of their *Histoire des animaux*. They revised plates (table 8b) and dissected corpses from the menagerie at Versailles or from the Hôtel des invalides. Du Verney and Méry published several articles reflecting their dissections during the 1680s and 1690s, and the Academy debated their conflicting views about the circulation of the blood in the fetus. Homberg earned the gratitude of his colleagues for enlivening meetings with papers and demonstrations and with recollections from his travels.[35]

Practical Projects. Utilitarian problems continued to interest academicians, who investigated hydraulics (table 5), mapping, new inventions, and military technology.[36] The crown subsidized these studies modestly, however, in order to launch two new projects—writing a natural history of arts and crafts and designing a new typeface for the Imprimerie royale—that complimented Pontchartrain's program of publishing. At first Bignon and Pontchartrain founded a separate Compagnie des arts et métiers to undertake this double mission. But in 1699 the Academy absorbed the Compagnie and its work, and that year alone the crown spent nearly 4,000 livres on engravings of arts and crafts (table 10d). By 1700 when Pontchartrain resumed work on the meridian, he had revived the two Colbertian projects that had lapsed under Louvois—cartography and mechanics—and thus reestablished the Academy's utilitarian program.[37]

Summary. During the 1690s research expenditure primarily maintained buildings and equipment or continued older projects. What Pontchartrain paid for research bore little relation to what academicians published, which mostly represented work done earlier or outside the institutional structure of the Academy. But the Academy bore impressive fruit under this parsimonious management, with academicians publishing in mathematics, botany, astronomy, and other fields.

The Academy suffered from reduced funding and poor morale during the 1690s, but it was also undergoing basic changes in its very conception. Under Colbert academicians worked principally in teams on long-term projects begun with the assurance of continuity and support. But facilities were cramped, funding diminished, and collaborative research faltered; these trends emerged late in Colbert's protectorship and Louvois and Pontchartrain exacerbated them. Under the circumstances the institution altered. As individual research became more prominent, the functions of meetings changed: they lost their importance for proposing research, debating hypotheses, and demonstrating experiments; instead academicians used them to referee manuscripts for publication. Thus the costs of research during the 1690s reflect the two inconsistent ministerial policies of austerity and rejuvenation, against the background not only of foreign wars and domestic insolvency but also of a changing institution.

CONCLUSION

The seventeenth-century Academy cost its royal patron at least 2,000,000 livres for pensions, the Observatory, and research subventions. It also benefited from nearly 400,000 livres spent on it and other royal establishments jointly. In principle, the Academy cost on average about 63,000 livres a year, with pensions representing the lion's share at 32,400 livres; the Observatory was in second place at 21,440 livres, and research was the least costly at 9,100 livres a year.[38] In fact, from Colbert to Pontchartrain the Academy's budget declined markedly, and under Pontchartrain academicians saw their pensions become annuities. Both trends had deleterious consequences for academicians personally and for their research.

Royal funding influenced the nature of the Academy's research as much as it did the character of the institution itself. Access to royal funds gave the Academy an advantage over other scientific societies, because it could mount ambitious collaborative projects. Yet the result was to limit the institution's scientific vision. Academicians and their protectors believed that theory would improve practice and that accumulating data was the necessary preliminary to hypothesizing. But collective projects gained a momentum of their own, so that theorizing was sometimes neglected in favor of practical applications. Furthermore, dependence on royal funding made research more vulnerable to ministerial interference, and this could be damaging when the protector failed to appreciate scientific priorities.

Finances, however, tell only part of the story. They mostly reflect the expensive, collaborative projects, but academicians also pursued more modest research as individuals, albeit with moral and material support from the Academy. Indeed, Carcavi and Huygens boasted that members "did not make enquiry into any one subject in particular but every one took unto his examination what suited best with his own fancy and genius."[39] How may such freedom of choice be reconciled with the facts of corporate planning and ministerial control? Three general points resolve this paradox. First, academicians agreed among themselves about the important questions, and they were in sympathy about the general aims of the institution.[40] Second, because academicians believed that theoretical science should have practical benefits for society, they shared with their patrons various utilitarian expectations of the Academy. Third, the distinction between official and individual projects enabled an academician to work on several problems at once; team members cooperated on descriptions of plants or dissections, but an academician might also pursue specific interests such as the circulation of sap or the nature of hearing. This flexibility was available more readily in natural history than in astronomy, where the hierarchy of the Academy restricted the choices of some academicians, especially the students, who principally assisted others. Nevertheless, academicians used the company as a resource for work that interested them, and thus patronage for official projects also protected the individual projects of academicians.

To understand the Academy, however, it is necessary to look behind the scenes, to explore what royal funds actually bought, to observe the institution at work. The Academy's research on plants exemplifies many characteristics of the institution as a whole. It reveals conflicts between individual and collaborative projects, between theoretical and practical expectations, and between academicians and their protectors. Because botany was in flux, academicians were often uncertain of themselves: they had to adjust to the failure of their theories, test new instruments, and explore new analogies. Their research also offers a glimpse into the relations of this elite and somewhat secretive institution with the larger scientific and lay community. Thus, botanical research, a neglected but important aspect of the scientific revolution, may serve as a barometer of the institution as a whole.

PART III
Botanical Research at the Academy

CHAPTER 6

The Natural History of Plants: Rival Conceptions

The Academy was the realization of the House of Solomon, an instrument of royal propaganda, the hub of a sociopolitical network of influence. But it was more. It was a catalyst for thought whose facilities, customs, and corporate self-consciousness influenced the work its members performed. The institution cannot be understood apart from its scientific research and writings, for these were its ostensible raison d'être. Whatever other functions the Academy had, a satisfactory scientific performance was a necessary condition for its survival.

The Academy's interests are so broad as to elude detailed analysis in any one study. Concentrating on a single facet of its research, however, clarifies both the discipline and the institution. Botany offers a particularly rewarding case, for under the influence of chemistry, microscopy, anatomy, and physiology, it was changing. By academicians' own standards, moreover, much of their botanical research was a failure, and the historian can often learn more from failures than from successes. Since some of the Academy's work on plants attracted the concern of its protectors, it clarifies the direct and the indirect effects of patronage. Finally, with respect to organization of research and modes of reasoning, studies of plants resemble other projects at the Academy and thus offer a key to the institution as a whole.

CHANGING WAYS OF THINKING
ABOUT PLANTS

Academicians thought of research on plants as being of two types: descriptive natural history (*l'histoire*), which will be discussed in chapters 6 through 8, and explanatory natural philosophy (*la physique*), which will be discussed in chapters 9 through 12.

The concept of natural history goes back at least to Aristotle, whose *Historia animalium* lays out what its author knows about animals, organizing that knowledge into such categories as number and type of limbs, mechanisms for eating, and mode of reproduction. Aristotle offers a series of generalizations, each modified by exceptions and strengthened by comparisons. Subsequent natural histories always contained these two elements: they enumerated the pertinent facts and they generalized in order to organize the facts. Aristotle's book introduced his treatises on the functions of the various parts of the body, reproduction and generation, and other aspects of a natural philosophy of animals. It was the necessary preliminary to causal analysis. But the blend of assumption and generalization in Aristotle's work, as in the natural histories of later authors, betrays a pattern of causal thought embedded in the method itself.

Surveying the two-thousand-year-old tradition of natural histories, Bacon tried to clarify their uses and limits. Perhaps in revulsion against the magical or superstitious element found in many of them, he declared that a good natural history should present fact shorn of explanation. A natural history of the world would enumerate all observable phenomena, category by category (for example, winds, heat and cold, plants, animals, and minerals). Only when savants had compiled this information could they ascertain the underlying causes of phenomena, that is, examine the natural philosophy of the world. Given the immensity of the first task, an ideal Baconian approach would make it difficult ever to reach the second stage.

Some natural histories, consistent with Bacon's recommendation, did little more than illustrate and describe flora and fauna without explaining their behavior or nature. Bauhin's *Pinax*, for example, the most comprehensive seventeenth-century guide to plants, described the external appearance, cultivation, and uses of each plant. Zoologists, tempted by analogies between human and animal behavior and impressed by the lessons of comparative anatomy, went further. In his ornithology, for example, Aldrovandi not only portrayed the skeletal structure and reproductive organs of birds but also discussed the development of the chick in the egg, the roles of male and female in reproduction, and the sexual mores of fowls. For Aldrovandi, knowing animals entailed knowing their anatomy and physiology and explaining their behavioral characteristics. Aldrovandi's and Bauhin's different methods show how the natural histories of plants and animals diverged at the beginning of the seventeenth century. Zoological research was prompted primarily by comparative anatomy, while the principal incentive to study plants was pharmacological, so that books on plants tended to be practical manuals.

In the course of the century, however, several influences altered ideas about how to study plants. The conceptions of natural history and natural philosophy changed, bringing botanical and zoological research closer in intent and method. First, the number of known plants grew quickly, making new compendia necessary. Second, among the species discovered in the new world were sensitive plants (fig. 1), which challenged the old Aristotelian distinction between plants and animals, for they moved when touched. Third, aesthetic appreciation of plants and gardens as objects of beauty was developing, and with it a desire to collect botanical illustrations. Insofar as this change represented a taste for plants on their own merits — and not as symbols or simples — it also represented a new way of thinking about plants that would affect botanical studies. Fourth, by redefining the differences between humans and animals, mechanistic theories created a greater incentive to test assumptions about plants and animals — especially the view that plants and animals were analogous in many respects — by searching for the limits of their similarities. Fifth, savants tried to put their causal accounts of the universe on a new footing. Astronomers sought a new celestial mechanics and developed a mathematical key to the language of the universe, while students of terrestrial phenomena turned to chemical and mechanical explanation to account for animal and vegetable processes. Sixth, as new scientific fields emerged and the interdisciplinary character of scientific inquiry was placed on a firmer footing, traditional fields were redefined and theories or methods that developed in one area were applied to another.[1]

All of these factors influenced botanical research, which emerged in the late seventeenth century as a more independent field of study. From the appearance of Bauhin's work until the 1660s, no major treatises on plants had appeared. But during the 1660s Robert Hooke included plants in his *Micrographia,* Johann Daniel Major suggested that sap circulated in plants like blood in animals, and Robert Morison and John Ray began a new assault on the problem of classification. By the 1670s Nehemiah Grew was publishing anatomical studies of roots and stems, and Marcello Malpighi took up these inquiries in the 1680s.

The Academy was thus founded at a time when botanical research was in flux, exhibiting at once conservative and innovative elements. The Academy's own projects reflect both tendencies. Academicians settled quickly on publishing a definitive natural history of plants, to be produced as a team effort. This study was old-fashioned and stressed descriptions,

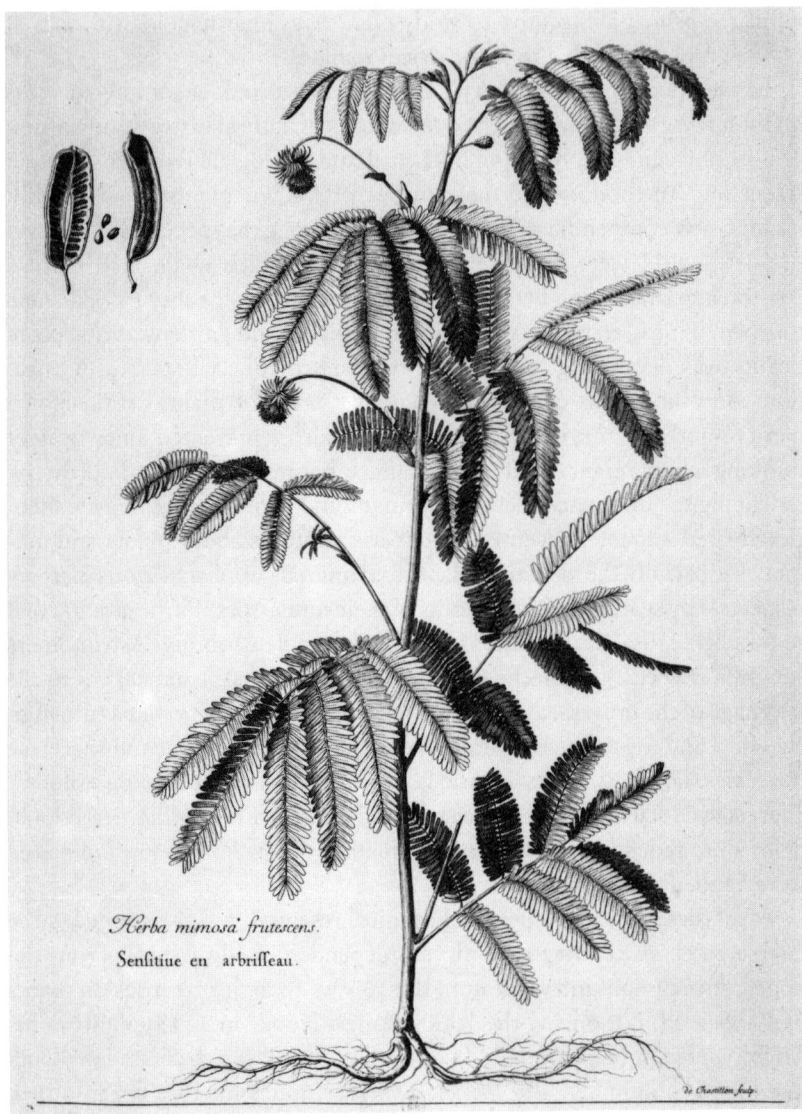

Fig. 1. Herba mimosa frutescens/Sensitive en arbrisseau. (From *Estampes;* drawn and engraved by Chastillon; photograph courtesy of Bibliothèque Nationale, Paris.)

lists of synonyms and sources, explanations of medical uses, tips on cultivation, and illustrations, all reminiscent of Bauhin. At the same time, however, academicians introduced new elements such as the chemical analysis of plants. The blend of old and new features, the research methods chosen, and the roles of the patrons and of the institution all contribute to the story of the Academy's natural history of plants. By and large, that story is one of failure. Academicians recorded hundreds of pages of botanical notes and compiled more than twenty volumes of notebooks from chemical experiments. They drafted assorted chapters, prepared more than three hundred engravings, and wrote several books. But they never published the natural history as planned.

The Academy's natural history of plants failed for four reasons: intellectual, natural, accidental, and institutional. The project was intellectually ambitious, its scope and style innovative but overly inclusive. Academicians never focused their inquiry adequately but studied plants rare and common, medicinal and edible, French and foreign. They disagreed about how to describe plants—where to start, what to emphasize, and how to balance bookishness and observation—and they were sometimes undecided whether a specimen corresponded with a plant already described in the literature. Natural obstacles, such as obtaining and cultivating uncommon plants, also taxed their ingenuity. Since they did not keep a herbarium of dried specimens, they could not always check their descriptions or illustrations. Such intellectual and natural problems were not peculiar, however, to academicians. Anyone writing a natural history of plants faced these difficulties, and academicians' solutions and mistakes resembled those of their contemporaries.

The Academy's project was distinctive in certain respects. It was to include chemical analyses of the plants. It was the work of an institution, not an individual. It suffered from editorial rivalry, uncertain funding, and theft. The patronage that made the project possible also undermined its completion. These special features of the project contributed to its failure. Academicians were able to isolate the most treacherous intellectual problem—chemical analysis—from the project, but they could not prevent the accidents and institutional problems that damaged their natural history. To understand how all these factors affected the project, it is necessary to know why the Academy decided to prepare a natural history of plants at all, how innovative the project was, what was the institutional character of members' efforts, and how patronage influenced their work.

PROPOSALS FOR A NATURAL HISTORY OF PLANTS

Between 1550 and 1700, the number of known plants quadrupled while the number of botanical compendia declined.[2] If only to take account of discoveries, a new natural history of plants seemed a necessity in the late seventeenth century. In 1674 John Ray pointed out the absence of a "general History of Plants." He complained that in order to have the available botanical lore in a single work, it would be necessary to combine the publications of Bauhin, Columna, Alpin, Cornut, Parkinson, Margrave, Morison, and Boccone.[3] Most of the authors he cited were no longer active. But when Ray wrote, members of the Royal Academy of Sciences in Paris had already committed themselves to just such an undertaking as he described. One of their earliest plans was to publish a general natural history of plants.

Huygens was the first to propose that the Academy publish a natural history. The project he described was Baconian and all-encompassing, intended to investigate weight, heat, cold, light, color, magnetic attraction, the composition of the elements, animal respiration, and growth in metals, plants, and stones.[4] The Academy would assign topics to its members, who would report weekly on their findings. This stilted proposal helped convince Colbert and his advisers to found the Academy, but Huygens's colleagues modified his ideas and adopted two specific projects for the anatomists, botanists, and chemists. These were the natural histories of animals and of plants.

It was Huygens's friend Perrault who suggested in January 1667 that the Academy publish a natural history of plants.[5] Perrault too was influenced by earlier models, especially Bauhin's *Pinax,* which Nicolas Marchant had already begun to revise. Perrault tried to define the field and to differentiate the kinds of research required for a comprehensive study of plants. He identified two ways of studying plants: "pur Botanique et Risotome" and natural philosophy. The former studied the "histoire" of plants by "botanizing" or "herborizing," that is by collecting plants and roots and studying their external characteristics and medical applications.[6] The latter Perrault defined, in the Theophrastean and Baconian tradition, as the inquiry into the causes of medical properties of plants or of vegetable reproduction and nutrition. For such research he envisaged chemical analysis, microscopic observation of seeds and shoots, tests of theories about propagation and generation, and studies of whether sap circulates like blood.

Perrault's plan of research for the natural history was more bookish than experimental. It was at once grandiose and modest. The Academy would treat all known plants in a comprehensive publication containing descriptions, illustrations, and a topographical index, but would take its information from existing literature. Because classification was problematic, Perrault proposed that academicians choose an existing system or dispense with one entirely. A catalogue of all known names of plants would be useful, but ancient names and descriptions could not always be correlated with modern plants. The compendium would be illustrated from watercolors painted by Nicolas Robert for the duke of Orléans rather than from life.[7]

Like Huygens, Perrault had a traditional conception of natural history. He referred to the ancients but not to the Americas, and indeed the Academy looked primarily to Europe and the Near East for unusual species, leaving American flora to the Minim Charles Plumier. As a physician, Perrault also stressed the medical merits of the project, urging academicians to correct and expand materia medica. The Academy's task was to collect useful information as efficiently as possible and make it available to the public.[8] This literary approach especially suited a society that had existed formally for only one month, as yet possessed no laboratory and little apparatus, and still used meetings to plan or debate. Perrault was also uncertain about the extent of royal patronage. A modest proposal, firmly rooted in work already begun, stood a chance of succeeding and might stimulate a more broadly conceived project, if only Colbert would authorize it.

By criticizing Robert's paintings, Perrault ultimately justified expanding the Academy's project. He showed that aesthetic criteria did not meet scientific needs. Most of the paintings did not show roots or indicate the relative size of a plant, defects that would trouble a scientist but not a connoisseur. Perrault's remedy was to add the missing roots and to depict leaves, fruits, and seeds in blank corners of the page (fig. 2). Should the Academy be allowed to commission illustrations from life, however, Perrault recommended that the artist portray the plant life-sized or provide a scale and show the important parts of plants; when the appearance of a plant changed markedly as it grew, the artist should depict both the young and the mature plant.[9]

Perrault's colleagues were not content with studying books but wanted to study nature. They would start with the literature and go beyond it. Indeed, since European botanical literature referred to only a small proportion of known plants, the greatest need was to study the new or "rare" plants. Here, too, the Academy was influenced by work begun by Guy de La Brosse and the duke of Orléans, especially since Marchant had worked

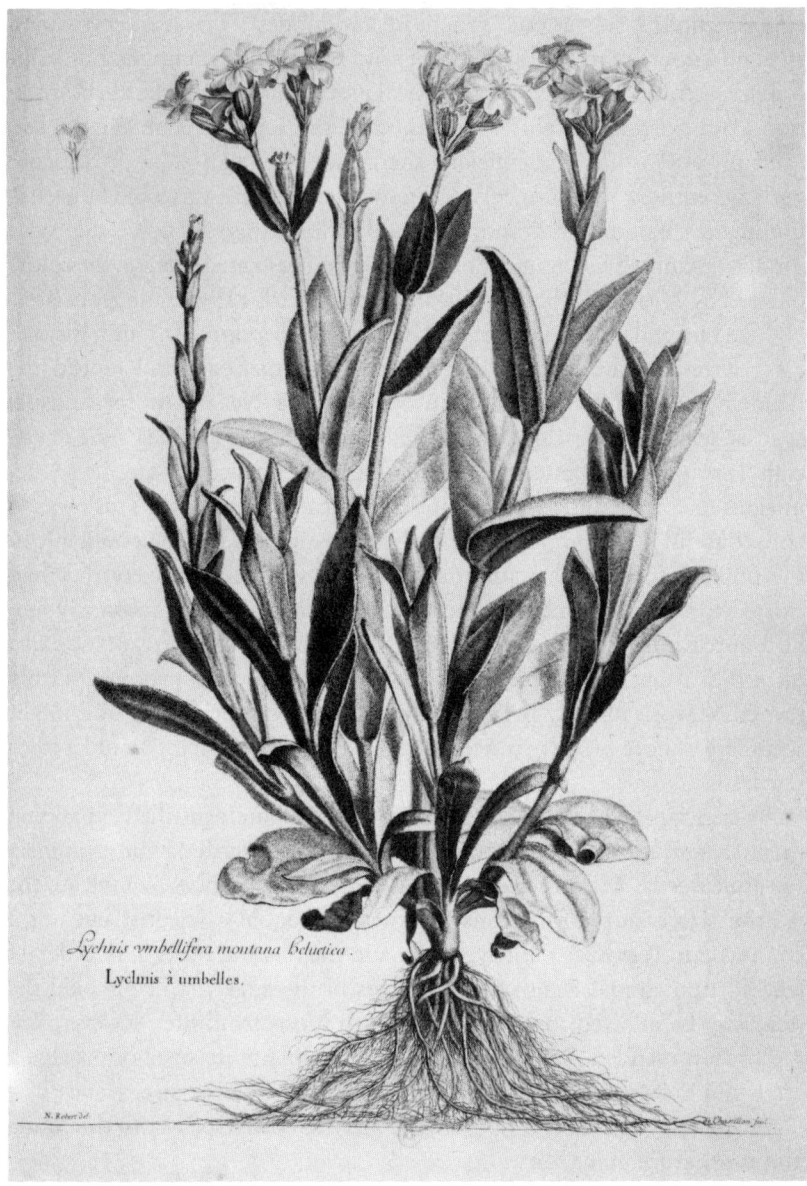

Fig. 2. Lychnis umbellifera montana helvetica/Lychnis à umbelles. (From *Estampes;* drawn by Robert, engraved by Chastillon; photograph courtesy of Bibliothèque Nationale, Paris.)

for the latter at Blois. But the Academy also added an experimental twist to Perrault's natural history by incorporating chemical analyses designed to explain what they were describing. This was Duclos's contribution to the project.

As director of the project, Duclos quickly put his own stamp on it.[10] To the basic elements—engravings and descriptions of plants—he added a classification according to Theophrastus's system.[11] He also expanded the descriptions to compensate for the faults of the illustrations. Descriptions of the size, parts, and products of plants would form a catalogue of characteristics that would distinguish one plant from another. He stressed for example the "carriage" or appearance (*le port*) of the plant in the earth, that is, whether a plant was tall (*eslevée*), or rested its branches on the earth without sending out roots from them (*couchée*), or rested its branches on the earth and sent out roots from them (*rampante*), or leaned (*appuyée*). Duclos also demanded precise descriptions of the root, trunk, leaves, flowers, fruits, seeds, and natural products such as gums, resins, or liquids.

In a more radical departure, Duclos added chemical analysis to the work. He planned to distill plants in order to mention in the descriptions the consistency, color, smell, and taste of distillants. He hoped to discover the chemical constitution of plants by analyzing their distillants, by testing a decoction of sap or juice in various solutions, and by studying crystals formed by condensed juices.[12] Because Duclos believed that chemical explanation of organic matter was fundamental, he made chemical analysis an integral part of the natural history of plants. Under the guise of description, therefore, he introduced inquiries that bordered on natural philosophy. In so doing Duclos set the natural history on a new and difficult footing that made the project controversial and delayed its completion.

Duclos also thought the natural history should have a regional bias, focusing on French flora. Thus, he recommended that the common French names be in the list of synonyms of plants:

> And because we plan to write this natural history in the French language, it would be good to be informed about the names which the common people in the major French provinces give to each plant, so as to add them to the names used in other languages.[13]

The natural histories of plants and animals, like nearly everything academicians wrote, were published in French. The Academy intended to reach above all a French audience. First and foremost, that meant the king, the ministerial protectors, and the persons to whom they distributed the book, as a group probably unfamiliar with Latin. Duclos's interest in popular

French names for plants was controversial, however, and his proposal allied him with the "moderns" against the "ancients" and with the "realists" against the "purists." The same controversy raged in the Académie française, which ruled out of its dictionary neologisms and technical language, the very vocabulary that Antoine Furetière struggled to learn for his own dictionary. Within the Academy of Sciences Duclos had many allies, for academicians coined words so that they could write about plants in their own language, and in the 1670s French names were added to some of the plates.[14]

Duclos's request that the Academy study only French plants was less palatable, despite its advantages. Valuing experience over authority, Duclos wanted descriptions to reflect direct observation, which was possible only when the plants were near at hand. He believed that the provincial flora of France were insufficiently appreciated in scholarly circles. He wanted the Academy's project to be manageable, and he worried that plant species varied when transplanted. Academicians, however, resisted this restriction. In the late seventeenth century gardeners were proud of the exotic flowers and fruits they could cultivate, and Louis XIV's nurseries and orangerie were famous for defying climate and seasons. Connoisseurs and savants alike wanted to expand their knowledge of flora and fauna, not limit it to what was native to France. Although Duclos's proposal was formally approved, the Academy never stopped cultivating, describing, and illustrating rare plants and never limited its projected book along geographical lines.[15] The resulting lack of focus impeded the project.

The Academy accumulated proposals and smoothed over disagreements. Its corporate procedures and plans grew by accretion and ignored inconsistencies. The successive adoption of Perrault's and Duclos's proposals exhibits this tendency well: in some areas they agreed, in others they disagreed, and the Academy simply glossed over any problems of coherence between them. Where Perrault and Duclos were in harmony, they reflected a centuries-old approach to studying plants, with Perrault emphasizing illustrations, Duclos text, to convey information. Duclos also improved on Perrault, whose language had sometimes been vague. But Duclos wanted the natural history to concentrate on French flora and to include chemical analysis, which Perrault regarded as more appropriate to the natural philosophical studies of plants. In any case, when the Academy formally approved Duclos's recommendations in 1668, a basic framework for the natural history existed. The designers were Huygens, Perrault, and Duclos, but the research and writing were almost entirely the responsibility of Nicolas Marchant and Bourdelin.

Duclos quickly lost control of the project to Dodart, who entered the Academy in 1671. His statement of principles appeared in the *Mémoires pour servir à l'histoire des plantes*, published in 1676 with Marchant's *Descriptions de quelques plantes nouvelles*. The books were an inconsistent introduction to the project, for Dodart discussed chemical analyses of plants at length, but Marchant omitted them from his descriptions.[16]

Dodart accepted many of Perrault's and Duclos's criteria for describing plants; he also learned from Marchant's experience of writing descriptions. He reaffirmed that the Academy's goal was to publish a description of every known plant. He agreed with Perrault and Duclos that the function of any description was to enable the reader to distinguish one plant from another. As a result, he limited descriptions to the parts of plants that served this purpose, or that helped to discover the uses of the plant, or that revealed "some particular industry of nature." When its surroundings affected the appearance of a plant, these also were to be indicated. Because the botanical vocabulary of the French language was limited, Dodart warned, academicians would coin words or borrow them from the vernacular.[17]

More important, the Academy was now able to commission its own drawings from life. Many of the illustrations in Marchant's *Descriptions de quelques plantes nouvelles* were copied from the duke's watercolors. But Dodart's *Mémoires des plantes* announced that the king's patronage would henceforth suffice to obtain engravings of the highest scientific standard; the Academy's artists would refer to the watercolors only if Marchant could not grow certain rare plants. The Academy would take the utmost pains to obtain accurate and detailed pictures. The engravers drew delicate parts or very small plants with the help of a microscope, and they used etchings (*eau forte*) rather than line engravings (*taille-douce*) to suggest shades of color (figs. 1–12). Instructions to the artists reflected Perrault's and Duclos's recommendations. Thus, illustrations would indicate the actual or relative size of each plant and portray the appearance of the plant in the earth (figs. 3, 4). They would also include a picture of the young plant, "whenever it first appears in a shape different enough to make it difficult to recognize" (fig. 5).[18]

Familiar with his colleagues' views, Dodart adopted them selectively, usually siding with Perrault. He rejected all known systems of classification, washing the Academy's collective hands of the problem that exercised Morison and Ray and whose solution later brought international fame to Tournefort.[19] Dodart shared Perrault's interest in testing methods of propagating plants and wanted to disprove the Theophrastean claim that plants could be propagated from their saps alone. Both Perrault and Dodart also

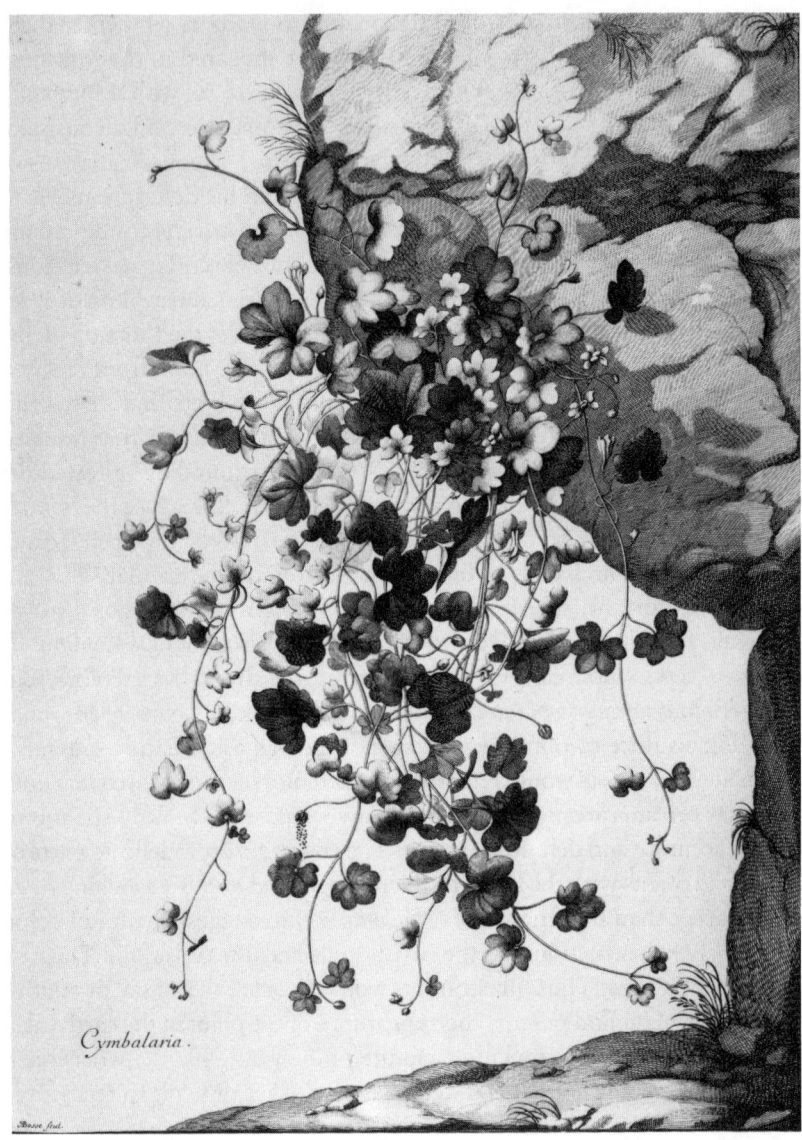

Fig. 3. Cymbalaria. (From *Estampes;* engraved by Bosse; photograph courtesy of Bibliothèque Nationale, Paris.)

Fig. 4. Melo vulgaris/Melon. (From *Estampes*; engraved by Robert; photograph courtesy of Bibliothèque Nationale, Paris.)

Fig. 5. Sanicula, sive Diapensia/Sanicle. (From *Estampes;* drawn and engraved by Robert; photograph courtesy of Bibliothèque Nationale, Paris.)

hoped the Academy could displace superstitions with observations, in that way teaching the public and raising the standard of knowledge about nature.[20]

Despite such similarities of approach and interest, the *Mémoires des plantes* was in certain respects Dodart's personal statement about the natural history. First, he affirmed the Baconian principle that one should not "condemn as false something that has not succeeded for oneself, but [instead] simply report the methods and results of one's experiments."[21] Dodart presented the Academy's raw findings, without hypotheses or conclusions. Duclos and Perrault, in contrast, expected the natural history to go beyond mere reports of experiment and difficulty. Second, Dodart introduced plant physiology into the natural history by including an explanation of how a plant grew, perhaps as an analogy with natural histories of animals that reported the development of the chick in the egg. He remained purist enough to exclude nutrition or the movement of sap from the natural history, but the cultivation of plants lent itself to analyses of germination and soils.[22] Whether or not Dodart intentionally challenged traditional definitions, no other academician had included these subjects in the natural history.

In ten years four academicians — Huygens, Perrault, Duclos, and Dodart — designed research on plants. They were inspired by earlier natural histories and by a well-established dichotomy between natural history and natural philosophy.[23] As work began in earnest, old distinctions were eroded and the fate of the project was irremediably altered by the addition of chemical analysis. Before considering the effects of that decision, however, the progress of research in all its aspects — cultivation, description, illustration, and chemical analysis — must be reviewed.

RESEARCH FOR THE NATURAL HISTORY

Academicians began work soon after Perrault proposed the natural history, but they needed plants for study. Perrault called for an "Academic Garden,"[24] and by the 1670s Nicolas Marchant had commandeered part of the vast and unused territory of the Jardin royal for the Academy. This plot became known as the *petit jardin* and was formally recognized as belonging to the Academy.[25] In it Marchant and his son cultivated seeds from all over the world, collected by friends, acquaintances, and colleagues.[26]

After cultivating a plant, the Marchants described it, gave it to the illustrators, and if there was enough supplied it to Bourdelin for analysis.[27]

Jean Marchant preserved unusual specimens in his *cabinet* at the Jardin royal.[28] The final description, however, was the collective business of the Academy.[29] While the Marchants read their preliminary drafts aloud at meetings, academicians looked at the plant and proposed improvements; during the spring of 1668 Nicolas Marchant and Duclos debated correct descriptive style.[30] The Academy soon approved a division of labor that persisted for several decades. The Marchants grew plants, not just rare ones, in order to explain their cultivation and development.[31] Bourdelin analyzed plants chemically,[32] while Dodart and the Marchants compiled comprehensive nomenclatures of each plant and wrote descriptions.[33] Finally, the artists drew and engraved the plants, working from life whenever possible. These patterns of work survived changes in the directorship, making the natural history of plants a team project that reflected the ideas and work of several academicians. It was too ambitious an undertaking for any one savant.

Illustrations occupied a prominent place in the natural history and in publicity for the Academy. Engravings of rare plants accompanied Marchant's *Descriptions de quelques plantes nouvelles* and Dodart's *Mémoires des plantes;* they were printed in an expensive folio format that would appeal to the king and encourage him to continue publishing the Academy's work. With the same purpose in mind, academicians showed Louis drawings of plants when he visited the Library in 1681.[34] Illustrations were essential because even the clearest description could not identify a plant so well as a picture of it. John Ray had likened "a history of plants without figures" to "a book of geography without maps," and regretted that engravings were beyond his means; Robert Morison impoverished himself to illustrate his text.[35] Royal funding gave the Academy a distinct advantage over savants who lacked patronage, because the Academy could spend substantial sums on illustrations.

Three engravers—Abraham Bosse, Nicolas Robert, and Louis Claude de Chastillon—shared the work, which cost more than 25,000 livres (table 8) from 1668 through 1699.[36] Like so much of the Academy's program, this too suffered from Louis's wars, and Colbert stopped paying for engravings in 1681, despite the pleas of academicians.[37] Louvois reinstated funding for the Academy's illustrations and Pontchartrain continued it, concentrating his resources on Tournefort's *Élémens* and finally publishing all of the Academy's engravings of plants in the early eighteenth century.[38] But many of the plants that Bosse, Robert, and Chastillon drew were never engraved.[39]

Academicians tried, not altogether successfully, to hold Bosse, Robert,

The Natural History of Plants

Fig. 6. Gentianella alpina verna magno flore. (From *Estampes;* engraved by Bosse; photograph courtesy of Bibliothèque Nationale, Paris.)

and Chastillon to a new standard of scientific accuracy. They supervised the artists closely, comparing drawings and engravings with descriptions and actual plants.[40] They were especially critical of Bosse's work, finding not one of the flowers of Cymbalaria (fig. 3) accurate,[41] protesting the superfluous branch and pot in the picture of Gentianella (fig. 6),[42] and deriding his anthropomorphic Mandragora mas (fig. 7) as "a ridiculous affectation."[43] Chastillon misrepresented the proportions of the leaves to the plant in depicting the mimosa (fig. 1).[44] Microscope and loupe were called for to correct certain features.[45] Extraneous details such as butterflies (fig. 9)[46] and birds marred some illustrations. Chinese characters (fig. 12) betray an engraving taken from an illustration rather than life. Many pictures lacked roots, seeds, and proper scientific names.[47] For reasons of economy, the size of engravings was reduced under Louvois,[48] but for reasons of accuracy, a picture of the seed, flower, and other parts of plants was added to some illustrations (figs. 1, 2).[49] These problems slowed the work, and the engravings as published in 1701 were far from representing the standards of the Academy.

Bourdelin performed the chemical analyses, which the entire Academy reviewed and Dodart interpreted. As soon as the laboratory was minimally equipped in June 1668 Bourdelin began distilling plants,[50] which remained his major occupation until his death in 1699. Indeed, Bourdelin analyzed more plants than were described or illustrated and his reports became a regular fixture at meetings.[51] It took large quantities—a hundred *livres* according to Dodart's estimate—to analyze a plant thoroughly.[52] Plants that did not grow around Paris[53] were cultivated in the Jardin royal or were purchased.[54] From Bourdelin's findings Dodart tried to discern the basic composition of plants. Interpretation depended on minutiae, and Dodart struggled unsuccessfully to extract from overly plentiful details the generalizations that would justify Bourdelin's labors.[55]

Academicians collaborated in planning and researching the natural history of plants, and this cooperative spirit was a matter of pride. Four members designed the research, four carried it out, and many others contributed at assemblies. Bourdelin, Nicolas and Jean Marchant, and Dodart were the principal collaborators. Although Perrault proposed the project in 1668, he worked primarily on the natural history of animals, which he had suggested at the same time. But it is odd that Duclos, who directed the project in the 1660s and early 1670s, contributed very little afterwards to its completion. The explanation does not lie wholly in Duclos's preference for other work, such as his study of mineral waters, which distracted him from the natural history of plants.[56] Rather, his disassociation from the natural history was involuntary, the result of a struggle over who would edit the project.

EDITORIAL RIVALRY

Duclos's nemesis was the young Dodart, physician and protégé of Perrault, whose association with the Academy began in 1671.[57] Dodart quickly took a position of responsibility and leadership. He was instrumental in reinstating the practice of keeping minutes and in reviving the Academy during an early slump.[58] Within five years he had become director of the natural history of plants, and his *Mémoires des plantes* reveals his control rather than Duclos's. With Perrault's protection, his influence transcended his lack of seniority in the Academy.

Dodart rose at the expense of Duclos. Director of the botanical project and leading theoretician of chemical research at the Academy in the 1660s, Duclos found his authority diminished during the 1670s. The minutes chart this decline. In his heyday from 1667 until 1669, Duclos read an average of more than three substantial papers a year—on topics ranging from coagulation and solvents to a detailed analysis of one of Boyle's books—filling roughly five hundred pages of minutes. During his decline in the period from 1675 to 1683, Duclos presented an average of fewer than two papers a year, and these fill at most twenty-odd pages in the minutes. In the 1660s Duclos dominated chemical and botanical planning with his long-range proposals, the status symbols that were the preserve of members who controlled the facilities and supervised others. Thereafter, this became Dodart's prerogative. Duclos conducted only his personal research by the mid-1670s, while Dodart supervised some of the work in chemistry and directed the natural history of plants. Why did this happen?

Dodart's interests and relative youth made him a plausible replacement for Duclos as director of the natural history of plants. Duclos's papers focused on experimental or theoretical chemistry, while Dodart was fascinated with all natural phenomena.[59] Dodart supervised the natural history and chemical analysis of plants energetically.[60] Duclos, however, was preoccupied with his books on mineral waters and on alchemical subjects. Since Duclos was at least thirty-five years older than Dodart, his flagging energy and waning interest in all but his favorite projects make Dodart's assumption of the natural history of plants even more understandable. Yet Duclos did not happily relinquish his responsibility to the new junior colleague. Thus Dodart's interests and qualifications do not explain a succession that was forced rather than amicable.

Fig. 7. Mandragora mas/Mandragore. (From *Estampes;* engraved by Bosse; photograph courtesy of Bibliothèque Nationale, Paris.)

The Natural History of Plants

Fig. 8. Mandragora flore sub caeruleo purpurascente, C.B. / Mandragore. (From *Estampes;* artist unknown; photograph courtesy of Bibliothèque Nationale, Paris.)

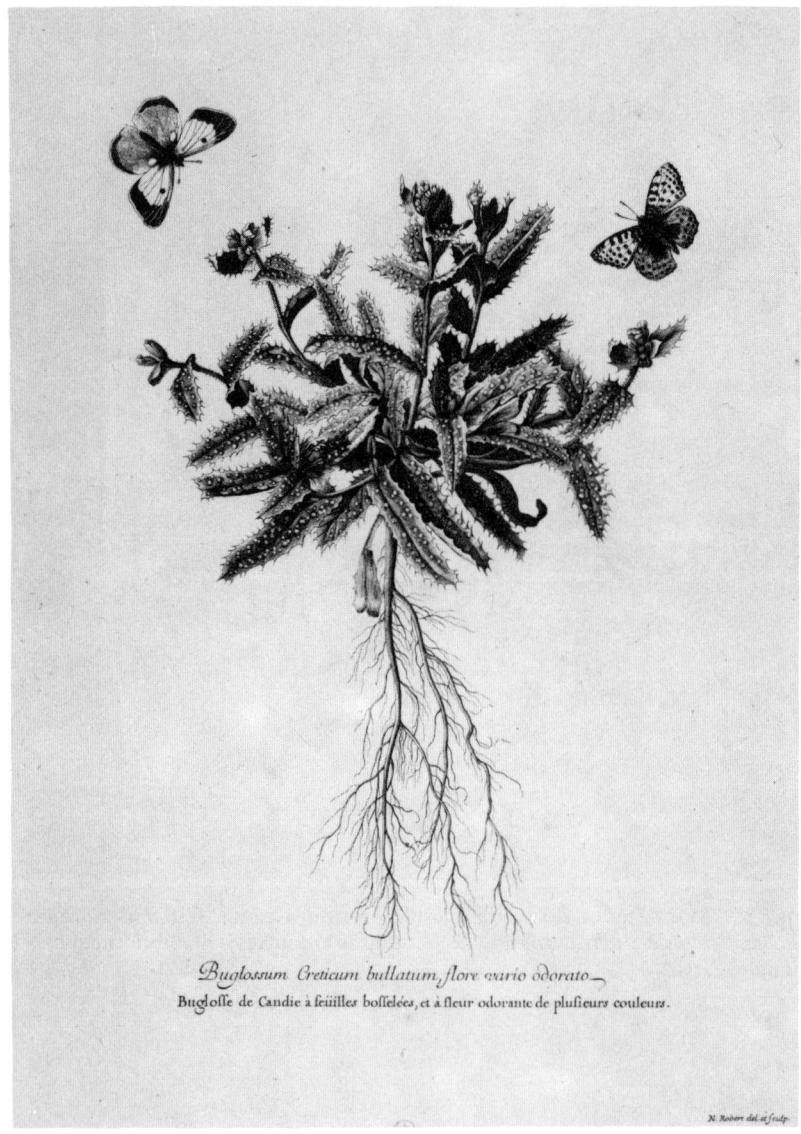

Fig. 9. Buglossum creticum bullatum, flore vario odorato / Buglosse de Candie à feuilles bosselées, et à fleur odorante de plusieurs couleurs. (From *Estampes*; drawn and engraved by Robert; photograph courtesy of Bibliothèque Nationale, Paris.)

Duclos's espousal of Platonist and Paracelsian views and his lifelong alchemical study complete the explanation. He made no secret of these interests, which he presented to his colleagues in several papers during the 1660s. For a while the Academy tolerated Duclos's pursuits. Later it feared embarrassment should his leanings become associated in the public mind with the institution itself, and academicians went so far as to refuse Duclos permission to publish one of his books.[61]

Duclos deeply resented Dodart's usurpation and counterattacked by maligning his editorial, scholarly, and collegial integrity. He accused Dodart of writing badly and reproached him for ignorance and careless reasoning. He unfairly denied that the Academy asked Dodart to write the *Mémoires des plantes*. Most important, Duclos criticized Dodart for misrepresenting the Academy. Dodart, he claimed, attributed ideas improperly to the Academy, represented his own views as those of his colleagues, portrayed the opinions of a few as if all academicians accepted them, and misrepresented theories he did not share. Duclos claimed that Dodart failed to collaborate with other academicians who had directed the research. The truth was that Dodart had simply rejected many of Duclos's views. Finally, Duclos was appalled because Dodart's book elaborated methodological issues instead of presenting conclusions about the nature of plants. At the heart of their disagreement was an argument about the purpose of analyzing plants chemically: Duclos had anticipated substantial insights into the nature of plants, but Dodart found the analyses more beneficial for medicine.[62] Duclos's animosity, therefore, had both personal and professional aspects; the latter, which focused on the purposes of chemical analysis, will become clearer in the following chapter.

Dodart certainly used his editorial power to alter the project, and his *Mémoires des plantes* was a very different book from anything Perrault or Duclos had conceived. It enumerated the obstacles to carrying out Duclos's instructions of 1668. In contrast, Marchant's *Descriptions de quelques plantes nouvelles* simply ignored them. Duclos had hoped to establish from the chemical analyses a theory with practical applications, but Dodart declared such efforts fruitless and advocated a more pragmatic use of Bourdelin's findings.[63] It is not surprising therefore that Duclos perceived the book as a disavowal of his views.[64]

CONCLUSION

Dodart claimed that the *Mémoires des plantes* was the first stage of a more comprehensive study of plants and a showpiece of collaboration.[65]

Both claims were only partly correct. The Academy never published the natural history of plants as conceived, and its cooperative research foundered on personal and substantial quarrels. Several problems imperiled the project. There was no perfect correspondence between descriptions and illustrations, with some plants described but not engraved and others engraved but not described. Funding was inadequate. Academicians disagreed about the style and content of descriptions, had to invent a botanical vernacular, and lacked simple criteria for selecting plants. Despite repeated revisions, illustrations remained inadequate and descriptions did not reflect the Academy's recommendations. The surviving notes reveal disagreements and delays in correcting problems.[66]

Still another issue—the chemical analysis of plants—divided academicians and jeopardized the natural history. Academicians' expectations and difficulties form the subject of the next chapter, which explains why they persisted with such recalcitrant research.

CHAPTER 7

Justifying the Chemical Analysis of Plants

The most controversial aspect of the Academy's natural history was the chemical analysis of plants. This introduced an element of causal explanation that Perrault believed was inappropriate. Even Duclos and Dodart, who approved chemical analysis, disagreed about its usefulness in the project. The results were difficult to interpret, and academicians did not know what precisely they were seeking as causes. Exacting but of uncertain merit, the chemical analyses of plants were debated by academicians throughout the remainder of the century.

The method of analysis was distillation. There was no initial dispute about this choice, for it was the traditional way to define the composition of mineral waters and to extract from animals and vegetables certain ingredients for medicaments. Chemists at the Jardin royal had given public demonstrations of distillations in their course on chemistry, and treatises by William Davison, Christophe Glaser, and Sébastien Matte La Faveur described the methods used there.[1] Employing similar procedures, Bourdelin distilled plants for the Academy until his death. Interpreting his data and improving his method were the concern of several other academicians.

The Academy's chemical research has traditionally been dismissed as a waste of resources, thus falsely obviating the need for a closer examination of its institutional role.[2] The number of academicians involved and the amounts of time and money devoted to the project show that the Academy regarded this work as important. Although academicians argued among themselves or admitted that they could not interpret their findings, they still

89

persevered with the research and refined Bourdelin's techniques of distillation. Personnel, method, and goals forced the Academy to persist with a project many members found unrewarding.

THE CONTROVERSY OVER DISTILLATION

Distillation—the process whereby plants were placed in a receptacle and heated to obtain liquid and solid products—was the obvious choice for analyzing plants. It was also known to be flawed. Academicians had to justify their choice in the face of well-aired criticisms, but the shortcomings of distillation were most forcefully presented to them by their own research. Its defenders argued not only against contemporary chemical literature but also against objections raised within the Academy itself.

Duclos was the first academician to explain how to distill plants. He described in detail how to extract the chemical constituents of plants, that is, "their distilled waters, their acrid, sulphurous, acidic, and mercurial spirits, their oils, and their fixed or volatile salts."[3] In explaining how distillation worked, Duclos used the kind of old-fashioned teleological language that Perrault and Mariotte sometimes ridiculed: the heat of the fire, he argued, made an impression on the plant and then rarefied it; rarefied matter would rise, but some matter was more disposed to rise than others.[4] Duclos's subsequent dismissal as director of the project, however, meant that others had to defend the method he had chosen, and they did so in very different language. Dodart argued the case in his *Mémoires des plantes,* Mariotte used the results in his *Végétation des plantes,* Homberg tried to exonerate Bourdelin, Tournefort studied Bourdelin's method and findings, and Fontenelle summarized some of the arguments for distillation when he wrote the *Histoire.*

Two analogies seemed to warrant the distillation of plants. Distillation could be considered the equivalent of dissection (with the fire serving as the knife) or as the counterpart of digestion (with the still replacing the stomach).[5] These were variations of ideas that had been current since Paracelsus.[6] Nicolas Lémery, John Ray, and Nicaise Le Febvre, among others, had advocated "anatomizing" plants by distilling them; Le Febvre also emphasized that distillation would show how the heat of the stomach acted on the food it digested.[7] But the analogy with the stomach also drew attention to some shortcomings of distillation: a fire could not transform plants the way the stomach could, did not extract the same nourishing substances, and required higher temperatures.[8] These analogies reveal not only academicians' assumptions but also their motives. They wanted to anatomize plants in order to understand the secrets of their structure and to know what caused their effects on humans.

Even more revealing than such justifications are the Academy's debates about the shortcomings of distillation. One difficulty was that the products were not necessarily extracted in their purest forms. Because chemists could not always tell when the distillants changed in nature from one substance to another, in any single distillation they would get mixed substances as well as relatively pure ones. Dodart expected that this problem was not serious, because even mixtures would reveal information about the composition of a plant. Obtaining pure distillants was not the original goal: the Academy wanted to discover the constitution of plants.[9]

Another problem was that distillation was destructive. Analyzing plants destroyed the very components that produced the effects—nutritive, gustatory, poisonous, or medicinal—academicians sought to understand. Dodart responded that such effects did not necessarily result from "the union of all the principles [that is, chemical constituents]," and, anyway, that effects "which depend on several of these principles joined together often depend on the dominant principle." He did not deny that the fire itself might pass through the apparatus and mix with the plants, but he replied that even so distillants differed from one another. So long as the fire and the vessels were the same, any variations must derive from the plants and not from the distillation.[10]

The most common objection was that, since all plants released the same constituents, these could not account for diversity among plants.[11] Dodart, in reply, pointed out subtle differences in the proportions and strengths of the constituents. He hoped that some of the "more ordinary effects" of plants might thus be explained and that, with more experiments, unusual effects might become understandable.[12] Mariotte, in contrast, granted the premise on which the objection rested but was more interested in why all plants had the same basic constituents. He concluded—using a thought-experiment that resembled an actual experiment described by Helmont and Boyle—that all plants had the same "gross and sensible" constituents because they received their nourishment from the same sources, earth and water.[13]

Where academicians sought diversity—in the products of distillation—they found uniformity. Where uniformity was essential—in the replicability of experiments—they found diversity. They recognized the importance of being able to reproduce experimental results. Bourdelin weighed the plants he distilled and the products he extracted from them; he recorded the exact conditions of each distillation, including, as best he could, the temperature of the fire. But even with all his precautions, the results of iterated experiments might vary, in some cases dramatically. Dodart tried to minimize this discrepancy by saying that chemists were entitled to ignore small variations and should take only major ones as significant.[14]

Unexpected variations in the results of similar experiments aggravated still another problem—the unmanageability of the data. Even by 1676, when Bourdelin had been distilling plants for only eight years, academicians found the amount of information compiled from his distillations so vast as to defy their analytical skills. The responsibility of interpreting Bourdelin's data fell on Dodart, who did his best to extrapolate a few generalizations from them.

The most threatening objection, however, was that the fire created new substances instead of merely separating substances that already existed in the plant. This view was widely accepted and had been asserted by numerous English scientists throughout the century.[15] Some academicians feared that this was indeed happening, and Dodart had to acknowledge certain disadvantages of distillation in refuting this view. As Fontenelle later pointed out, something as violent as fire must alter the constituents of a plant, especially the fixed salts, which were obtained by lessives only after calcination.[16] Mariotte suspected that distillation might fix volatile salts and make fixed ones volatile; the fire could even create a poisonous substance from a nutritious plant or form new chemical unions from the plant's constituents. On the whole, however, Mariotte believed that fire did not produce the constituents found in plants, because all of them could also be obtained naturally without recourse to fire.[17]

This problem worried academicians, who searched the data for reassurance. Even Duclos changed his mind about the effects of distillation. In 1668 he had believed that the fire assembled similar elements and separated dissimilar ones when heat excited motion in the substance being distilled.[18] By 1676 he came to believe that fire changed a plant's material virtues without making its formal and specific virtues better known.[19] Homberg later wrote that fire united some parts of a plant to form oil.[20] Against such views, Dodart argued that a fire did not often create new products, although he admitted that it might change the structure of the basic particles that compose plants and that some elements might escape through the vessels.[21] Dodart tried to define the nature and limits of any changes that fire could produce and asserted that any loss from the vessels was inconsequential with respect to both weight and character.[22]

Academicians criticized the procedure they had selected, and they disagreed about continuing to use it. Whatever doubts existed when the project started were not assuaged as it progressed; rather, Bourdelin's research brought to light still more problems. As a result, his colleagues considered abandoning or refining the method, sought a more effective one, and in the meantime changed Bourdelin's procedures.

THE METHOD OF DISTILLATION

Bourdelin's first technique was to distill a plant and collect the distillant in a single container. He then subjected the product to further operations in order to separate it into spirit, oil, salt, phlegm, and earth.[23] This was plant distillation as Le Febvre had taught it at the Jardin royal, and Duclos recommended the same procedures to the Academy in 1668.[24] Duclos described how to change the temperature of the fire, explained that the ashy residue (the *teste-morte* or *charbon*) in the receptacle containing the plant was to be calcinated and lixiviated to extract salts, and recommended that various distillants be tested with color reactors similar to those he used for mineral waters. The distinguishing feature of this method was that the distillant was collected in one container, to be separated and analyzed later. Forty-two plants were examined this way in 1670.[25]

In 1670, shortly before Dodart joined the Academy, Duclos's method was abandoned for one that obtained more varied products. The new procedure changed the recipient (the glass receptacle that collected the distillant) every time the heat of the fire changed. Bourdelin varied this second method over the next three decades, while other academicians tried to improve it.[26] By the time Dodart wrote the *Mémoires des plantes,* more than one hundred plants had been analyzed this way.

Dodart described this new technique, which he in fact revised, in some detail. He named the vessels used, told how to regulate the fire, discussed the substances obtained, and described how the ashes were treated. Everything was distilled in a glass or earthenware retort, to which was attached either a *balon à tétine* or a *balon sans tétine,* that is, a recipient with or without an udder-like protrusion. Organic matter was placed in the retort, the recipient was attached, and the retort was placed over a fire. Bourdelin regulated the fire and changed the recipient carefully.

> We start the fire so slowly that it can scarcely heat the retort. We increase it slightly until some liquid passes into the receiver, and we keep the fire in this state. We increase the heat only when scarcely any more liquid comes out. We increase it slightly degree by degree during a period of fourteen or fifteen days, and we make it as hot as possible. We empty the receiver, not only whenever we

increase the fire, but more often, and we keep all parts of the distillant separated.[27]

Distillation continued until the fire had reached its maximum temperature and no more liquid would come out. Then the ashes remaining in the retort were removed and treated. As many as fourteen different distillants might be extracted from the plant, in this order: sharp (*acres*) spirits; essential oils, given by aromatic plants; sulphurous spirits; simple waters; waters with a hidden taste of acid or sulphur; acid spirits; mixed spirits; urinous spirits, either with or without acid; volatile salts; black oils; fixed or saline or lixivial salt; and earth.[28] These products were tested with color reactors and by other means to classify them further. Each watery liquid was characterized as either "insipid, acid, sulphurous, urinous, or mixed." All the insipid liquids were combined and set aside, then all the acid liquids were combined and set aside, and so on. Once all the products had been identified and organized, each was examined for its weight and other observable properties (*propriétés sensibles*).[29]

This new method was not an invention of the Academy, but academicians applied it more rigorously than did their contemporaries. Glaser, for example, also changed recipients during distillation, but not so frequently, and as a result he did not obtain so many different distillants.[30] But Glaser and Dodart had different purposes. Glaser simply wanted to extract certain substances that he could use as medicaments,[31] whereas academicians wanted to identify all the constituents of plants.

By the 1690s, when Tournefort studied Bourdelin's research, the chemist had abbreviated his procedures. He removed the branches and juices from a plant and crushed it before distilling it. Then he put five *livres* of the plant in a tinned cucurbit, covered it with a glass head, and placed it in a water bath or a steam bath for two to three days, with the fire going day and night. Bourdelin next tested the liquid products with his repertory of indicators to determine whether they were acid or alkali. Next he distilled the dry residue in a retort with a large balloon or recipient, increasing the fire gradually. After twelve or fourteen hours he put the distillant in a glass alembic and attached a new recipient to the retort. He increased the heat of the fire and collected further distillants, separating them with a large glass funnel.[32] By this time, the chemist was no longer regulating the fire and treating the *teste-morte* as he had in the 1670s, and distillations lasted only a few days instead of a fortnight. The changes perhaps reflect his declining stamina.

Bourdelin's procedures never satisfied academicians, who suggested either embellishing or replacing distillation. Dodart was frankly overwhelmed by the data and asked Bourdelin to focus his work. By distilling

Justifying the Chemical Analysis of Plants   95

more selectively, he would avert interminable research. Thus Dodart abandoned a Baconian search for every possible phenomenon. Instead he adopted a more carefully designed program that chose the objects of inquiry according to some preconceptions. Dodart's stamp was felt on the Academy's choice of plants for distillation thereafter.[33]

Chemical analyses, like dissections of animals, required painstaking work and could be dangerous or unpleasant. Just as a slip of the knife might cause an infection (like the one that killed Perrault, who cut himself while dissecting a camel), so distillants were risky, for chemists identified many of them by taste. Rotting corpses and distilled plants stank. Anatomists treated decaying flesh with *eau de vie,* and Bourdelin treated plants by digesting (that is, heating without boiling), fermenting, or macerating before he distilled them. Unfortunately, this treatment altered them.[34] Dodart wanted to assess any changes caused by prior treatment, but other academicians tried to overcome any effects. Perrault thought this could be accomplished by distilling macerated or digested plants over lower heat for a longer time.[35] His idea was to compensate for the diminished force of the fire by increasing the duration of the distillation, a principle of substitution that he derived from mechanics.

The quest for a more satisfactory method of analysis continued well after the *Mémoires des plantes* appeared, but with few new ideas. By mid-November 1678, Bourdelin was on the defensive. He may well have been resisting pressure to disband his distillations. Borelly reflected on Bourdelin's recalcitrant research in the 1680s (as Homberg would do in the 1690s), probably as a result of a ministerial request. He stressed ways of rectifying distillants and designed a furnace for extracting substances from the *testes-mortes.* Above all he favored solvents for analysis. Some academicians had high hopes for his work. La Hire, for example, wrote to Huygens that Borelly "is searching as hard as he can for new ways of testing the liquids extracted in analyses." The chemist had discovered "something very curious," but La Hire's ignorance of chemistry prevented him from explaining Borelly's discovery.[36]

For years after Dodart published the *Mémoires des plantes,* academicians debated distillation. They were so dissatisfied that they nearly abandoned it. Researchers could not be certain that their methods were adequate or that their results were meaningful. Instead of rejecting distillation, however, they refined the process.

WHY DISTILLATION?

Given the pervasive skepticism about distillation by fire, why did academicians not discard it in favor of alternative methods? They could have tested the natural juices of plants with color reactors, observed the crystals formed by plant juices, studied vegetable dyes, or used solvent analysis.[37] Duclos, Dodart, and Perrault had discussed the first three of these techniques, while Borelly and Duclos promoted extraction by solvents. But two academicians—Bourdelin and Dodart—saw to it that the Academy continued distilling plants, in spite of shortcomings and alternatives.

Bourdelin's influence is surprising, because his role in the institution was so circumscribed. Of all the academicians involved with the natural history of plants only the two Marchants had as little power as Bourdelin. After the 1660s, his contributions to meetings were confined almost entirely to reporting on his distillations. His early papers on chemical research were ignored by the Academy, and his notebooks record experiments made according to the instructions of Duclos, Dodart, Borelly, and others. Yet if he could not initiate research, he could veto it, and he was markedly reluctant throughout the century to use any method of analysis other than distillation. Dodart suggested that soils be lixiviated instead of distilled, but Bourdelin continued distilling them, and when Borelly criticized him for this, Bourdelin stopped analyzing soils altogether. Both Duclos and Borelly wanted to use solvents, but again Bourdelin resisted. Since no other academician was willing to devote all his time to analyzing plants, animals, and minerals chemically, Bourdelin was able by default to perfect his chosen technique.

Dodart, too, favored distillation, and as director of the natural history his opinion carried weight. Distillation seemed appropriate for two reasons: it was a universal method which permitted comparison of all plants according to a single standard,[38] and it promised insights into how food nourished and medicines cured the body. Bourdelin's analyses hence seemed promising to Dodart's own research, and because the natural history could not proceed without Dodart and Bourdelin, their advocacy was decisive.

The most touted but controversial alternative to distillation was solvent analysis. Duclos had originally laid out a narrow sphere for solvent analysis in 1668. Distillation by fire, he argued, was best for separating the chemical constituents of most substances. The exceptions were "fixed substances and those which cannot be burned." These required "dissolving menstruums which break up the mass and render the constituent parts separable." Any substance that a fire could not distill required analysis with solvents. Pure earths, metals, glass, chalk, and minerals were all "fixed" in varying degrees; solvents offered the only hope of analyzing them.[39]

Duclos's interest in the subject had Paracelsian origins, and he supplied the recipe for what he claimed was the true alkahest or universal solvent.[40]

Solvent analysis was one of the issues that alienated Duclos from Dodart. The two argued about solvents in the early 1670s. When Dodart came across Duclos's recipe for the universal solvent, he mocked it as worthless for analyzing plants. In January 1675 he derisively asked the chemist to consider whether the solvent might shed light on the "marvelous effects" attributed to plants. Duclos's recipe, Dodart maintained, was as enigmatic as those of Paracelsus, Helmont, or Deiconti. Even if it was possible to make a universal solvent, it "would not help us understand the nature of plants any better, because each plant would be reduced by the operation of these solvents to a state" in which it would be indistinguishable from any other plant so treated. He derided universal solvents as being as useless as the theories of signatures and temperaments.[41]

This exchange occurred after Duclos modified his view. He now believed that solvent analysis offered

> a much better method than that of the fire since a solvent does not alter things, but leaves them as they are and reduces them to their constituent principles while preserving their virtues and their specific properties, something the fire cannot do.[42]

Furious at Dodart's attack, Duclos criticized "the author of the project who always speaks in the name of the Company without being so charged" for having characterized "universal solvents as vain and useless." Dodart embarrassed the Academy, he claimed, by representing it as mistrustful of methods recommended by "famous chemists."[43] Ironically, it was Duclos who discomfited his colleagues by publishing his alchemical *Dissertation sur les principes des mixtes* in Amsterdam after a committee of academicians had advised against its publication.[44]

Duclos's alchemical interests made him an unconvincing proponent of solvent analysis. Borelly, however, was untainted by Paracelsianism and he too favored solvents over distillation. Like Duclos he collected reports about their use, and the year after Duclos died Borelly proposed that all kinds of solvents be prepared. Perhaps he hoped to convince his wary colleagues that solvent analysis did not necessarily depend on alchemical precepts.[45] But his death in 1689 left the field to Bourdelin and Dodart.

Why did the Academy continue to analyze plants? Members recognized the shortcomings of distillation and its results baffled them, but they mistrusted solution analysis more. Why did they persist? The answer does not lie merely in the persuasiveness of the method's proponents, who brushed aside problems as due to imprecise observations. Rather, the steadfast analysis of plants by academicians in the face of apparent failure results from the high premium they placed on the basic goals of chemical analysis.

THE GOALS OF CHEMICAL ANALYSIS

Academicians had many reasons for analyzing plants. They hoped to find support for a particular theory of matter, to discover the nature of plants, to ascertain the medical and nutritional uses of plants and their products, to distinguish among the parts and types of plants, and to determine the limits of the method itself.[46] Over a period of thirty-three years, more than half a dozen different goals were enunciated by the eight or nine men concerned with analyzing plants. What induced academicians to justify their research with such varied reasons? Were there differences of opinion, or did opinion change gradually during three decades? The sources indicate that some academicians did disagree about the aims of this research and that their attitudes often changed as the research unfolded, but that they never totally abandoned certain fundamental expectations.

Perrault was the first to articulate goals. Chemical analysis had two objects for him. First, he hoped to obtain some experimental support for the corpuscular theory of matter. Perrault believed that the shapes of salt crystals were related to the shapes of corpuscles, an idea shared by Lémery and Homberg.[47] Although Mariotte later agreed that chemical analysis might prove that corpuscles existed, this view never caught on in the Academy.[48] Perrault's second goal struck a more sympathetic chord among his colleagues: he wished to identify what caused the properties of plants, that is, what made some nutritious, others medicinal, and still others poisonous.

Perrault's ideas anticipate the three major goals that motivated academicians until the end of the century: to identify the constituents of plants, to improve medicine, and to understand how plants nourish humans. The Academy's chemical analyses of plants promised both theoretical and practical results, with the latter contingent on the former. Duclos, for example, hoped to describe the "constitution" of plants,[49] while Dodart wanted to uncover "what plants are" and thought that chemical analysis might reveal the intimate structure in plants that produces their effects.[50]

The main purpose of analyzing plants chemically was to discover their constituents.[51] But by the mid-1670s, frustrated by distillation, some academicians became disillusioned about the prospects of understanding the nature of plants.[52] Instead they emphasized more practical purposes, such as improving medicine, without the benefit of an improved theory. Rather than try to put a practical art on a firm theoretical basis, they would operate pragmatically. Instead of deducing the effects of plants from general constituents, they would simply test specific distillants.[53] Bourdelin's reports often prompted discussions of remedies that could be made from the plant in question. Dodart scrutinized Bourdelin's notebooks for any pharmacological benefits, and Homberg told Bignon that he expected to find some medical uses for the distillants that Bourdelin had identified.[54] Dodart also proposed feeding poisonous plants to animals and dissecting the victims to trace the action of the poisons. He even considered carrying practical inquiry to the extreme, reversing the order of the inquiry: he suggested that pharmacological discoveries might clarify what plants were in themselves, that causes could be inferred from their effects. The difficulties of such an approach, however, were daunting.[55]

The third major goal—understanding nutrition—was Dodart's particular interest.[56] Indeed, the experiment for which he is probably best known stemmed from this quest: Dodart weighed himself before and after his Lenten fast, measured his daily intake of food and liquid, compared it with what he excreted, and concluded that the additional weight loss was due to transpiration.[57] Seizing the opportunity to get comparative information when Roemer traveled to England in 1679, Dodart asked his colleague to find out how racehorses were fed and trained, to look into the training and eating habits of men and women who were long distance runners, to find out how patients were fed in hospitals, to discover whether oatmeal was mixed with cucumber or fruit, and to let him know the eating and drinking habits of the Scots and Irish.[58] Furthermore, Dodart hoped that chemical analysis would clarify the food chain linking soil, plants, animals, and humans.[59] Finally, Bourdelin distilled various fruits, grains, and green vegetables for Dodart in the hope of identifying what made them wholesome. But these analyses did not reflect what was already known about plants. Dodart noticed that nourishing fruits, like peaches and apples, seemed to contain only water and yielded little oil during distillation. Because these distillants could not account for the food value of the fruits, however, Dodart concluded that there must be a fixed oil in peaches and apples that only the stomach could extract.[60]

After 1675, when it was clear that the primary goal of understanding the nature of plants would not swiftly be achieved, academicians devoted more attention to the second and third goals. They also posed more specific questions, about the salts and oils in plants and about the chemical differences between various parts of plants. As a result, Bourdelin no longer tried to analyze every possible plant with utmost thoroughness; instead he selected particular plants or distillants for particular purposes, often those suggested by his colleagues.[61] Dodart believed that in addressing such small questions academicians had made the best of things: while they could not, for example, explain why acidic and sulphurous substances differed, they had at least contributed to knowledge about the two.[62] Furthermore, by concentrating on simpler problems first, academicians might establish a basis for examining the more complex issues.[63]

In summary, the Academy's chemical analysis of plants changed in the 1670s. In the first half of the decade, as before, the principal reason for analyzing plants was to determine their chemical constituents. In the second half and thereafter, Dodart's two practical interests—nutrition and medicine—dominated chemical analysis. When academicians sought to identify the nature of plants, they were propounding an unanswerable question, given their methods and knowledge. This failure consequently forced them to pose more limited, manageable questions and to refine further their methods of analysis. These strategies, however, did not solve the quite different problem of how to present unsuccessful work to the public in a favorable light.

PUBLICITY AND DISCRETION

The natural history of plants was plagued by uncertainty. Academicians, therefore, continually modified their goals and research procedures. Neither perfectionists nor pedants, academicians were realistic experimentalists. The blend of tradition and innovation in their project, the too general nature of their first goal, and the unsuitability of distillation for their work, all disrupted their research. So did rivalry among colleagues.

At the very time when the Academy had decided to publish its results, its members were raising the most serious objections to the project. Dodart was dissatisfied about chemical analysis: "Since it scarcely seems that the distillants obtained by the analyses show us what plants are and what they can do, we must at least learn from the analyses what can be done, by any method whatever."[64] This justified persistence but allowed only a small hope that more general conclusions might be reached.

Dodart was desperate because he was editing the *Mémoires des plantes* for publication. Some engravings and descriptions of plants were ready, but Bourdelin's research evaded all efforts at interpretation. Yet Dodart had to present the Academy's work in the best possible light. After all, the Company was not ten years old, and savants in England and elsewhere awaited its publications eagerly but with skepticism. Everyone knew of the generous royal funding, academicians' pensions, and the institution's grandiose plans, but there had already been rumors of dissension. The public would judge the fledgling society by its publications. The Academy's natural history of plants seemed to meet a scientific need, and its chemical analyses made it somewhat innovative. But in 1675, the year when he was writing a first installment of the natural history of plants, Dodart was worried.

The *Mémoires des plantes* reflects Dodart's ambivalence about analysis, but it puts the best possible face on the Academy's work. Dodart addressed the problem directly. In the preface he invited the public to send information to the Academy. In the text he laid out Bourdelin's methods and results, the original goal and its more realistic modifications, and the difficulties encountered. Defending the Academy, Dodart pointed out that its laboratory had extracted several new substances from plants. Furthermore, he asserted that even if the Academy could not demonstrate "what is in each plant," then showing at least what plants are good for

> constitutes an important aspect of the History of Nature, and should add considerably to materia medica, as will be seen in the rest of this work. That is the sole certain usefulness which the Company anticipated from this research, leaving the rest to the conjectures of the Natural Philosophers.[65]

Dodart adroitly defended the Academy's failed chemical analysis. Its accomplishments, he argued, were well within the proper limits of natural history, while its failures belonged to the realm of natural philosophy and thus lay outside the scope of the project. Finally, he stressed the practical applications of the Academy's work.

The *Mémoires des plantes* was a clever smoke screen meant to make a good impression on the public. It emphasized the most plausible aspects of the Academy's work. But a careful reader would have realized that academicians still hoped that distillations might reveal the composition of organic matter. Indeed, Dodart's views were often confused and inconsistent because he was trying to do justice to the more ambitious goals of the Academy without making it look foolish.

CONCLUSION

The Academy had many reasons for asking Bourdelin to analyze plants. Principally, it hoped to discover the basic chemical constituents of plants, to develop new medicaments, and to understand what makes plants nutritious or poisonous. These purposes, along with other, secondary goals, explain why academicians persisted with this research on plants. Even though they worried that what they did might be fruitless, their multiple goals made them flexible and optimistic. They could justify distillation on medical grounds, for example, and hope that it would also explain the constituents of organic matter. They continued because what they sought was so important, because alternative methods seemed even more doubtful (to all but Duclos and Borelly), and because they thought they could perfect the one method in which most of them had any confidence at all.

For academicians and contemporaries like Grew and Boyle, chemistry was pivotal because it contributed to natural history, natural philosophy, and medicine. They hoped that chemical analysis would uncover the basic constituents of living matter and perhaps corroborate the corpuscularian theory. Their hopes dashed, academicians had to adopt more limited, practical goals; at worst chemical analysis might help generate medical reforms.

Both editorial rivalry and intellectual disputes undermined the project. Academicians disagreed about its goals and conduct, and several problems stemmed from the attempts to make the natural history innovative and to give it a theoretical foundation. Yet these obstacles were not fatal to the project, which failed for still other reasons, while Bourdelin's work was endorsed in a way that no one had anticipated.

CHAPTER 8

Ministerial Intervention and an Unexpected Outcome

Once the *Mémoires des plantes* appeared, Dodart's control over the natural history of plants grew stronger for a time. His view that the project should be published in installments was adopted, and he chose the plants to be distilled. Throughout the 1680s he directed the project, although he could never get it published. In the 1690s Homberg and Tournefort supplanted Dodart, just as he had replaced Duclos. These new members reinvigorated research and molded it to their own interests while Dodart's responsibilities as *médecin ordinaire* to the king eroded his participation in the Academy; indeed by 1699 he even needed a special dispensation to receive his pension because he missed so many of the Academy's meetings. When Homberg and Tournefort entered the Academy at the end of 1691, the one investigated Bourdelin's research at the behest of the ministerial protector, while the other saw how to use Bourdelin's research himself and preempted Dodart's plans. But they were not the first to thwart Dodart, whose project was imperiled by the Dutch Wars, by highwaymen, and by an illness of the king.

THE LOST SECOND INSTALLMENT

Dodart wanted to publish essays on plants annually, or at least in regular installments. He thought that the second volume of his natural history of plants should set out generalizations and exceptions and describe individual plants.[1] Hoping that the public would share his interest in explaining nutrition, he planned to write not about the rare plants that intrigued Nicolas Marchant but about homely vegetables that formed part of the diet, such as "coriander, lettuce, wild and domesticated chicory, watercress, etc." (figs. 10, 11).[2] This concern with nutrition greatly expanded the work. From 1676 to April 1678, Bourdelin analyzed more than four hundred and fifty plants and animals for Dodart. In the spring of 1679, Jean Marchant sowed more than four hundred different seeds from France and abroad, "with the object of describing them for the general history of plants."[3]

Between August 1680 and mid-June 1681, Dodart wrote the second part of the natural history of plants and prepared three other treatises for publication. But he fell victim to a dreadful scholarly accident. All of his treatises—the second part of the natural history of plants and the works on medicine, natural philosophy, and plants—were stolen from him. Du Hamel had to explain the loss to Colbert:

> ...all these treatises which ought to have composed a good volume were stolen from him on his way into Paris, where he was bringing them in order to complete them for the printer; and since all the efforts which he has made to recover them have been futile, he has been obliged to redo the two most important of these treatises, and to collect from his papers anything that he can find that will help him rewrite the other works.[4]

The highwaymen were probably so disappointed with their worthless booty that they threw it away, but Dodart had to spend the rest of the year rewriting the stolen treatises.[5] By mid-December 1681 Du Hamel could report that among the books ready for publication were Dodart's "Deuxième partie du projet de l'histoire des plantes" and "Analyses des plantes," and Marchant's "Environ 200 descriptions de plantes gravées."[6] The first two were apparently reconstructions of the stolen books. By December 1681, therefore, despite the theft of Dodart's manuscripts, the Academy was ready to publish three volumes pertaining to the natural history.

Why were these works never published? It cannot have been disputes about distillation that delayed publication, for by separating the "Deuxième partie" and the "Analyses des plantes," Dodart had insulated the uncontroversial aspects of the Academy's work from contamination by chemical analyses of uncertain merit. Blame for the failure to publish lies partly with the academicians and partly with their patron. Many of Marchant's descriptions were inconsistent with the plates or with earlier treatises, and discrepancies had to be resolved before publication. Financial reasons also delayed publication. Funding for the Academy diminished in the late 1670s, and in the 1680s Colbert still refused to release funds for

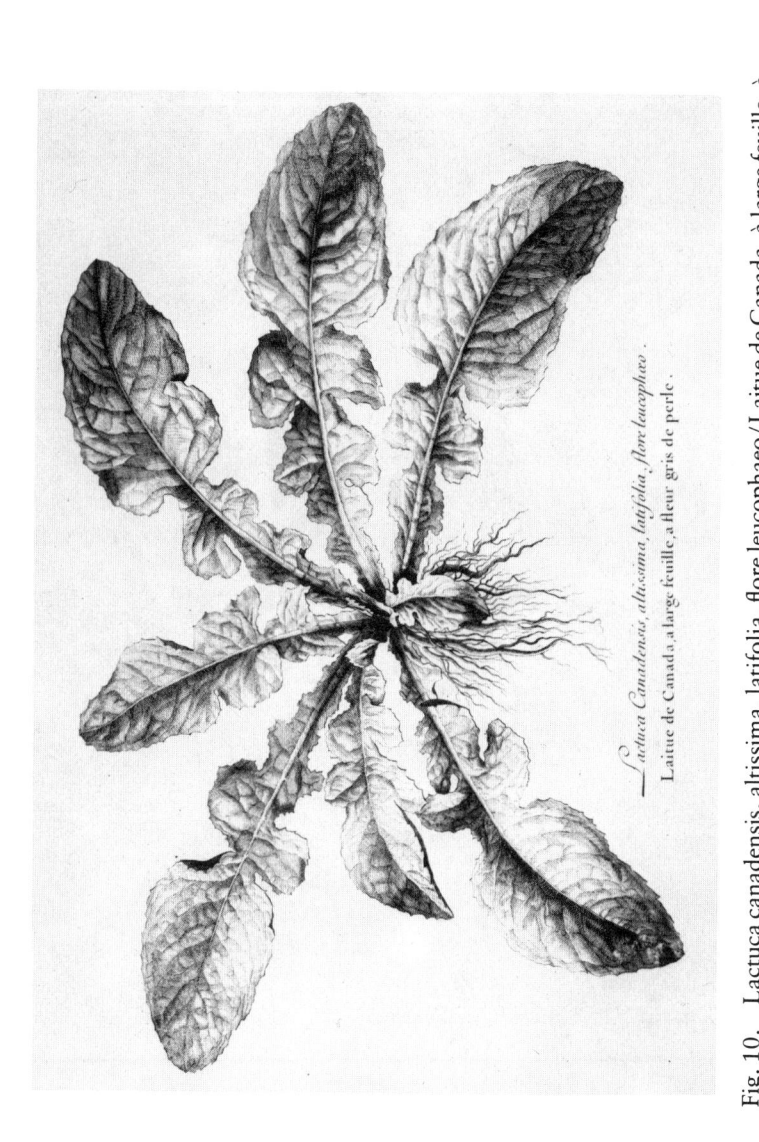

Fig. 10. Lactuca canadensis, altissima, latifolia, flore leucophaeo/Laitue de Canada, à large feuille, à fleur gris de perle. (From *Estampes*; artist unknown; photograph courtesy of Bibliothèque Nationale, Paris.)

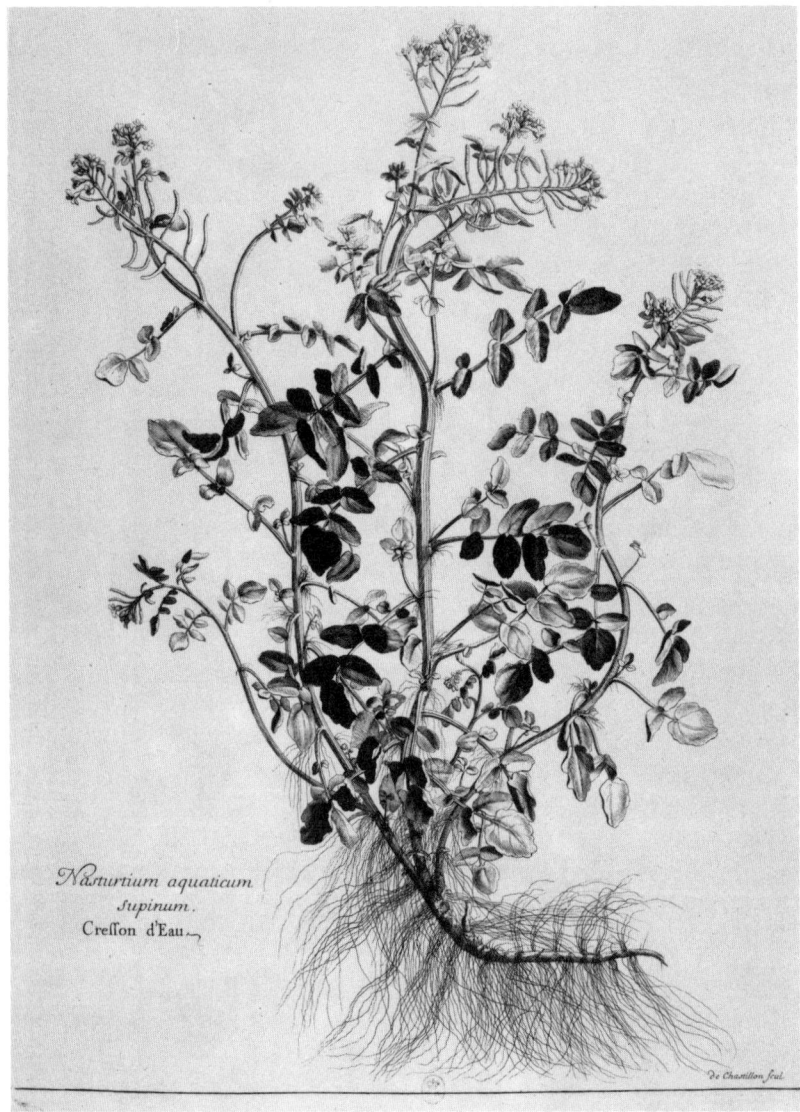

Fig. 11. Nasturtium aquaticum supinum/Cresson d'eau. (From *Estampes;* engraved by Chastillon; photograph courtesy of Bibliothèque Nationale, Paris.)

engravings despite persistent appeals from academicians. When the king visited the Academy at the Bibliothèque du roi in December 1681, academicians presented him a list of treatises ready for publication, to no avail. But these refusals must have been couched in encouraging terms, for academicians continued their work, and morale in the early 1680s was better than it had been a decade earlier. They lived on promises and persevered.[7]

Even the transition from Colbert's to Louvois's ministry did not at first disrupt work on the natural history of plants. During the first thirty months of Louvois's protectorship, work continued apace, especially on engravings, which Louvois had reinstigated. Louvois encouraged academicians to ready their work for print. His ministry marked closer ties between the Academy and the Jardin royal, especially with Fagon and Tournefort, whom Fagon took under his wing. Fagon chose plants for Chastillon to draw for the Academy's "Mémoire des plantes," and Louvois sponsored Tournefort's herborizations in the Iberian peninsula.[8] In the winter of 1684 the Company planned to work on roots, seeds, and woody parts of the plants that had been engraved, while Bourdelin continued his analyses.[9] In 1685 and 1686 Dodart and Marchant concentrated on engravings and descriptions.[10] In 1685 Du Hamel believed that the Academy would soon publish a volume of the natural history of plants, for which many engravings were completed.[11] But after 1686, the grand botanical compendium was mentioned in the minutes only as a project of the past.

MINISTERIAL INTERVENTION

What explains this abrupt abandonment of the natural history of plants? Academicians had overcome editorial rivalry, discouragement about chemical analyses of uncertain merit, theft of manuscripts, and parsimony prompted by the Dutch Wars. Despite all of these discouragements, they researched and wrote for publication. But in the mid-1680s a new and more dangerous impediment arose. What jealousy, controversy, theft, or economy had not accomplished, ministerial interference did. The Academy's botanical work was injured by misguided enthusiasm on the part of its patron.

When a patron's own interests interfere with the conception or execution of a creative project, the quality of work suffers. Interference from a patron who does not understand the technical language and skills or the theoretical assumptions of the work can be especially damaging.[12] This is precisely what happened. In 1686 Louvois intervened, upsetting the delicate balance between theoretical and practical expectations. Academicians had hoped that their natural history of plants would benefit medicine and add to basic knowledge about the nature of the world. They disputed among themselves as to whether sound theory was a basis for or an outcome of practical advance. But no academician advocated choosing between utilitarian and abstract goals; rather they debated the precise relationship between practical and theoretical knowledge. Into this scholarly discussion was injected a ministerial command to obtain practical benefits at the expense of theoretical research.

Louvois took an interest in natural history and preferred it to the other sciences partly because he thought it promised the quickest utilitarian results. He became especially impatient when the king's life was endangered by a serious illness that his physicians had been unable to treat effectively.[13] Looking for a scapegoat, resenting an Academy that could not save its monarch's life, needing more funds for Louis's wars, and expecting scientific inquiry to lead inexorably and swiftly to improvements in the quality of life, Louvois lost patience. He told the Academy how to do its work.

At the meeting of 30 January 1686, Henri Bessé de La Chapelle, Louvois's spokesman in the Academy, read a short paper to academicians. Willfully insulting and openly disdainful of the Academy's chemical research, La Chapelle warned academicians that they must move in new directions. He exhorted academicians to eschew "curious" research, which he called "a game" or "an amusement of the Chemists," and to apply themselves instead to "useful research that has some connection with the Service of the King and of the State."[14] La Chapelle recommended that academicians study medicine, improve and republish Duclos's book on mineral waters, try to desalinate sea water, and analyze wines. He also discussed the relationship among medicine, natural philosophy, and the natural history of plants:

> The other research more appropriate for this Company and which would be more to the taste of Monseigneur de Louvois is anything that can illustrate natural philosophy and serve medicine, these two things being practically inseparable, since medicine takes consequences and profit from the new discoveries of natural philosophy.[15]

La Chapelle believed that the Academy's chemical analyses were worthless because they belonged not to the practical realm of botany and medicine but to the abstract and theoretical realm of natural philosophy. The Academy's natural philosophers controlled chemistry and the natural history of plants, and they stood accused of subverting these subjects by seeking the underlying causes of things.

While La Chapelle agreed that practical results followed from theoretical discoveries, he noted that chemical research at the Academy seemed to make little progress in either direction. For that reason he warned the Academy to concentrate on empirical matters. In singling out Duclos's book on French mineral waters for praise, La Chapelle cited a work that perfectly blended science, medicine, and chauvinism to please a royal patron interested in practical accomplishments. Furthermore, Louvois and La Chapelle unerringly criticized the most vulnerable of the Academy's projects, the one that had stirred personal and theoretical controversy and whose publications were few by comparison with its cost. Whatever Louvois's motivation—Louis's nearly fatal illness, some behind-the-scenes influence, or impatience with a project much discussed but showing few practical results—his instructions were clear. Academicians were to spend more time on medical research. He would continue to support the natural history of plants only so long as the Academy adjusted its contents to his expectations.

Since Louvois never repudiated his spokesman or his instructions, he must have been content with La Chapelle's speech.[16] Other academicians perhaps agreed, for only eighteen months later La Hire explained La Chapelle's role to Huygens, to whom he sometimes complained about Louvois's policies. Louvois, he said, had "entirely committed the care of our academy" to La Chapelle, who "does us the courtesy of attending our meetings and of communicating to us his good ideas (*belles lumières*) in the sciences."[17] La Chapelle's place in the Academy was assured and his harangue of January 1686 was heeded by academicians, up to a point.

La Chapelle's speech is a curious blend of criticism and advice, of general statements about method and specific suggestions for research, of familiarity with and ignorance about the Academy's chemical research. Many of his suggestions were superfluous. Duclos, Dodart, and Bourdelin had already studied the tastes of plant distillants, analyzed earths, and tried to desalinate sea water.[18] Academicians had always sought medical applications of their work and planned to include the medical uses of plants in the natural history. La Chapelle named specific vegetable and mineral substances—mercury, antimony, quinine, laudanum, poppy, tea, coffee, and cocoa—for study,[19] but academicians had already examined them. Perhaps, in his awkward double role as ministerial spokesman and as colleague of the academicians, La Chapelle was trying to soften Louvois's criticisms implicitly by proposing work he knew his colleagues had already performed. Louvois's motives are clearer than La Chapelle's, but the effect of the speech, no matter what its genesis, was unmistakable.

La Chapelle's talk had marked repercussions. First, academicians no longer presented papers about the natural history of plants. Second, the chemists were more carefully supervised and had to draft new research proposals. Both Borelly and Dodart immediately suggested projects incorporating Louvois's requirements, and Du Hamel turned over the chemical proposals to La Chapelle in April 1686.[20] During January 1688 Borelly was asked to keep a notebook of his chemical experiments, and he presented his findings to the Wednesday assemblies.[21] Third, chemical analyses emphasized the potentially useful natural products of plants, such as gums and materia medica. Fourth, medical remedies became an even more common topic of discussion at meetings. Fifth, fewer plants were described or engravings verified than in previous years.[22] Sixth, only two other papers on plants were read; one was a letter sent to academicians about a deformed pear, and the other was Sédileau's report on the insects that caused galls in the bark of trees in the royal orangerie.[23] Thus the only botanical paper produced by an academician from 1686 until 1690 concerned a disease of Louis's orange trees, not disinterested inquiry but institutional flattery of the patron. The years 1688 and 1689 represent the nadir of botanical research in the seventeenth-century Academy. Only Bourdelin continued his normal research, perhaps because while working at home he could isolate himself from ministerial pressures. When La Hire reported the Academy's activities to Huygens in 1690, he did not mention any work on plants.[24] By the time Pontchartrain took over the Academy from Louvois, botanical research at the Academy had very nearly collapsed.

With the decline in pure botanical research, academicians emphasized the nutritional, medical, and industrial uses of plants and their products. In 1688, 1689, and 1690, they assessed a coffee-flavored beverage brewed from roasted rye, tasted the milky juice in common chicory, considered a remedy for hemorrhoids, and compared methods of treating wood to obtain good charcoal.[25] Only in 1690, when Louvois's health and interest in the Academy were declining, did La Hire revive the study of plant vegetation and Dodart and Marchant again read descriptions of plants.[26] Their example was lauded by Gallois and, oddly, by La Chapelle, both of whom urged a study of roots.[27] Despite this flurry of activity, Louvois's reprimands about curious research and chemical games had both an immediate and a lasting detrimental effect on the natural history of plants; the project temporarily came to a halt and never recovered its full vigor.[28]

The natural history suffered from fiscal exigency and internal difficulties both intellectual and personal, but ministerial manipulation was decisive.[29] Indeed, theoretical botany revived when Pontchartrain became protector. In 1692 Dodart planned botanical research and read a paper on the structure of the bud of a tree, while La Hire demonstrated that fig trees produced flowers, and Tournefort and Marchant described plants for the Academy's new monthly publication.[30] Observations sent from the Far East fired enthusiasm and broadened the scope of inquiry, so that the Academy inspected drawings of hitherto unknown Chinese plants (fig. 12) and studied the root of ginseng.[31] Dodart reported the growth of leaves and shoots on an elm felled fifteen months earlier. Tournefort examined an unusual mushroom, the flowers of *Apocynum maius syriacum rectum cornuti,* and the contraction of fibers in certain plants.[32] Morin de Toulon discussed a plant called *tartunaire* in Provençal, and La Hire *fils* described how a vine attaches itself to walls.[33] La Hire and Sédileau studied orange trees, and Jean Marchant reported on his and his father's emendations of Bauhin's *Pinax*.[34] In 1697 Dodart pointed out that the base of a tree's crown was always parallel to the ground in which the tree grew.[35] No longer did botanical presentations emphasize utility.[36]

Finally, Tournefort and Homberg agreed that the natural history of plants should be published, and they cooperated for a time with Dodart, Marchant, and Bourdelin. By 1692 Dodart and Marchant were once again reading descriptions of plants to assemblies almost as frequently as they had before 1686.[37] They also compared plants with engravings, while Bourdelin analyzed plants and Homberg and Tournefort studied his findings.[38] Homberg was initially enthusiastic and his manuscripts confirm that the Academy expected to publish the natural history of plants during the 1690s.[39]

In the number and variety of botanical activities and in the balance of pure and applied research, the 1690s were a period of marked improvement. Ministerial appointments spurred this revival. Unlike Louvois, who named relatively unknown scientists to poorly paid positions and added no chemists or botanists to the Academy, Pontchartrain selected well-known and respected scientists — Boulduc, Homberg, Charas, and Tournefort — and paid some of them decently. At least as important, he and Bignon allowed academicians to control their research, insofar as the royal treasury could underwrite it. Finally, in appointing Tournefort, Pontchartrain injected into the Academy a savant with a forceful intellect and powerful backing who would usurp the natural history of plants and shape it to his own purposes.

Fig. 12. Aubépine (plante chinoise). (From *Estampes;* drawn and engraved by Chastillon; photograph courtesy of Bibliothèque Nationale, Paris.)

A NEW EDITOR

Pontchartrain's choice of Tournefort had obvious merits. Tournefort shared many interests with Jean Marchant, Dodart, and Bourdelin, and he collaborated with other colleagues on varied research.[40] He agreed that it was important to correct errors in traditional botanical literature and to develop the medical uses of plants.[41] Like the Marchants, Tournefort advocated collecting information from all countries on the medicinal uses of plants; he believed that the task merited royal patronage and promised new cures for dangerous diseases.[42] His descriptions of plants reinforced the efforts of Dodart and Jean Marchant.[43] He studied Bourdelin's chemical analyses of plants and examined the chemical constituents of soils. He also conjectured about the constituents of mixed bodies and the medicinal properties of plants. These ideas had been the hope and despair of Duclos, Dodart, and Mariotte before him. Along with his colleagues, Tournefort tempered expectation with doubt, for he was skeptical of ascertaining the "the primary qualities" or the "configuration of the parts" of plants and soils.[44]

Even Tournefort's *Élémens de botanique* complemented the Academy's other botanical research. It was an intellectual prolegomenon to the Academy's natural history, although its cost delayed publication of the latter.[45] Tournefort classified plants and rationalized their nomenclature, did not include in his engravings "pictures of the entire plant," and omitted the "virtues" of plants from his descriptions.[46] Thus the *Élémens* laid the groundwork for the Academy's natural history of plants, which Pontchartrain's and Bignon's policies seemed to revive.

Tournefort promised in the introduction to his *Élémens* that the compendium would soon appear:

> The Royal Academy of Sciences, which has made Botany one of its principal activities, will soon furnish to the public some papers about the natural history of plants, with illustrations, descriptions, and analyses, all worthy, if one may dare to say it, of the magnificence of the King, and which will demonstrate just how far the science has been perfected.[47]

Yet the work he predicted so confidently in 1694 was not to appear. Within four years Tournefort himself sabotaged it. His next major book, the *Histoire des plantes qui naissent aux environs de Paris* of 1698, superseded the Academy's natural history of plants. Thus the unexpected outcome of three decades of research was the usurpation by a new member of prerogatives clearly staked out by his colleagues.

After 1694 the Academy decided not to publish its grand natural history of plants. Homberg retracted some of his optimistic assessments of Bourdelin's analyses, and Tournefort's *Histoire des plantes* announced that the Academy's natural history would never appear in the form originally conceived. Granting the importance of correcting previous botanists, Tournefort nevertheless deprecated any plan to begin "a general history of Plants on the basis of new expenses." This caution was sensible during the 1690s, but coming from the author of the costly *Élémens,* it must have struck some of his colleagues as unseemly and self-serving.[48]

Tournefort's *Histoire des plantes* supplanted and transformed the Academy's natural history. Unlike his *Élémens,* which solved a problem that academicians had disregarded, the *Histoire des plantes* drew on the botanical work of Dodart, Bourdelin, and the Marchants. Tournefort's book grew out of his lectures at the Jardin royal and also out of the Academy's work. It credited the Marchants with supplying plants for analysis, used Bourdelin's research to delineate the medical uses of plants, and included alternative plant names such as the Marchants and Dodart had painstakingly collected. Tournefort, however, did not make their intentions his own. Instead he focused on plants in the Paris region that were medically useful. In so doing, he made an unwieldy mass of data manageable, and he captured the sentiment of academicians that the medical implications of Bourdelin's work were its most viable dimension. He also revived Duclos's idea that the Academy emphasize French flora. Eighteenth-century academicians viewed his book as the sequel to Dodart's *Mémoires des plantes,* while Tournefort declared it to be the first of several regional natural histories of French flora.[49]

In effect Tournefort divided the Academy's project. In appropriating the description and analysis of useful local flora to himself, he left foreign flora to Marchant. Marchant worked until the end of his life on the old project, but after publishing 319 engravings in 1701 he got no further with its publication.[50] When Reneaume and Terrasson tried to revive the natural history of plants in 1709, their conception of the work was considerably different from his.[51]

Tournefort redefined the Academy's natural history of plants and welded a new alliance between the Academy of Sciences and the Royal Garden, with the latter dominating. His role in the Academy during the 1690s resembled Dodart's during the 1670s: both entered the Academy with influential mentors, assumed direction of a project that seemed to be floundering, and published specific parts of the Academy's research. Their books appeared under their own names and reflected the research of other academicians, although Dodart overstated and Tournefort minimized the contributions of his colleagues. Tournefort isolated Marchant's work but published some of the Academy's results in forms that its original proponents could not have foreseen. When at last the engravings appeared as *Les plantes du roi* in the eighteenth century, they brought to an elegant close the series of publications that represented the seventeenth-century Academy's failed natural history of plants.

CONCLUSION

The Academy's corporate character affected its scientific inquiry, by providing continuity of goals across the lifetimes of individual members and by encouraging division of labor within research teams. These advantages favored ambitious projects. But the natural history of plants never came to fruition as planned because it was plagued by serious problems. The goals of research were unrealistic given the methods available. Chemical analysis was difficult to interpret. Relations among academicians were not harmonious. Funding was erratic, and manuscripts were stolen. Ministerial intervention on the side of practical rather than theoretical research damaged morale.

Neither the structural benefits of collaboration nor the relative independence of strong-minded individuals entirely surmounted these problems. Dodart and Tournefort faced a difficult choice. If they were true to the original intentions of the project, these would become a straitjacket. But if they changed its goals, reported research selectively, and published their own views rather than those of their co-workers, they alienated fellow academicians. Both chose the latter course.

By contrast the Academy's work in the natural philosophy of plants required little team effort yet achieved a modest success. It was carried out by academicians who worked independently of one another and did not always submit research plans to the institution or its protectors but instead read finished papers to their colleagues preparatory to publishing them. It enjoyed minimal material support from the institution but at least was free from ministerial interventions.[52] In the natural history, academicians emphasized experiment and observation over theory, but the uncertain theoretical implications of that work undermined the project. In the natural philosophy, theory and experiment were more effectively wedded. If the natural history of plants tested the Academy as a company that produced science, the natural philosophy of plants revealed it as a company that reviewed what its individual members had produced. For the natural history the institution provided labor, materials, experiment, observation, and analysis; for the natural philosophy the institution served primarily as a referee of ideas. Academicians' failures in the natural history revealed tensions between the Academy and its patrons in the seventeenth century; academicians' achievements in the natural philosophy helped establish the pattern of the Academy's activities in the eighteenth century.

CHAPTER 9

Analogical Reasoning: The Model

In the late seventeenth century botanical thinking derived its inspiration from four sources: natural phenomena; the more or less traditional ideas that defined and constituted the field itself; ideas borrowed from other disciplines; and newly invented scientific apparatus.

The stimulus provided by natural phenomena was particularly dramatic in the early modern period because, with the discovery of the new world and the more assiduous exploration of the old, the number of plant species known to Europeans quadrupled. As a result, fifteenth-, sixteenth-, and seventeenth-century botanical literature reflects a delight in the superabundance of nature. Many botanists focused on the natural resources of the discipline, trying to record and describe all known types. Academicians contributed to this enterprise by preparing their natural history of plants.

Traditional botany provided part of the conceptual framework for students of plants, who distinguished flora from other living things in an Aristotelian manner and modeled their treatises in style and content after those of distinguished predecessors. But by the middle of the seventeenth century there was a marked shift in botanical writing away from herbals and toward specialized treatises on classification, regional flora, plant anatomy and physiology, and exotic specimens. When academicians looked backwards, it was often to disprove survivals from earlier literature that they regarded as superstitions. Insofar as they were influenced by botanical literature it was by natural histories of the recent past rather than by physiological treatises by contemporaries such as Grew or Malpighi.[1]

In the late seventeenth century savants challenged and expanded the traditional ideas. They did so through cross-disciplinary borrowings of ideas and instruments, which provided new interpretations and phenomena for contemplation. Theories adapted from other disciplines, especially from animal anatomy and physiology, stimulated botanists to reinterpret the structure and behavior of plants. Instruments such as the microscope and air pump helped them see plants in more detail and from new perspectives. As a result of such inspirations, academicians and their contemporaries were swept up in a compelling new explanatory momentum.

Both descriptive and explanatory botanical writing drew from all four sources—the plant, the field of botany, borrowed ideas, and new apparatus—to some extent. As a result, by the eighteenth century the concepts of the plant and of botany were transformed. Natural history focused primarily on naming, describing, and classifying plants, while natural philosophy studied the processes related to the plant's life cycle, drawing heavily on theories and equipment developed in other contexts.

At the Academy the natural history was inspired primarily by natural phenomena and by ideas from traditional botany. When academicians drew on ideas from other sources, they relied on chemistry. Oddly, the natural history of plants excluded anatomy, even though this was the mainstay of the Academy's natural history of animals. Instead academicians applied any anatomical study of vegetables to their physiological theorizing. Their natural philosophy, in contrast to their natural history, depended for its inspiration principally on other disciplines, especially zoology, and on new instruments. In the end it altered the very idea of the plant itself.

Academicians focused their natural philosophical research into plants on three questions: does sap circulate in plants the way blood circulates in animals; which accounts better for plant physiology, chemical or mechanical explanation; and do the air pump and microscope clarify how plants reproduce and grow? The present chapter describes analogical reasoning and the Harveian model that caused academicians to ask the first question; the following chapters address their answers to all three questions.

THE NATURE AND FUNCTIONS OF ANALOGICAL REASONING

Analogical reasoning has played an important role in the development of the sciences. As a means of explaining the unfamiliar in terms of the familiar or of subsuming one field under the laws governing another, it has didactic and heuristic value; fertile analogical theories may enrich both the borrowing and the lending disciplines. The stimuli to reasoning from analogy may be both general and specific. In the seventeenth century, Galileo, Descartes, Newton, and others had predisposed savants to "a unitary conception of natural forces," and discoveries in many individual scientific disciplines were so impressive as to become paradigmatic for other fields as well. Thus botanists were beguiled in particular by advances in animal physiology to essay zoological methods and theories in their own domain.[2]

Two characteristics distinguish late seventeenth-century analogical reasoning about plants and animals from earlier comparative theories. First, unlike the traditional theories of sympathies and antipathies, the new analogies provoked further tests; insofar as the model itself was experimental, so was the analogical hypothesis. Second, savants spurned the technological models and mathematical standards that many had previously favored and chose instead models and standards from the life sciences themselves. That was possible principally because the Harveian theory explaining the motions of the heart and blood offered a seductive model. Consequently, much analogical reasoning in botany reveals a double trend: toward increased experimentalism in botany and zoology and toward greater self-reliance within the biological sciences.

The most ambitious and elusive of botanical analogical leaps in the seventeenth century was the hypothesis that sap circulates in plants as blood does in animals. First propounded in 1660 by Johann Daniel Major, only one generation after William Harvey published his *De motu cordis,* the idea caught on quickly in England, France, and Italy. Although Nehemiah Grew and Marcello Malpighi are probably the best known adherents of the theory, they were not alone in exploring it systematically. Members of the Academy were the first to push the analogy to its limits. Claude Perrault and Edme Mariotte defended the idea at meetings of the Academy during the summer of 1668, with Nicolas Marchant demonstrating Mariotte's experiments and Samuel Cottereau Duclos opposing the hypothesis. In 1679 and 1680, Mariotte, Perrault, and Duclos published revised statements on the subject, and later in the 1680s other academicians, especially La Hire, tried to repair the analogy.[3] Their efforts shed light on how the Academy fostered the biological sciences, on the dramatic changes in botanical research in the late seventeenth century, and on the merits of analogical reasoning.

Before examining the theory of the circulation of the sap as developed in the Academy, a double foundation must be laid by establishing the nature of analogical reasoning and by describing the features of the Harveian model that inspired academicians. Only then can academicians' elaboration of the hypothesis be assessed.

Analogy is a means of comparing two things. By identifying similar traits in both objects, scientists infer the existence of a causal mechanism affecting both. Several types of analogy used by scientific savants have been identified by historian-philosophers of science. Claire Salomon-Bayet distinguishes "lazy-universal" from "experimental or observational" analogies; Georges Canguilhem differentiates mathematical analogies from explanation by reduction; and Mary Hesse compares formal and material analogies.

Salomon-Bayet's distinction is addressed specifically to the early modern period. Taking Paracelsianism as the exemplar, she defines lazy analogy as a mental habit of ancient origin that simply assumes untested the sympathy or antipathy of all parts of the universe. By contrast, experimental or observational analogy is open to verification and correction. Experimental analogy subsumes particular objects or phenomena under general theories either by applying the laws or theories of one discipline to another or by employing a model; the first method is more fertile than the latter, which has didactic but not explanatory power.[4]

Canguilhem focuses on analogical reasoning in biology, contrasting deduction (or the use of mathematical models) and explanation by reduction (or the use of mechanical analogies or analogical models). The former is less naive but also less useful than the latter for biology, which is not always susceptible to expression in mathematical language.[5] Canguilhem's discussion also clarifies Salomon-Bayet's distinction between analogy from theories and analogy from models. Two examples suggest the two kinds—structural and functional—of analogy: the stirrup and anvil after which the bones in the ear are named, and the ancient irrigation system which inspired the Greek concept of the motion of blood. Unlike Salomon-Bayet, who insists that a fertile analogy must prompt experiment or observation, Canguilhem allows that in biology analogy from models can be an alternative to experiment, because models permit "the comparison of entities which resist analysis."[6] Both Canguilhem and Salomon-Bayet agree that models may aid the quest for laws.

Mary Hesse is interested in both the physical and the biological sciences and, unlike Canguilhem and Salomon-Bayet, systematically analyzes analogical reasoning as a logical tool appropriate in all sciences. Hesse distinguishes between material analogy and formal analogy. Material analogy is both substantive and predictive, whereas formal analogy is neither, since it is simply a one-to-one correspondence between different interpretations of the same formal theory. Thus Hesse's analysis of material analogy is relevant to seventeenth-century botany.

Material analogy is a means of comparing different organisms or phenomena by pairing and comparing their individual traits. The model is the organism or phenomenon that is already understood; the explicandum is the organism or phenomenon that needs to be explained. A systematic comparison of their traits will determine whether the explicandum can be understood in terms of the model. Thus, if the traits in the model resemble those in the explicandum, and if the traits in the model constitute a causal mechanism, then the same causal mechanism may be inferred in the explicandum.

Hesse provides a schema for such comparisons, listing the paired traits in two columns, one for the model and one for the explicandum. Paired traits are subject to "horizontal" comparison, while the theory linking a set of traits as the causal mechanism in the model provides a "vertical" connection. The closer and more numerous the horizontal similarities between pairs, the more likely that the vertical causal chain of the model may be inferred to exist in the explicandum.[7]

Material analogies may have explanatory power if three conditions are met. First, the model and explicandum must have something in common beyond the analogy in question, that is, there must be "pretheoretic" similarities between them. Second, their horizontal similarities must be substantial. Third, there must be a causal connection among the traits in the model.[8]

Hesse's analysis clarifies Canguilhem's and Salomon-Bayet's distinctions. Thus the most fruitful dichotomy is not between models and the application of one discipline to another, as Salomon-Bayet argues, nor between mathematical models and explanation by reduction, as Canguilhem would have it. Rather, there are empty and productive analogies, and the latter may be either mathematical or material; but only material analogies with pretheoretic similarities may have predictive power.

Analogical reasoning has various advantages and shortcomings. As a form of induction, it is prey to all the shortcomings of the inductive method. But analogical argument enjoys a special role when observation or experiment are inadequate.[9] In such a case, however, it is inconclusive not only for all the usual inductive reasons but also because it rests on an incomplete identification of similarities. Nevertheless, analogy offers a means of selecting a hypothesis, because it draws attention to comparable properties that may betray causal similarities.[10]

As differentiated from experiment, analogy permits comparisons between traits or phenomena that cannot be analyzed.[11] Just as an experimenter uses theory to suggest predictions that do not proceed from tests and observations, so the savant uses analogy to suggest hitherto unsuspected causes or theoretical entities.[12]

Analogy resembles theory, because it offers a way of subsuming a pattern of behavior under a set of laws, and because it holds out the promise of generating explanations and predictions. Canguilhem explains how analogies differ from experiment and are like theories:

> What validates a theory is the possibility of extrapolation and prediction which it permits in directions which the experiment, keeping to its own level, would not have indicated. Similarly, models are judged and tested one against another by the completeness of the accounts they give of the properties to which they direct attention, in the object of study, and also by their aptitude for revealing properties hitherto unnoticed. The model, one could say, predicts.[13]

In explaining this similarity of analogical reasoning to theory, Max Black's study of metaphor is useful. A metaphor is a comparison whose thrust is indefinite. So long as the scientist does not know "how far the comparison extends" and tries to push the analogy to its extreme, unexpected theoretical implications may emerge, for "it is precisely in its extension that the fruitfulness of the model may lie."[14]

Good analogies, therefore, supplement and encourage experiments and, like theories, suggest what, unobserved, might have remained unsuspected. It sometimes happens that an analogy, like a metaphor, changes the way both the model and the thing being explained are viewed. This may be due to the impoverishment of the model,[15] or to changes in the meanings of the concepts associated with the model and explicandum; sometimes "the two systems are seen as more like each other."[16]

Analogical thought has influenced biological thought in both positive and negative ways, depending on how the model is selected. For example, when models are more appropriate than experiments, analogy can stimulate alternative observations or anticipate evidence that is inaccessible given experimental capabilities. But overreliance on mechanical and technological models has injured biological analysis, by emphasizing structure at the expense of function. Thus, Greek and Latin anatomical nomenclature suggests the appearance but not the function or causal mechanism of the anatomical part. Such analogies cannot "show the identity of the general laws of the two fields of phenomena which are brought together" and thus are causally insignificant.[17]

The risks and rewards entailed in the choice of model are illustrated in two theories of the motion of the blood. Both Harvey and the ancients explained the motion of the blood analogically. The ancients compared it to the unidirectional supply of water to irrigation channels and hence argued that blood was absorbed by the body and had to be continually replenished; their theory was conceptual, not experimental. Harvey replaced the notion of irrigation with the idea of circulation within a closed circuit, an idea that was compatible with his experimental findings.[18] Models chosen from technology on the basis of structural resemblance are likely to lead to an explanatory cul-de-sac when applied to the biological sciences. But in the seventeenth and eighteenth centuries savants began to use biological models more often, with positive results for the life sciences and especially for plant and animal physiology.

Analogical reasoning can serve as the prelude to comparative method. This is particularly important in analogies between plants and animals, which often resemble one another functionally, but have different structures. Either similarities or differences may be emphasized, but pretheoretic resemblances determine whether the comparisons are valuable.[19] The important point for comparative method is to test, not assume, any similarities. As will be seen in the following chapter, some academicians used analogy in precisely this fashion; starting with a comparison between plants and animals, they tested the analogy experimentally, admitting structural dissimilarities between plants and animals and investigating how plants accomplished equivalent functions in the absence of equivalent organs. In such cases, analogy identifies crucial dissimilarities and becomes a preliminary step to understanding causal differences.

Analogies are full of pitfalls for the unwary. Even scientists who rigorously examine the traits of two organisms for dissimilarities need to guard against claiming too much for an analogy. In the eighteenth century, Albrecht von Haller complained that analogy had become a substitute for experiment. This was the flaw that had weakened medicine, he believed, because "the great source of error in physics" was due to "physicians, at least a great part of them, making few or no experiments, and substituting analogy instead of them."[20] Analogies should inspire experiment, and although they may extend it, they should not be a substitute for it. The overenthusiastic savant runs the risk of conferring "a representational value on a model" and of letting the model become axiomatic instead of being only the lender of a mechanism.[21] Analogies are frail inductive tools. When the identification of similarities or dissimilarities is incomplete, then the analogy is inconclusive.

In summary, analogies have more than a didactic value to scientists. They offer a means of selecting hypotheses. As a weak form of induction, they can be useful when only a few instances are known or when only sparse observational data exist. They should inspire, may supplement, and will sometimes replace experiments. They resemble theories in suggesting new experiments, subsuming the behavior of an object under a general rule, predicting, and explaining. They are fertile because, like metaphors, their extension is indefinite; when scientists push analogies to their extremes, they may discover what would otherwise have eluded them. The best analogies pay compound interest on what they have borrowed, by changing the way both the borrowing and the lending disciplines are perceived. Finally, failed analogies have a particular use: by drawing attention to crucial dissimilarities in the things being compared, they stimulate a search for the causal mechanism. Failed analogies, therefore, are the beginning of comparative method and show where further research will be necessary.

Improved analogical reasoning helped advance the biological sciences. The choice of model was important. What was needed was a broadly conceived, well developed theory, one with adequate observational and experimental supports to win adherents, and sufficient specificity to avoid sectarian splits. Chemistry was appealing, and was regarded by Paracelsus, Helmont, Boyle, and others as a source of knowledge about the basic and unchanging components of the universe, but it was beset by doctrinal divisions.

In the absence of an approved general explanatory theory, savants had recourse to smaller-scale theories with a more limited explanatory range. William Harvey's writings on the movements of the heart and blood offered just such a theory, and as his hypothesis became acceptable scientists in other fields borrowed it. In the Academy and elsewhere, botanists formulated the hypothesis that sap circulates in plants. Their analogical leap is an important case study for several reasons. It is an example of borrowing from within the biological sciences and illumines the relative importance of structural and functional analogies in botany. It indicates whether savants tested analogy experimentally and how they responded to its limitations. Finally, it reveals that when the model failed, academicians used analogy as a stimulus to comparative method and thus came to ask an important botanical question.

THE CIRCULATION OF THE BLOOD

Three theories of circulation competed in France after 1628: the hypotheses of William Harvey, Jean Riolan, and René Descartes. Harvey claimed that blood made a complete circuit of the body, that the heart pumped it into the arteries, that blood then passed to the veins and returned to the heart, and that blood nourished and heated the body. He believed that the heart, veins, and arteries were "constructed for that purpose with extreme foresight and wonderful skill," and thus that their structures revealed their functions.[22] Descartes and Riolan accepted the idea of circulation. But they disagreed with Harvey about important details, challenging, for example, his estimate of the speed with which blood circulated. As a result, they proposed alternative explanations of the motions of the heart, and they retained certain elements from ancient theories about the motion of the blood. French botanists who wished to formulate an analogical theory of the circulation of sap had, therefore, three models from which to choose. None, however, was perfectly compatible with vegetable anatomy and physiology. These models must be clear if the theories of the circulation of the sap are to be understood.

Harvey's theory was the most important and was widely accepted in scholarly circles by the late 1660s. It was novel in several respects: it unified the venous and arterial systems, described the pulse as a mechanical effect of the heartbeat, calculated the quantity of the blood, and characterized the circulation as a closed system in which all blood returned to the heart without being consumed. Harvey challenged standard notions about the hierarchy of bodily organs. He maintained that the blood was more important than the heart because it preceded the heart in the development of a fetus. He also claimed that the heart was formed prior to the brain and liver and was thus more important to life than either of those organs.

Harvey also retained certain traditional views. He believed, for example, that the purpose of circulation was to nourish and warm the body by generating heat and spirits necessary for life.[23] This made a circulatory motion necessary, in his view, not only so that all the parts of the body "may be nourished, warmed, and activated by the hotter, perfect, vaporous, spirituous and, so to speak, nutritious blood," but also in order to repair the blood which "may be cooled, coagulated, and be figuratively worn out" in its travels. The heart held a special, beneficial position in the body, for it was the "source or the centre of the body's economy" and could restore blood "to its erstwhile state of perfection. Therein, by the natural, powerful, fiery heat, a sort of store of life, it is re-liquefied and becomes impregnated with spirits and (if I may so style it) sweetness."[24] The Harveian model was experimental and mechanistic, but as such passages reveal, it was also teleological and vitalist.[25]

To prove his theory, Harvey tested and confirmed three assumptions. First, "the blood is continuously and uninterruptedly transmitted by the beat of the heart... into the arteries" in large quantities that cannot be made up by intake of food. Second, the pulse of the arteries drives the blood into every part of the body, in greater quantities than necessary for nutrition, and in such amounts that a rapid circular motion must be assumed. Finally, "the veins themselves are constantly returning this blood from each and every member to the region of the heart." As proof, Harvey cited his measurement of the amount of blood passing through the body in a half-hour; his experiments with ligatures of blood vessels; and his description of the structure and function of valves in the veins. Once the three suppositions were confirmed, Harvey could state "that the blood goes round and is returned, is driven forward and flows back, from the heart to the extremities, and thence back again to the heart, and so executes a sort of circular movement."[26]

Although Peiresc and others in France defended Harvey's theory from the beginning, Riolan and Descartes both proposed alternative theories of the circulation of the blood. Riolan, a respected anatomist and member of the medical faculty of Paris, did not object to the idea of circulation in itself. But he found Harvey's formulation distasteful because it challenged Galen and undermined some of the theoretical bases of traditional medical practice. Riolan also mistrusted Harvey's assumption that the anatomy of animals may resemble that of humans.

Riolan argued that the blood traveled through the arteries and veins to the extremities of the body and returned to the heart two or three times a day. Not all blood returned to the heart, however, because some of it was assimilated into the body. Although the normal route of the blood was away from the heart in the arteries and to the heart in the veins, when the veins of the arms and legs threatened to become empty the blood in the veins of the trunk could flow backwards to prevent a void. Thus Riolan maintained conventionally that blood ebbed and flowed in the veins and that it was consumed as nutriment by the parts of the body.

Riolan also calculated the amount of blood in the heart and the entire body and the quantity of blood that passed through the body in one hour. But he disagreed with Harvey. Riolan did not believe that the heart propelled the blood, as Harvey had shown. Instead he claimed that the blood kept the heart in motion, as a stream moves the wheel of a water mill. In

Riolan's view, blood prevented the heart from drying out, while the heart reheated the blood and replenished it with spirits. Although Riolan agreed with Harvey about that function of the heart, he contradicted him in insisting on the primacy of the liver.[27]

Descartes's theory was closer than Riolan's to Harvey's. Thus Descartes accepted the full circulation of the blood through the body, but he rejected Harvey's theory of the motion of the heart. Arguing that physiological phenomena resulted from chemical processes, Descartes claimed that when the wet blood reached the hot heart it vaporized and expanded. This stretched the heart. As the blood cooled, it condensed, and the heart contracted. This alternate vaporizing and condensation accounted, in Descartes's system, for the heartbeat and pulse.

Both Descartes and Riolan agreed with Harvey that there were anastomoses connecting the arteries to the veins. Descartes incorrectly gave Harvey credit for discovering them, although Harvey had simply assumed they existed. Descartes accurately summarized three of Harvey's proofs for the circulation, namely the argument from ligation, the argument from the function of valves in the veins, and the fact that all blood in the body can exit from one cut artery. But he did not stress Harvey's estimate of the amount of blood that passes through the heart in an hour. Like Harvey and Riolan, Descartes believed that circulating blood carried heat and nutrition to the body. Like Harvey, he argued that blood was not itself a nutriment but carried food. In order to explain how the body obtained this food, Descartes drew on an analogy with sieves, which permit small particles to pass but retain larger ones. Descartes believed the heart repaired and renewed the blood. To explain the motion of the heart, Harvey and Riolan cited mechanical models—a pump and a mill—while Descartes, the mechanist, derived his explanation from chemical processes.[28]

Harvey first published his theory in *De motu cordis* in 1628. It quickly found defenders and detractors in France. Although it was banned from the Parisian medical school, lecturers at the Jardin royal disseminated the theory, and by the 1660s a Harveian school had established itself in France, counting among its members Claude Tardy, Jean Pecquet, Jacques Mentel, Pierre Guiffart, Jean Martet, Jacques Chaillou, and Pierre Betbeder. Several defended the Harveian theory of circulation in vernacular treatises about such topics as the lacteal veins, the lymphatic vessels, chyle, and the preparation of blood. Their affiliations reveal that medical faculties had become receptive to the theory of circulation and that even physicians educated by faculties hostile to the theory might adopt it. Tardy was physician to the duke of Orléans and doctor regent at the Parisian medical faculty. Chaillou practiced medicine at Angers. Martet was a master surgeon and royal anatomist in the faculty of medicine at Montpellier. Mentel had been educated in medicine at Paris in the late 1620s and early 1630s. Pecquet corresponded with Harvey and became one of the original members of the Academy. Most of Harvey's defenders published in the vernacular, perhaps hoping, as Guiffart put it, to reach a less dogmatic audience. Harvey's French proponents stressed his quantitative findings (although they gave different figures) and his experiments with ligatures, but they relied far less on analogies to explain the theory than had Harvey.[29]

ANALOGICAL REASONING IN THE HARVEIAN MODEL

Harvey prefaced his *De motu cordis* with a plea for analogical reasoning. Indeed, in developing his theory in *De motu cordis* and *De circulatione sanguinis,* he discussed about a dozen analogies, using most of them to support his own theory. Some analogies he borrowed from the Greeks, some from political theory. Some were biological, and some were mechanical.

Harvey indulged in lazy or macrocosmic analogy only twice. In one case he made the commonplace observation that circularity is a natural phenomenon, citing Aristotle's view that "the air and rain emulate the circular movement of the heavenly bodies," that the condensation and evaporation cycle is a kind of meteorological circularity, and that the sun's circular motions cause storms.[30] Harvey's purpose here was to justify a loose use of the terms "circle," "circularity," and "circulation." Moving from the truly circular revolutions of heavenly bodies—Harvey was no Keplerian—to the figurative circularity of the condensation-evaporation cycle, Harvey suggested that in this figurative sense, repetition constitutes circularity, which is thus akin to rejuvenation.

The second instance of macrocosmic analogy evolved into a biological analogy. Harvey wanted to show that "blood permeates from the right ventricle of the heart through the parenchyma of the lungs into the vein-like artery and the left ventricle." To show that such a passage is possible in nature, Harvey reminded the reader that water seeping through the earth "gives rise to streams and springs." Two other examples—sweat passing "through the skin" and urine "through the parenchyma of the kidneys"—show that Harvey employed a lazy analogy to demonstrate the general possibility of seepage in nature. But he chose a biological comparison with sweat and urine to illustrate seepage in the body.[31]

A clearer instance of causal biological analogy exists in Harvey's explanation of the two motions of the heart, which "occur successively but so harmoniously and rhythmically that both [appear to] happen together and only one movement can be seen."[32] He cited three analogies. Two were technological (involving comparisons with geared machinery and flint-lock firearms). The third and most extended comparison was with swallowing, and Harvey used this biological analogy to make his causal point.

Harvey anticipated later seventeenth-century biologists in taking "as a model of the living thing the living thing itself."[33] He did not hesitate, however, to use mechanical models — such as the pump, the glove, and the filling of leather bottles — to draw causal inferences, starting from the premise that similar motions have similar causes.[34] Thus, in careful hands even technological models could serve as causal analogies for the biological sciences.

Harvey's analogies reveal various causal assumptions and traits of argument. First, he explained biological processes chemically. Second, several models, such as the image of the "carefully planned and ingenious arrangement of ropes on a ship," were solely didactic. Third, in scrutinizing the analogies of his opponents, Harvey demanded rigorously close comparisons, pushing the analogies of others to their limits and ridiculing inept comparisons (like the notion that blood flows like water in the seas) by reductio ad absurdum.[35]

In summary, Harvey used analogies in three ways. He proved that a process (such as permeation) was possible in one structure because it was already known in another. He taught by likening a phenomenon to a more familiar sight (such as a gun, a machine, or the ropes on a ship) whose workings were either well known or obvious to the observer. Finally, he justified the loose use of the word "circulate." Reasoning from macrocosmic models was relatively insignificant for his argument. While Harvey demanded that his opponents' analogies be accurate with respect to both behavior and causation, his own models were principally a source of general inspiration or a means of teaching. Whether heuristic or explanatory, they came mostly from outside the biological sciences. His causal analogies (the comparisons with swallowing, the leather bottle, the glove, or the fermentation of wine) simply assumed that similar phenomena resembled one another because they had similar causes. Finally, Harvey's analogical reasoning was not systematic but rather ad hoc or episodic. Thus, analogy rarely inspired him to apply either the methods or theories of another discipline to his own; it was unusual that likening the heart to a pump led him to measure the flow of blood through the body or that chemical comparisons led him to draw theoretical inferences.

CONCLUSION

Harvey's theory of circulation unified the heart, veins, and arteries in a single system. He believed that his theory had utilitarian implications and could explain some "events fundamental in practical medicine," such as "the suppression or cause of hemorrhage, sloughing and gangrene, the assistance derived from ligature in castration or the removal of tumours."[36] The explanatory power of the theory encouraged Harvey and his proponents to use it also to improve general knowledge of physiology, both animal and vegetable. His work offered more than a theory to botanists; it was also a model of experimental method and analogical reasoning.

By the 1660s the idea that blood circulates was well established in France, save in a few ultra-conservative circles. Although Harvey's theory had triumphed, it remained controversial and some savants were loyal to the competing views of Riolan and Descartes. Circulatory theory depended principally on observation, quantitative analysis, and an assumption that structure and function were closely related. But it was also indebted for its inspiration and exposition to analogical reasoning. It was only reasonable, therefore, for natural philosophers to apply circulation theory to plant physiology. While German and English savants were the first to raise the possibility, the French academicians soon surpassed them by testing systematically how well the model applied to the vegetable kingdom.

CHAPTER 10

Analogical Reasoning: The Theory

The hypothesis that sap circulates in plants just as blood circulates in animals was not the first attempt to liken plants to animals. Animals had served as models for explaining plants since ancient times, and seventeenth-century savants equated seeds to eggs, named the parts of plants after the parts of animals, and compared the structures of plants and animals.[1] Such ideas became productive in the late seventeenth century, however, because botanists relinquished the Aristotelian distinction between animal and vegetative souls, which separated flora and fauna into separate causal categories, and because they sought causal mechanisms instead of causative "faculties."[2] What had formerly been lazy analogical thought that merely affirmed the unity of living things became instead a guide to the empirical investigation of causal mechanisms. Thus, Harvey's theory reanimated botany at the end of the seventeenth century because it offered a heuristic model: it suggested experiments and practical applications that might ensue from a transplanted theory, and it stimulated the search for a causal mechanism.[3]

THE CIRCULATION OF THE SAP

In the summer of 1668 Claude Perrault and Edme Mariotte defended the hypothesis that sap circulates in plants as blood does in animals. They identified five ways in which plants resemble animals: plants have two sorts of vessels, corresponding to veins and arteries; there are two sorts of sap, and these are the equivalents of venous and arterial blood; sap is nutritious for the plant, just as blood nourishes the animal; the root manufactures sap just as the liver produces blood; and sap circulates frequently and quickly through the plant, replenishing itself with water from the leaves and being recooked in the root, just as blood circulates and is refreshed during its circuit through the body. Like Harvey, they had a hierarchical concept of the organism and emphasized "the control and stewardship" of one part of the body over the others, and like Riolan they incorporated traditional physiology—for example, the idea that the blood was itself a nutriment—into their theory.[4]

The debate that the Academy sponsored marks the first systematic effort to apply circulatory theory to plants. Similar ideas were current outside the Academy in the 1660s: Johann Daniel Major had suggested the analogy, Timothy Clark had written about a circulation of the liquid in sensitive plants and had searched with a microscope for structural equivalents of valves, and Nicaise Le Febvre had compared the functions of sap and blood. In the 1670s and 1680s Nehemiah Grew and Marcello Malpighi impressed the botanical world with their systematic studies of plant anatomy and physiology.[5] But it was Mariotte and Perrault who first pushed the analogy between blood and sap to its limits.

The debate of 1668 represents one of the Academy's most productive efforts at refereeing research. It began with the conflicting claims of Mariotte and Perrault for priority, and ended amicably by recognizing that their independent judgments had coincided. The exchange of evidence and opinions in 1668 influenced not only the two protagonists but also Duclos. Originally drawn into the debate to review the evidence, Duclos opposed the theory during the summer of 1668 but supported it in 1680. Finally, Perrault, Mariotte, and Duclos published their views about the theory.

The essential scientific traits of Perrault and Mariotte are exemplified by their work on the circulation of sap. Perrault was theoretical. He conjectured, offered plausible arguments, and identified the need for experimental support. In citing experiments, however, he rarely used the first person, and all the experiments cited in his 1680 book were actually performed by Mariotte, Huygens, Duclos, and Bourdelin at the Academy, not by Perrault himself.[6] In contrast, Mariotte was emphatically experimental, and even his initial inspiration that sap circulates was prompted by an experiment.[7]

Proving the circulatory hypothesis required academicians to address the three issues that Hesse has identified as crucial. Academicians had to establish the pretheoretic similarities between plants and animals that would make the analogy materially plausible; here they relied on lazy analogies and on a functional resemblance. Next they had to push the analogy to its limits, testing for traits in plants that would correspond to those in animals; here academicians identified crucial dissimilarities that ruled out the relations of causality they originally anticipated. Finally, they were faced with the problem of crucial dissimilarities: plants were not comparable to animals in several significant respects, and in particular they lacked any internal motive force that could pump the sap as the heart pumped the blood. Because this latter dissimilarity could not be resolved, the analogy as a whole failed, and Mariotte and Perrault were left with a most difficult and important botanical question: how does sap rise in the first place? For an answer they turned ultimately to disciplines other than botany or zoology.

PRETHEORETIC PLAUSIBILITY

Botanists in the Academy first had to show that the circulation of sap was plausible. Like Harvey, who reminded his readers of other circular motions in nature and claimed "as much right to call this movement of the blood circular as Aristotle had to say that the air and water emulate the circular motion of the heavenly bodies," Perrault compared the circulation of sap to the cycle of condensation and evaporation. Harvey and Perrault both dwelt on the differences between living creatures and other things, Harvey stressing the connection between heat and life, Perrault the "natural connections" that unite the various parts of the body. Mariotte emphasized another of Harvey's themes, the connection between movement and life. Harvey affirmed that motion is necessary to generate the heat that is associated with life and pointed out that blood clots when it does not move, and Mariotte observed that still liquids stagnate and become corrupt or death-like.[8]

Pretheoretic analogy or plausibility rested here on lazy analogy. Plants and animals were living creatures that depended for life on the motion of liquids that transported vital heat to all their parts. These were seventeenth-century truisms about the nature of life. Better pretheoretic support came from the specific context in which circulation of the sap was proposed. Perrault and Mariotte believed that this was above all a question of nutrition.

In 1667 Perrault had introduced the theory of circulation as a way of explaining how plants are nourished. In the following year he cast his entire analysis in the context of whether plants and animals are nourished similarly. When Mariotte explained what aspects of blood he meant to compare with sap, he started with the reception of chyle by the lacteal veins in the mesentery and their transmission of this food to the venous blood. Following this non-Harveian interpretation of the lacteal veins,[9] Mariotte described the full circuit of the blood and then a hypothetical circuit of the sap:

> Probably the ends of the roots imbibe liquid from the earth and carry it into the body of the root. From there it passes into small vessels in the stem; and then it is distributed to the branches and ends of the leaves. The remainder is carried along different small channels to the root to be perfected by a type of cohobation and in order to become a well-digested sap, appropriate for the nourishment of flowers and fruits.[10]

When Mariotte published the theory, he called his treatise "On the Vegetation of Plants," and he embedded the circulatory theory in a broader discussion of the chemical composition of plants, their germination and growth, the origins of vegetable nutrients, and the effects of plants on other living creatures.

These were the general grounds on which circular motion of sap was plausible. In addition to the vague principle that motion is necessary to generate life-sustaining functions, academicians cited the more specific need for nourishment characteristic of all plants and animals. A functional rather than a formal resemblance between plants and animals stimulated the analogy. But to confirm it Mariotte and Perrault still had to show that sap actually circulated and to find structures in plants that resembled the circulatory organs of animals.

PUSHING THE ANALOGY TO ITS LIMITS

Academicians sought to prove that sap circulated, first, by showing that their general theory of growth necessitated circulation and, second, by developing experimental evidence that demonstrated the descent of sap. Scientists at the end of the seventeenth century explained growth mechanically as resulting from the pressure of blood or sap against vessels in the extremities of animals and plants. That is why, explained Perrault, the French use the word "pousser" to speak of growth. Since all parts of plants and animals grow larger, all must be subject to this pressure. From a mechanistic theory of growth, it followed that sap must push downward as well as upward, because roots and the tops of plants grew, downwards and upwards respectively, in proportion to one another.[11]

Experimental evidence for circulation was sought in several ways. Everyone assumed that sap rose from the root to the top of the plant. The novelty was in showing that it descended again to the root. In 1668 Mariotte had already performed most of the experiments that proved descent of sap from the tops of plants toward the roots. He cut stems and observed that sap flowed in both directions. When he planted seeds upside-down, or placed them with the leaf-end in water and the root-end exposed above the water, the seeds grew. When he cut the filaments on roots, they bled. Uprooted chives, placed in water with only the shoots immersed, survived and grew for a fortnight. Most of the early experiments immersed seeds, plants, or parts of plants upside-down in water. Usually the plant survived and grew for several days. From these experiments, Mariotte and Perrault concluded that sap did move toward the root, that they had found two kinds of sap (yellowish and whitish in color, or thick or thin in consistency) in most plants, and that leaves could absorb water.[12] Proving experimentally that sap descends to the root hence led to a promising observation, that there were two kinds of sap, and to an unforeseen consequence, that leaves had an important role in the nutrition of plants.

If leaves could absorb food, they endangered the analogy with animals, because an animal ingested food through a single mouth. Mariotte tested leaves in water and observed the sap in their branches. He concluded that leaves not only absorbed water but also carried more water than did the root to the vessels containing yellow sap. He also noticed that drops of water formed on the leaves of plants under a glass bell. Assuming that these were dewdrops, seeing that the plants remained healthy, and believing that no other water was available to the plants, Mariotte concluded that the leaves absorbed enough water to sustain the plant.[13] Mariotte, the experimentalist, showed that plants, unlike animals, could take their food through two orifices. Perrault, the theoretician, tried to reconcile the new evidence to the analogy with animals. Here Harvey's own theory offered an explanatory model, for he had noted that medicaments applied externally to one part of the body entered the bloodstream and traveled to the entire body. Perrault, therefore, suggested that leaves absorbed liquids the way the skin of a dog absorbs the heat of a fire, or the skin of a butcher the fat of the meats he is handling. Mariotte affirmed the consequences of the experiments, while Perrault drew on folk wisdom and lazy analogy to reconcile the behavior of plants and animals.[14]

If the descent of sap proved the circulation of sap, the existence of two kinds of sap in a plant provided another likeness between blood and sap. Harvey had shown that there were two kinds of blood: one, going out to the body from the heart, that was "hotter, perfect, vaporous, spirituous and, so to speak, nutritious," and one whose nourishment and heat were exhausted and which returned to the heart "cooled, coagulated, and . . . figuratively worn out."[15] For sap to be comparable to blood, therefore, rising sap should nourish and falling sap be weak and useless. Perrault asked whether there was a thick sap equivalent to arterial blood and a watery sap equivalent to venous blood, and Mariotte found plants that contained two different saps, yellow or white, thick or thin.[16] The difficulty was to show that one was more nutritious than the other. Citing trees tapped for their sap in the spring, Perrault argued that falling sap cannot be nutritious, or trees would die from the loss of so much of it. Academicians analyzed the saps chemically but could not agree on the results.[17] Although academicians compared the two sorts of sap with venous and arterial blood, they asserted resemblances more effectively than they proved them.

The analogy also constrained academicians to find separate vessels for carrying sap up and down the plant. In 1668 Mariotte could not determine whether there were two kinds of vessels. Perhaps mindful of Harvey's insistence to Riolan that one vessel could not accommodate simultaneously the flow of two liquids in opposite directions, Mariotte inferred from the two kinds of sap that there must be two sorts of vessels. Perrault, on the contrary, defended Riolan's stance, and tried to show by analogy that a vessel might permit both upward and downward flow, if the two liquids were sufficiently different.[18] By 1679, Mariotte had studied the anatomy of plants more systematically and with a magnifying glass. He described the appearance of stems and the arrangement of fibers, channels, and spongy matter in them.[19] But he still could not positively identify different vessels for different saps. In 1680, Perrault argued from Mariotte's 1668 experiments and from general information about trees that some plants had separate vessels and were therefore like "perfect" or higher animals. But he cautioned that the absence of separate vessels could not disprove circulation. Some plants were like insects, which do not have separate vessels. Perrault argued that the parts of plants must be able to differentiate cooked from raw sap, perhaps because of the disposition of their pores; in other plants the double bark or the bark and the pith might serve as separate conduits for different saps.[20]

Equally serious was the lack of a mechanism for controlling the direction of flow. Mariotte and Perrault were uncertain whether they could identify valves or equivalents that prevented the rising sap from falling prematurely.[21] Among academicians, La Hire agreed with Robert Hooke's sentiment that valves seemed "very necessary for conveying the juice of trees up to the height of sometimes 200, 300, and more feet; which he saw not how it was possible to be performed without valves as well as motion." La

Hire claimed to have found valves in canes and reeds.[22] Allowing expectation to prejudice observation, he argued from similarity of function to resemblance of structure and insisted that he had found valves where none existed.

Skepticism about the existence of valves or separate conduits required academicians to consider whether sap flowed in different directions in the same vessel. Grew believed this was so.[23] Perrault tried to explain that it was possible by using an analogy with water vapor, which can rise through oiled paper but cannot penetrate it again once the vapor has condensed. He offered a second analogy, with two sponges, one soaked in water and the other in oil, each of which absorbed liquid of its own type when placed in a mixture of oil and water. Both arguments presumed that the parts of plants were designed to accept one kind of sap and reject the other.

Each structural comparison endangered the analogy. The closest resemblance was between blood and sap, liquids that were of two kinds and that made a complete circuit of the body in question. Organs and blood vessels, however, challenged the ingenuity of savants. Academicians tried to compare the root to the heart, the soil in which a plant stood to the intestines, but even with a microscope they could not identify two distinct sets of vessels. The most serious difficulty, however, was that since plants lacked a heart, they had no pump to drive the sap.

SOLVING THE PROBLEM OF CRUCIAL DISSIMILARITIES

Failure to find equivalent organs weakened the analogy seriously. The absence of what Hesse calls horizontal analogues meant that the causal mechanism was missing in plants. Academicians were reluctant to discard the circulatory hypothesis altogether, once they had found evidence that sap descends. But purported equivalents, putative valves, and two-way vessels were inadequate to explain the principal problem of the circulatory analogy, the very issue Perrault had identified with his opening words in 1668: how does sap rise in the first place?

One escape from the failed analogy was to compare plants with "lower" animals, whose anatomical organization was less differentiated. Such a move was consistent with the Harveian model, for Harvey had noted that valves are "not present in all animals" and are not "made with equal skill in all the animals in which they are present." Harvey had also allowed for circulation in animals that lacked hearts, and plants seemed similar to what Harvey had called "plant-animals." These were creatures such as oysters and earthworms, which had only rudimentary hearts or no hearts at all, because they were too small, too cold, too soft, and "too little differentiated in their structure." Harvey admitted that they did not need "a propulsive organ to transmit food to their extremities." Their bodies were limbless and homogeneous. In them, ingestion "and expulsion of food is an in and out movement produced by contraction and relaxation of the body as a whole." They need no heart because "they use the whole of the body as such and an animal of this sort is in effect nothing but a heart."[24]

Valveless veins and heartless creatures, therefore, offered two escape routes to scientists who used Harvey's theory as a model for plants. A heart, valved veins, and pulsating arteries were not necessary for circulation in all animals, since the entire organism might serve as a propulsive mechanism, driving nutrients and excrement through itself. But plants differed from the plant-animals: they tended to be larger, had a more differentiated structure, and were stationary. Botanists who took the view that the entire plant could propel the sap upward would have to look for an external motive force. To show how sap rose at all, therefore, academicians had to look beyond anatomical or structural identities. They had to seek nonbiological forces operating outside or within the plant. Botanists still sought a pump, not organic but figurative, that could impel sap in a direction contrary to its natural downward flow.

In the absence of a biological causal mechanism, academicians turned to two explanatory modes: chemical and physical. The former operated inside the plant and was part of the normal physiology of a living creature. The latter could be either external or internal, depending on what kind of phenomenon was cited. Chemical explanation was consistent with the Harveian model, for the digestion of nutrients was thought to be a chemical process that resulted in effervescence and rarefaction of the digested substances. The physical explanation, on the contrary, relied on concepts of air pressure and capillary action unknown to Harvey. Both chemical and physical mechanisms were invoked by Perrault and Mariotte, who thereby diluted the biological model with nonbiological explanatory mechanisms. But the physical model itself was uncertain, since scientists in the late seventeenth century were unclear about the causes of capillary action. Thus the analogy with the circulation of blood, which had a visible and organic causal mechanism, led botanists to adopt as a causal mechanism a mysterious phenomenon that had only recently been investigated. When the analogy between the motions of blood and sap could not be sustained, academicians related the rise of sap to capillary action and air pressure.

When those explanations seemed inadequate they cited the chemical phenomena of effervescence and rarefaction.

EXPLAINING THE RISE OF SAP

The concept of capillary action was regarded as novel during the 1660s. Robert Hooke believed the phenomenon had first been observed by French scientists:

> An Eminent mathematician told me one day, that some inquisitive Frenchmen (whose Names I know not) had observed, that in case one end of a slender and perforated Pipe of Glass, ... be dipt in water, ... the liquor will ascend to some height in the Pipe ... tho held perpendicular to the plain of water. And to satisfie me, that he mis-related not the Experiment he soon after brought two or three small Pipes of Glass, which gave me the opportunity of trying it.[25]

Robert Boyle experimented with capillary tubes in 1660, discussed capillary action in 1671 and 1676, and tried to measure the force of capillary imbibition in a seed. Huygens knew of his work, saw demonstrations of capillary action at Rohault's house in December 1660, and attended a meeting at Gresham College in April 1661 where the phenomenon was discussed; he also owned a copy of "Boyle's article on the rise of water in small tubes and on other phenomena which we call capillary."[26]

Capillary action seemed to seventeenth-century scientists to explain several natural phenomena. Hooke listed these effects of capillary action:

> ... the Rising of Liquors in a Filtre, the rising of Spirit of Wine, Oyl, Melted Tallow, &c. in the Weake of a Lamp (tho made of small wire, threeds of Asbestus, Strings of Glass, or the like) the Rising of Liquors in a Spunge, piece of Bread, perhaps also the ascending of Sap in Trees and Plants, through their small, and some of them imperceptible Pores, (of which perhaps I may say more on another occasion) at least the passing of it out of the earth into their roots.[27]

He believed that capillary action was the result of unequal air pressure, with the air pressing heavier on the reservoir of water surrounding the thin glass tube than on that within the tube itself. Hooke reiterated the view in 1665 and stated that air pressure caused sap to rise in plants, basing his explanation on the analogy with capillary action.[28] This confusion between capillary action and the effects of air pressure persisted throughout the century, despite G. A. Borelli's argument in 1670 that capillary action was not due to air pressure.[29]

Because capillary tubes resembled the stems and vessels of plants, they were an obvious model. In the late 1670s Mariotte and La Hire used the analogy between glass tubes and the vessels of plants to explain how sap rose. Mariotte limited the effect of capillary action, however, to "the first entry of the water into the roots." This, he said "occurs by a law of nature similar to the movement of union of which I have already spoken; since whenever very narrow tubes touch water, it enters, and it even rises despite its natural tendency to descend." For capillary action to work in plants, three conditions had to exist: the water in the soil had to touch the plant, it had to have access to that part of the plant in which it could rise, and something had to cause the liquid to enter and rise in the plant. Mere contiguity seemed to be insufficient because if a glass tube were dirty or rubbed with tallow, water would not rise in it. Furthermore, the pores of plants had to be properly "disposed to allow the subtle parts of other bodies to enter." Finally, these subtle parts had to "be pushed by some principle of motion." This principle Mariotte referred to as "a movement of union" and as an effect "that is popularly called attraction"; he did not cite the pressure of air as Hooke had done.[30]

But there were problems inherent in comparing the rise of sap in plants to the rise of liquids in capillary tubes: the cause of capillary action was disputed and liquids did not rise so high in capillary tubes as they did in plants. Thus, whether or not capillary action was cited, savants turned to air pressure in order to explain the rise of sap in taller plants.

Huygens initially favored the view that sap rose because of air pressure, and in 1668 Perrault propounded an explanation that depended on multiple causes, including air pressure.[31] The theory that air pressure causes sap to rise was known to be defective, however, by 1679. Pierre Perrault and Huygens debated whether sap rose because of air pressure, as Huygens maintained, or because of attraction and nature's abhorrence of a vacuum, Perrault's view. Perrault cited the following as a decisive argument against Huygens's theory:

> How can we understand the sap that rises in trees? Can one say that air pressure causes it to rise between the bark and the wood, as in a pump? For that it would be necessary for the foot of the tree to rest in a reservoir of sap. Even if that were the case, this sap could rise only thirty-two feet, but there are trees that are more than one hundred and twenty feet tall.[32]

Pierre Perrault concluded that sap rose as a result of "attraction due to abhorrence of the void," a phrase that he and Mariotte used apologetically and as a manner of speaking not intended to impute emotions to inanimate or mechanical objects.

Since neither capillarity nor air pressure seemed sufficient to raise sap higher than thirty feet, some savants adduced chemical reactions inside the plant. Claude Perrault had argued all along that sap rose for many reasons, including the existence of appropriate passages in the plant, air pressure, external propulsion from the wind, and coction of sap. Influenced by his brother's objections, however, he modified this view when he published his *Circulation* in 1680. There he cited both air pressure and attraction due to fear of the void, but he also elaborated in Cartesian fashion his earlier chemical explanation: when sap was prepared, fermentation and effervescence reduced its concentration so that more sap rushed in to fill the potential void. To show that sap would flow into an area of diminished pressure within a plant, Perrault cited the effects of an air pump: "plants that are full of sap let it run when the air is being evacuated, and when the pressure of the air is diminished, the sap dilates and becomes less condensed than it was."[33] Mariotte developed a slightly different theory in 1681: sunlight evaporated sap in the upper regions of the plant, creating an area of lower pressure into which sap rose.[34] Such formulations anticipated Hales's conjecture that the loss of liquid due to transpiration pulled sap upward, continuing a process begun by capillary action.[35]

Tournefort presented a different eclectic theory in 1691. He believed he had found two different systems in plants, one in which sap rose by absorption and another in which it rose by capillary action. The vessels in most plants, he said, were soft, spongy, and composed of many small, empty bladders or pouches that were connected so that sap passed through them. He compared them to felt strips or cotton that filtered and conducted liquids. Not all vessels were spongy, however; the stems of water-plants were like cylinders pierced longitudinally with holes. These tubes carried sap, and Tournefort thought they resembled capillary tubes: "this structure seems to favor the sentiment of some natural philosophers who believe that the sap ascends in plants for the same reason that water rises in very narrow glass tubes."[36] Tournefort followed G. A. Borelli, who claimed that the dilation and condensation of air enclosed in plants caused sap to rise and that the spongy matter in plants facilitated that rise.[37] La Hire, however, challenged the Borelli-Tournefort hypothesis after observing that water did not rise significantly in absorbent materials such as sponges inserted in glass tubes or paper strips. The best result was a height of 225 *lignes* over a period of more than eighty-four days.[38] La Hire concluded that neither absorbent matter nor capillary action could account for the rise of sap.

Instead, he maintained that only hollow tubes in plants could transport sap, and he claimed to have found hinged, woody valves in them that enabled sap to rise.[39]

The problem of how sap rose elicited varied responses from academicians. Huygens proposed air pressure, while Perrault listed several interrelated causes, such as air pressure, wind, fermentation, and the different weights of raw and cooked sap. Once air pressure was ruled out as a sufficient cause, Mariotte and La Hire focused on capillary action, while Tournefort combined capillary action with a Borellian theory about spongy matter in plants. Later, La Hire adopted the view that the vessels of plants were valved. In each case, academicians used analogies. Since the biological model of valves was inapplicable, they drew mostly on chemical or physical explanations. But the limitations of all models forced most academicians to develop theories based on multiple causation.

CONCLUSION

Academicians used the hypothesis of a circulation of sap to search for causal mechanisms. By choosing the Harveian model botanists tried to replace the two principal modes of explaining living things — chemical and mechanical — with a biological one. When the analogy failed, academicians had three choices. They could force a biological explanation by insisting on false structural resemblances. They could fall back on either or both of the traditional modes of explanation. Or they could draw on new physical models that were only half understood. Since the Harveian model itself assumed chemical processes within the physiological and retained a technological model for the heart, recourse to nonbiological explanation was broadly consistent with the model.

Circulatory theory had both substantive and methodological shortcomings. Although a circuit of sap was established, structural resemblances alone between plants and animals could not justify a causal analogue. The effort at methodological equivalence fared no better, despite the experimental ingenuity displayed by Mariotte and La Hire, whose demonstrations of the direction of flow sometimes resemble Harvey's tests. Harvey's crucial experiments were done on living creatures, however, and his theory owed a great debt to his skill at vivisection, but all of Mariotte's and La Hire's dissections were of dead plants. Experiment is necessarily an artificial procedure that may distort its object, and it is least informative when it examines defunct organisms in order to understand physiological processes.[40]

Given these failures of the analogy, did it have any value at all for seventeenth-century botanists? First, the circulatory analogy had useful consequences despite being a weak form of a partially failed induction. In the absence of a compelling alternative, botanists found a partial analogy better than none at all. Although the causal resemblance failed, a circuit of sap was established. By calling that a circulation, botanists implied that plants enjoyed the digestive and perfecting processes characteristic of animal nutrition. While the circulation of sap could not be subsumed under the laws governing the circulation of blood, the term "circulation" reminded botanists of both the circuit and its function, if not its causal mechanism. Hence, the analogy with the movement of blood supplied a suggestive language to botanists, who retained some of the connotations that the word "circulation" had grown to have for natural philosophers.

Second, because academicians used the experimental and observational form of analogy, their analogy was both positively and negatively useful. It aided "the investigation of structure and the relation of structure and function" and helped reveal some "properties hitherto unnoticed."[41] It also led academicians to ask new questions and to propose different causes, causes that were testable. Because the extension of the metaphor was more limited structurally than had been foreseen, however, botanical investigation did not affect notions about the physiologies of animals or "plant-animals." Finally, those who did not insist on putative valves in plant vessels learned from the circulatory analogy that there was no organ in plants to make sap rise.

Third, when a model cannot lend its mechanism because of structural disanalogies, the principal value of analogical reasoning must be as an inducement to comparative method. In the case of plants and animals, this may be inevitable, for the two types of organisms enjoy similar functions but dissimilar structures. Analogies used experimentally can draw attention to these problems. But for analogy to work as comparative method, the researcher must not assume the identity of the two things being compared. Therefore, an analogical argument must start by elucidating similarities and differences, as Hesse has pointed out. In this case, the analogy with animals forced botanists to ask how plants accomplish certain functions without having the appropriate organs. By retaining the analogy while admitting structural dissimilarity, academicians moved from analogical to comparative method. That is, they used the analogy to locate specific resemblances and differences; they then tried to explain the differences by comparing the causal mechanisms of the two. In the case of the Academy's study of sap, this transition was incomplete. Some academicians persisted in pressing the structural comparison by searching for valves. No one seems to have questioned the functional analogy at all, so that the physiology of plant and animal nutrition was assumed rather than tested.

Academicians simultaneously escaped from and succumbed to the dangers of analogy. Although they experimented and acknowledged numerous and crucial dissimilarities, they also assumed fundamental resemblances without examining them. In matters of nutrition, analogy did substitute for experiment. Moreover, when La Hire was driven to find nonexistent valves, he let the model become axiomatic.

The Academy's circulatory analogy could be disproved by experiment and observation. Unlike Harvey's own analogies, which were didactic or were instances of similarity chosen to promote general plausibility, the analogy between sap and blood was falsified by significant dissimilarities. As Canguilhem has pointed out:

> A good hypothesis is not always that which leads rapidly to its own confirmation,... It is that which obliges the researcher, by dint of an unforeseen discord between the explanation and the description, either to correct the description or to reconstruct the schema of explanation.... [I]n biology the models which have the chance of being the best are those which halt our latent tendency to identify the organic with its model[.][42]

The hypothesis of the circulation of sap had merit in drawing attention to a central problem in botany, namely, how sap rose at all.

It further represents an early attempt to use biology itself as a model for biological explanation. Academicians helped transform botany by finding new explanatory models. But when the biological model failed to supply a causal mechanism, academicians resorted to nonbiological causes, a fruitful reliance that affirmed the interdisciplinary character of scientific explanation. It was not "the impossibility of explaining how the vegetable machine works solely by the laws of motion"[43] that drove botanists to zoology for inspiration. Rather an incomplete correlation between the zoological model and the vegetable explicandum forced savants back to nonbiological explanation. Seventeenth-century botanists and anatomists found that mechanics, physics, and chemistry were necessary weapons in their explanatory armory. Far from demonstrating a failure of self-image in the biological sciences, the resort to chemistry and physics exemplifies the cross-disciplinary fertilization so important for early modern science, whose practitioners were adept in many fields. Moreover, it reveals a nondogmatic use of analogy. Academicians proved some resemblances but also identified crucial dissimilarities that led them to identify an important scientific problem whose solution lay outside the original analogy. In so doing, they transformed analogical reasoning into comparative method.

CHAPTER 11
Chemical and Mechanical Explanation of Physiological Processes

The seventeenth-century upheaval in natural philosophy meant that savants were embroiled in countless debates. They disputed theories and challenged evidence, and they disagreed about the order in which investigations should proceed. Although in principle academicians argued for completing natural historical research before addressing natural philosophical questions, in practice they could not wait. In botany, they sought principles that would account for the life cycle of a plant and turned to three forms of explanation: chemical, mechanical, and biological, with the biological model itself depending on chemical and mechanistic explanation. Chemistry was pivotal because it was susceptible to both mechanistic and vitalist interpretations; like botany it was in the throes of redefinition. Academicians were uncertain just how far they could reduce chemical processes to mechanical causes, and their eclectic accounts of plant physiology reflect their quandary.[1]

GENERATION AND REPRODUCTION

Perhaps the most perplexing aspect of the life sciences during the seventeenth century was generation. The respective roles of female and male, the nature and function of eggs and spermatozoa, and the sexuality of plants were all at issue. When academicians considered the generation of plants, they focused primarily on determining which of two incompatible mechanistic theories—spontaneous generation or preformation—accounted more plausibly for certain phenomena. According to the theory of spontaneous generation, animals and plants could be generated from soil, air, or corpuscles directly, without the need for progenitors of their own type. Preformationists held that living creatures carried within themselves perfectly formed descendants in miniature, so that the process of generation involved little more than giving birth to these entities. New experiments, observations with the latest instruments, and theological objections undermined both theories, however, and at the same time clarified the role of seeds in plants.[2]

Perrault opened the Academy's discussion of germination and reproduction with his paper of January 1667; academicians subsequently studied seeds, earths, and the conditions for germination, examined vegetative reproduction, and debated preformationism. Always interested in testing the claims of the ancients, Perrault and Dodart were skeptical about Theophrastus's assertion that plants grew from their saps, and also about the contemporary chemical notion that salts extracted from plants acted as seeds.[3] Other academicians emphasized reproduction by seeds and by vegetative propagation. Here the natural history complemented their natural philosophical research. The Marchants described the location and appearance of the seed, Bourdelin analyzed seeds chemically, and Homberg discussed their properties.[4]

Most academicians were dubious about spontaneous generation. In 1667 Perrault suggested testing the theory by observing earth taken from so deep in the ground that seeds could not have penetrated there; this would determine whether plants could grow without seeds. In the 1690s Tournefort showed that it was best to assume plants had seeds; if seeds had not been seen, it was probably because they were small enough to elude observation. He identified them in several plants previously said to have been seedless, especially ferns. Examining with a microscope the dust-like dots on ferns, he saw that they were "tiny sacks, each of which contains a large quantity of seed." In one capsule Tournefort counted more than three hundred individual seeds, and he grew a fern from some of them. He opened capsules previously thought to be the seeds of *Lunaria* and *Polypode* and, using a microscope, showed that they too contained numerous tiny seeds. Where even he could not find seeds, Tournefort argued analogically: if a plant whose seed was unknown passed through the same stages of growth as a plant known to grow from seeds, then it was proper to infer that the first plant also grew from seeds. Thus Tournefort claimed that the maidenhair fern of Montpellier (*capillaire de Montpellier*) grew from a seed, because its shoots consisted of a leaf and a thin root, like those of other plants.[5]

Although Tournefort described his own observations in the context of Grew's, Ray's, and Morison's discoveries, he criticized his English counterparts for stopping short of full discovery.[6] He contradicted Morison, for example, who believed that mushrooms sprang directly from the earth. In a large fungus taken from the woodwork in the abbey house of Saint Germain, Tournefort identified as seeds some fine dust attached by delicate threads to the pores of the fungus. He then looked for their source, what he called the ovary of a plant, in the rough crust on the back of the mushroom but concluded that this could not be the ovary since there was no seed there. Since he believed that similar natural objects should display similar natural processes, Tournefort disagreed with Morison's explanation of why the fungus *Erysimum* was more common after the London fire of 1666. Morison claimed that the mushroom had grown spontaneously from soil that was altered by the fire, but Tournefort replied that changes in the soil simply encouraged the seeds to germinate.[7]

Both Mariotte and Tournefort examined the mechanisms of seed dispersal, especially in plants like wood sorrel, dittany, wild cucumber, and others that threw their seeds great distances. Their shapes and appendages seemed to enhance propagation. Mariotte observed that the seed of a moss called rampion was slim and thus could slip through the dense growth to the ground. Tournefort believed the spiral shape of dittany seeds helped them spring away from the plant. Seeds were known to travel great distances, sometimes because their hooks and hairs attached to animals. Mariotte also described tip-layering and recalled that one of the Marchants had shown him a clover growing in the Jardin royal "whose flower, when it began to dry, curved and grew into the soil, so that the seed formed there and the clover in effect planted itself."[8]

Driven by a Baconian quest for data and inspired by zoological models, academicians observed vegetative reproduction and checked their hypotheses experimentally. Above all, they used their findings to adjudicate between the theories of spontaneous generation and preformation, with the mechanists Mariotte and Tournefort on opposite sides.[9]

The theory of spontaneous generation had a long pedigree stretching back to the ancient atomists. Although rejection of the theory has been heralded as a seventeenth-century contribution to modern science, some of the best minds continued to accept that plants and animals could be generated spontaneously.[10] Mariotte, one of the principal experimentalists of the Academy, argued that plants could be propagated when two or three corpuscles hooked together in the air, water, or earth. Such a union could "give the first impulse to [a plant's] growth," as for example when grass grew on the site of a dried pond where no seed could have fallen.[11] Unlike Tournefort, who used circumstantial evidence to subsume the exceptional under a general inductive rule, Mariotte invoked a corpuscularian theory of matter to perpetuate the exceptional cases.

The theory of spontaneous generation waned in popularity during the seventeenth century, albeit more slowly among botanists than among zoologists. Thus discoveries concerning generation often became grist for the preformationist mill. Savants cited animalcules to refute spontaneous generation and to support preformation. Plants had long seemed to offer particularly strong evidence in favor of preformation, and here, exceptionally, an analogy from botany influenced the development of theoretical zoology.[12]

Opinion in the Academy was split. At the end of the century, Tournefort and Dodart spoke for preformation.[13] But Mariotte had earlier rejected it altogether after he, Dodart, and Jean Marchant studied bulbs of tulips, lilies, and narcissi in 1677 and 1678 without finding even one entire, mature plant in miniature.[14] Mariotte concluded that the mature plant did not exist in either bulbs or seeds. Other evidence also seemed to disprove the theory: the knots on a rosebush produced flowers in the spring but branches and leaves in the autumn, and grafts might take three years to flower. Nor could preformationism account for variations within a type of plant: apple trees, pear trees, and melons were so varied that to accept the preformationist view—that one seed could contain an infinite number of identical plants—entailed making unpalatable assumptions, such as that one plant produced varied seeds or that plant families did not exist.[15]

Mariotte repudiated preformation and also the related theory of *emboîtement,* which maintained that the parent organism contained its descendants, but that these were not perfectly formed. He argued instead that plants developed their parts and properties gradually, as the result of an interaction between the plant and its sap. Seeds contained "only the principal parts of plants" and all other parts were developed "in succession as a result of the way the first parts affected the sap."[16] Mariotte concluded:

> But it is not believable that this small composite of corpuscles contains all the branches of this Plant, its leaves, its fruits, and its seeds; and even less that in these seeds there could be contained in miniature all the branches, leaves, flowers, etc., of the Plants which will be produced ad infinitum after this first germination.[17]

Like his contemporaries, Mariotte could not adopt both spontaneous generation and preformation (or *emboîtement*); unlike most of them, his corpuscularian theory of matter predisposed him toward spontaneous generation.

Mariotte was the exception to a seventeenth-century trend. Most savants cited microscopic research to disestablish corpuscularian theories of generation in favor of preformation. Here observations, not alternative theories, were decisive.[18] With the microscope, savants saw a new world of seeds, spermatozoa, ova, cells, and other objects previously unimagined. Finding seeds where earlier they had gone unsuspected helped dethrone spontaneous generation. But for Mariotte, corpuscularianism was too useful a theory to abandon. It even seemed to be confirmed by the microscope, which showed tiny particles, invisible to the naked eye, that might well be corpuscles. Because of his theoretical bias, Mariotte's microscope became a weapon not against spontaneous generation but against preformation. The microscope was important but not necessarily decisive. A savant assessed experimental evidence within a framework of theological or philosophical assumptions before choosing between theories. In Mariotte's case, theological objections to spontaneous generation were less persuasive than the philosophical merits of corpuscularianism.

GERMINATION, MATURATION, AND THE ROLE OF EXTERNAL FACTORS

Zoology also inspired the Academy's experiments and theories about germination. Perrault recommended microscopic research to discover just how plants grew from seeds, and Dodart compared the Academy's studies of germination to studies of chicks in the egg.[19] Although academicians never rivaled the detail of Grew's or even Samuel Foley's studies of the anatomy or development of germinating seeds,[20] they claimed interesting analogies between plant and animal parts. Working with a germinating white bean squash and other seeds, Mariotte and Perrault described the filament connecting lobes to leaves as an umbilical cord. Mariotte also thought that seed lobes of beans, pumpkins, cucumbers, and melons resembled the yolk of an egg or the liver in being a source of food for the embryo plants.[21]

Mariotte explained the process of germination mechanically:

> It is thus probable that the principal parts of the germination of Plants are contained in their seeds, and that they are predisposed to form fibers and pores suitable for the filtration and the union of certain principles that pass there, as

if through channels or molds; from which the other parts are formed, such as fruits, seeds, and the beginnings of the second germination.[22]

His mechanistic and analogical explanation challenged traditional views of savants like Duclos. Where Mariotte dismissed a vegetative soul in plants on the ground that no one had ever found one, Duclos preferred vegetative souls to analogies between the parts of plants and the organs of animals. Duclos thought plants and soil had similar natures; unlike his colleagues he did not differentiate the various parts of plants by their function, saying rather that all parts of plants (and of soils) contributed equally to germination and maturation.[23]

Any account of germination had to weigh the roles of the plant itself and of external conditions such as sun, soil, air, and water. Perrault and Mariotte stressed the sun. Perrault conjectured that it cooked nutritious minerals in the rainwater. Mariotte demonstrated that sunlight was essential to the growth of plants by comparing seedlings grown under earthenware pots with seedlings grown under glass domes. He proposed two theories, one mechanical, the other chemical, to explain the sun's effect: either sunlight encouraged water to rise in the plant, and then water turned the plant green, or sunlight affected the chemical nutrients in the water.[24]

Soil, air, and water seemed to be the most influential external conditions for germination and maturation. Local conditions could encourage or inhibit growth, and plants either depended passively on what they took from the soil or transformed it to suit their needs. Mariotte and Perrault discussed the nutrients plants required from the soil. Tournefort thought that the earth's juices were the single most important stimulus of germination. They were prepared in the soil by agitation, moisture, heat, or cold, and shaped by the air and the pores of soil through which they passed.[25] Tournefort emphasized mechanical processes — motion, air pressure, the sieving effect of pores — but he also considered the chemical composition of soils.[26]

Academicians studied earths systematically in the 1670s. In 1675 Dodart proposed several experiments. He planned to extract "the salts, and if possible, some other substances from the different kinds of earths." He hoped to distinguish soils chemically, to differentiate among the salts in soils, to discover a connection between salts found in the earth and the salts of plants that grew there, and to identify the different proportions of salts in the same soil under various conditions.[27]

For Dodart and Borelly the purpose of research on earths was to understand what made soil fertile. But they could not persuade Bourdelin to study soils according to the methods they preferred. From the end of 1675 until the end of 1677, Bourdelin analyzed marls, clays, and other kinds of soils, distilling them as usual in the retort, some with and others without the fixed salt of saltpeter. He calcinated the *testes-mortes* and then tested the liquids extracted for their reactions to chemical solutions, just as he did with the products of distilled plants. Bourdelin found that earths released liquid with a sulphurous odor, vitriolated salts, volatile salt, and oil; the *teste-morte* of one earth was said to taste like common salt.[28]

Borelly promoted solvent analysis over distillation and objected to using the fixed salt of saltpeter in distilling earths.[29] He and Dodart preferred lixiviating earths "in order to extract all the salt and all the various substances together in their chaos"; afterwards the earths could be rectified and purified so that their separate parts might be identified. Salts obtained thus could be used in numerous experiments.[30] Any earth remaining after lixiviation could be tested further by solvents.[31]

After Borelly criticized him, however, Bourdelin simply distilled fewer earths, while never adopting Borelly's method.[32] Like the dispute over the relative merits of distillation and solvent analysis of plants, so the disagreement over the proper way of studying soils found Bourdelin and Borelly in opposition. Both disputes were resolved in Bourdelin's favor by his intransigence; Bourdelin would not abandon his preferred technique, despite the pressure exerted by his colleagues.

HOW PLANTS GROW AND HOW THEY ARE NOURISHED

Mariotte and Tournefort explained plant growth mechanically, Homberg chemically. Mariotte thought that as water evaporated, various "earthy, salty, and oily parts" remained to mix and unite with the plant, creating "the hardness and solidity of the branches." He argued that although plants could not select what they took from the soil, they could transform it and actually absorb more oil as they matured.[33] For mechanists like Mariotte and Tournefort, the circulation of the sap caused growth by putting pressure on the extremities of the plant. Rising sap stretched plant cells, causing the cells and hence the plants to grow; when a cell could stretch no longer, the plant withered and died. Tournefort thought this theory also explained the maturation and release of seeds. Observing the organs of plants — the ovaries of hellebore, aconite, and crown imperial, the fruits of spiny poppy, false dittany, toothwort, and the pods of the plant Caspar Bauhin named *Lathyrus latifolius* — Tournefort found that plants released their seeds when the fibers in the ovaries dried and contracted, and he argued that when the ovaries opened, air entered and helped the seeds ripen.[34]

Homberg, however, used Bourdelin's analyses to develop chemical explanations of how seeds and plants grew. Homberg concentrated on the maturation of seeds. Noting that unripe seeds yielded a lot of water, less oil, but more fixed salt than ripe ones, Homberg gave this account of their development:

> ...the organs of the young seeds contain only a watery and very fluid sap, which is not yet well digested; after these salty, earthy, and watery parts have mixed together more perfectly over time, they thicken and create this oil that forms little by little.[35]

To strengthen his claim, Homberg compared young seeds with ripe fruits, nuts, and olives; stored in a dry place for three or four months they too yielded more, and thicker, oil than when they were freshly plucked. Homberg reasoned that the young seed resembled ripe fruit and became oilier as it matured. He also distilled fetid oil with quicklime; this diminished the oil, changed its color, and produced a lot of water. Homberg argued that the phlegm, salt, and earthy matter of young seeds "together create over time the quantity of oil that is found in ripe seeds."[36] He believed he had separated the oily compound into the simple substances out of which nature had formed it. When rectifying oil with quicklime, for example, the quicklime separated fixed salt and earth from the oil; thus, Homberg reasoned, the oil must consist of salt, water, and earth.[37]

All academicians agreed that only chemical analysis could explain the origin and nature of plant nutrients. Boyle, Helmont, and others stressed water and deprecated soil as the source of nourishment,[38] and academicians investigated the role of rainwater and dew in the growth of plants.[39] But savants who did not accept Helmont's view that water was the ultimate source of matter had to identify the origin of the nutriments found in water. Academicians debated whether these came from the atmosphere or from the earth, a question that turned on salts.[40]

For Mariotte and Perrault, nutrients came intermediately from the earth but ultimately from the atmosphere when sulphur, saltpeter, and volatile salts fell in rainwater to the ground.[41] Duclos disagreed with his colleagues. In his view, the fertile "fatty and sulfurous salts" formed in the soil, not in the air. As proof he cited his analysis of the waters that condensed inside and outside a concave vessel placed on the ground: dew had more volatile salt than either the rainwater or the air vapors that collected on the outside. His arguments convinced Perrault, who conceded that the nutrients originated in the soil from living or decayed plants and animals or from whatever produced mineral salts in well water. Perrault thought such nutrients were cooked by the sun when they rose in vapors.[42]

Applying chemical analysis to a concept of the food chain, Homberg developed a theory about the different origins of the different salts. He noticed that most plants were composed of three salts: fixed lixivial salt, volatile urinous salt, and volatile acid salt. Bourdelin's analyses showed that fixed lixivial salts and volatile urinous salts occurred only in the distillants of plants, of herbivorous animals, or of animals that ate herbivorous animals. The third substance, volatile acid salt, was found in soils, including ones that had no vegetation, and in plants that grew in all kinds of soil. Volatile acid salts thus came from the earth. The other two salts—fixed lixivial and urinous volatile—were therefore manufactured in the plant.[43]

The Academy was divided on whether seeds and plants were active or passive. Did they simply absorb juices already prepared in the earth or did they transform what they took from the earth? Tournefort took the passive view. He believed that juices altered in the earth and became more or less suitable to various plants.[44] Mariotte, Perrault, Duclos, and Homberg took the contemporary view that seeds and plants changed what they took from the soil, but they disagreed about how this happened. Mariotte believed the liquid was simply a vehicle for minerals and that plants used the mineral residue only after the liquid had evaporated. Perrault and Duclos believed the plant transformed both liquid and minerals, and Duclos thought it did so by coagulation. Homberg believed plants imbibed one kind of salt from the earth but created two additional salts themselves.[45]

CONCLUSION

Seventeenth-century botany was eclectic.[46] Zoological anatomy and physiology lent it a vocabulary, an experimental repertoire, and some explanatory theories about physiological processes. Chemistry and mechanics accounted for generation and reproduction, germination and growth, and nutrition. Some academicians took these explanations as complementary, especially when they answered related but different questions. Thus, they might explain chemically how a plant assimilates nutrients and mechanically how a plant gets larger or dies. Mariotte, for example, cited both mechanistic and chemical theories in his discussions of plant physiology. Homberg and Tournefort, however, did not try to combine the theories—Homberg's primarily chemical, Tournefort's predominantly mechanistic—that they presented separately but contemporaneously at meetings of the Academy. Nor did the Academy as a whole encourage them to unify their hypotheses into a general account of plants. Academicians never tried to develop a treatise on plant physiology that would rival either the detailed and original works of Grew and Malpighi or the derivative, systematic essays of Régis. The natural philosophy of plants was too tentative a discipline to justify a collaborative project similar to the natural history of plants.

The Academy as an institution never encouraged members to develop a comprehensive theory of plants, to resolve their inconsistent interpretations, or to explore the implications of their piecemeal observations and conclusions. Nevertheless, it aided the development and expression of their theories in three ways. First, the Academy offered a forum for discussing and publishing these early studies of plant physiology. Second, the Academy's natural historical research supported its natural philosophical inquiries. Studies of seeds, the cultivation of plants, and chemical analysis—all ingredients of the natural historical project—fed theories about reproduction, the conditions of germination, and maturation. The official project provided not only data but also protection for the hesitant and unofficial natural philosophical studies. As a result, collaboration on the failed natural history of plants bore fruit in individual studies of plant physiology, with chemical analysis, more than any other tool of research, connecting the two kinds of inquiry.

Third, the Academy's interdisciplinary composition and interests encouraged members to study plants with the new instruments. As the next chapter will show, however, these instruments were theory-laden; they revealed novel details but also interposed a screen between observer and object. When academicians examined a plant with the microscope or the air pump, they saw not just a plant but also a link in an evidential chain that was not primarily botanical. Indeed, the cogency of their research depended on the theoretical certainty of the discipline that stimulated them to use the new apparatus. Thus, microscopic and pneumatic botany had different fates: the Academy used microscopes to support theories about physiology, but it used the air pump to clarify theories about air and the void.

CHAPTER 12

The New Instruments and Botany

New theories, experiments, and observations were the hallmarks of seventeenth-century science. They often depended on recently invented instruments quickly applied to the most varied research. Scientific apparatus such as Galileo's telescope and Torricelli's tubes helped revolutionize the way people thought about the universe. It also had theoretical implications that remained fertile and controversial for decades.

The bounteous yield of data and hypotheses provoked many debates. Exactly what could be learned from the new instruments was unclear, since both the phenomena observed and the inferences drawn were generally called into question. Hence, the four instruments that transformed scientific research in the seventeenth century—the telescope, barometer, air pump, and microscope—opened an uncertain world to their earliest users. The Baconian ideal of compiling an encyclopedia of discrete facts about the world could not suffice when such instruments demonstrated the controvertibility of data and forced savants to adjudicate between discovery and theory.

The new instruments were puzzling but irresistible. Savants and amateurs both were enthusiastic about them. By the 1660s, many people wanted to perform the latest experiments on the most modern equipment or to see demonstrations of controversial phenomena with their own eyes. Private scientific societies whose sponsors could not afford up-to-date apparatus found their inquiries fettered, and dedicated experimentalists became exasperated by dilettantish amateurs who preferred discourse to experiment. Many savants worked with artisans to design and construct novel devices or to improve the quality of essential ingredients such as glass. Curiosity and optimism about the new tools permeated scientific inquiry, and any self-respecting savant required a sometimes expensive range of equipment.[1]

The scientific institutions organized in the latter half of the century reflect this enthusiasm. As curator of machines and experiments at the Royal Society, Robert Hooke demonstrated the latest experiments and equipment at meetings. Members of the Accademia del Cimento had at their disposal the laboratories and instruments of the Medici princes and became known throughout Europe for their experiments, especially with the Torricellian void, which were widely imitated. The Academy of Sciences collected models of new machinery and bought instruments for astronomical observations, surveying, dissections, and other research. In the best seventeenth-century tradition, academicians were inventors as well as users of apparatus. Several developed surveying instruments for use at Versailles. Auzout devised a micrometer, and Roberval described a new balance, while Huygens designed clocks, microscopes, and air pumps and briefly thought he had discovered a new kind of barometer.[2]

The Academy employed instruments in its botanical studies to analyze plants chemically, to anatomize plants, and to clarify vegetable physiology. Of these activities, chemical analysis was least affected, because the only new instruments used were the aerometer (a device for determining the specific gravities of plant extracts) and the thermometer.[3] In their chemical analysis, academicians sought the constituents of organic matter and looked for patterns in their data. They had an idea, not a theory, and while the aerometer provided more information, it did not offer the interpretive key.

More provocative were the Academy's studies of plant anatomy and physiology with the microscope and air pump. Here academicians could relate their findings to a broader range of theories and analogies, because both instruments had already been deployed in other fields. The microscope and the air pump opened plant studies to several interdisciplinary influences: microscopy connected vegetable and animal anatomy, and the air pump linked plant growth to the properties and effects of air. The botanical applications of these instruments at the Academy draw attention to some lesser known aspects of seventeenth-century botany and clarify the impact of new inventions on scientific theorizing.

EARLY BOTANICAL MICROSCOPY AT THE ACADEMY

Microscopes had already been used to study plants by the time the Academy was founded,[4] but only a few academicians were interested in botanical microscopy. As a result, microscopic observation affected the Academy's natural history of plants only peripherally. Descriptions were meant to distinguish plants from one another, but not to include their anatomies or to clarify vegetable physiology. The engravers found microscopes useful, because the Academy wanted the smallest external details of each part of the plant to appear. Otherwise, academicians only occasionally used microscopes to study plants.[5]

Perrault recommended microscopy for research on germination. In 1667 he made these plans:

> Experiments on how plants grow will be made by considering the roots and seeds and examining them diligently with the microscope, both before placing them in the ground, and by taking them out of the ground at various times in order to consider the different changes which occur with respect to size, or to shape of their pores, to their saps, weight, color, odor, taste, and so on. Then one will consider what happens to their sprouts when they begin to grow, especially to those which are enclosed within large seeds, such as acorns, where one notices the root, the trunk, and the branches of the entire tree, which seems already formed and distinct before emerging between the two sections into which the acorn normally separates.[6]

Perrault hoped that microscopes would settle disputes over preformation and *emboîtement* and would ease comparisons between the growth of seeds and the development of the chick in an egg.

Although Perrault's suggestions were not officially adopted, some members studied plants with microscopes and magnifying glasses. Examining hemp thread (*fil de chanvre*), La Hire saw filaments which he compared to capillary tubes and claimed that sap passed up them to nourish the plant.[7] Dodart examined young shoots of wheat to find the tiny grains; he was testing preformationist theory.[8] In the 1680s some academicians considered studying plants and their distillants with the microscope.[9] Not institutional policy but individual interests incorporated microscopes into the arsenal of discovery and argument in botany.

Unlike most of his colleagues during the 1660s and 1670s, Mariotte routinely used hand lenses and compound microscopes. They were invaluable in his studies of vegetable physiology. His arguments against preformation and for the circulation of the sap depended on meticulous observations. He examined leaves, bulbs, seeds, and cut stems; his descriptions of bark, skin, fibers, vessels, spongy matter, and saps are the verbal equivalent of Grew's illustrations. In shrubs he identified "canals or pores" in the marrow of the cutting; a microscope revealed these pores to be "several small oval cells [*cellules*]" resembling honeycombs.[10]

Without a lens Tournefort could not have found the "seeds" of ferns or examined the growth, desiccation, and contraction of seed cases, observations that he cited against spontaneous generation. When he examined plant vessels Tournefort found resemblances to bones and muscles, and he claimed that the very vessels that carried sap eventually dried and became fibrous, stiff, woody, and capable of supporting the plant. Microscopy provided evidence for Tournefort's theories about the motion of sap and the growth and reproduction of plants.[11]

Huygens contributed incidentally to botanical microscopy when he brought a spherical microscope to the Academy in the summer of 1678.[12] Although his primary interest was animalcules, his colleagues examined a section of a fir tree, some pollen from a lily, and the marrow of a fig tree.[13] But Huygens's apparatus held little interest for the botanists, and besides him only the astronomers Picard and La Hire used it to study plants, or rather their pollen. Picard compared the shapes of pollens from different plants, without conjecturing about the nature or function of that "dust" or "flour." He merely commented on the shapes, colors, and structures of pollen.[14] Huygens and his brother Constantyn went beyond Picard's simple comparisons of appearance. They considered internal structure and the connection between pollen and the activities of bees. Huygens observed that the "dust" of crocus flowers and the dust on bees' feet looked the same, and he argued that pollen adhered to the feet of bees, who made wax from it. When his brother expressed surprise that pollen stored for two months still contained a liquid, Huygens replied, "What you say of the liquid in yellow powder confirms again what I said, that it served to make wax."[15] Although he hypothesized about accidental uses of pollen, Huygens never applied his observations of pollen to any theories about plant physiology. No one in the Academy took more than passing notice of his or Picard's findings.

Only one of the academicians who normally studied botany, La Hire, used Huygens's spherical microscope to examine plants. This surprising lack of interest among the Academy's botanists was due to the difficulty of using the spherical microscope and to its limited range of applications. Huygens noted that many of his colleagues saw animalcules only with great effort, while others never saw them at all.[16] Proper use of the lens and careful mounting of the object were crucial for success.[17] Huygens, Roemer, and Hartsoeker developed a way of mounting several objects on a rotating wheel, an invention to which no Frenchman contributed, Huygens was quick to point out.[18] But even when every precaution was taken, the instrument was of limited use for studying plants, because it was not yet possible to cut plant sections so thin as to be transparent or to fit between slides. Few parts of plants, therefore, lent themselves to study with Huygens's apparatus.[19]

Despite their inherent defects and the difficulty of using them, spherical microscopes impressed Parisian scientific savants. Huygens was as gratified by the reception of his microscopes as by the inability of Parisians to make the lenses. He reported that "the curious" were "astonished by the great effect it makes"; Locke had heard of "the extraordinary goodness" of Huygens's microscope. Protestations of interest no doubt outnumbered clear sightings of animalcules, because the skepticism that had earlier greeted the telescope and air pump was now less tenable; by 1678 few amateurs would have chanced embarrassment by challenging Huygens on the basis of negative evidence. Instead, Huygens's enemies contented themselves with ghost-written articles disproving Huygens's claims of priority.[20]

Microscopy was always ancillary to other ways of studying plants at the Academy. Academicians who habitually studied the natural philosophy of plants used a convex hand lens and sometimes a compound microscope to observe the details of plant anatomy and to support various hypotheses about plant physiology; they applied the readily available microscopes to subjects that had long interested them. Huygens and Picard, on the other hand, were more engrossed by the novelty of the instrument than by its botanical applications; they never pursued plant microscopy exhaustively because they were more interested in other subjects. Finally, academicians who studied the anatomy and physiology of plants preferred theoretical issues to exhaustive description. For them microscopy was one technique among many, and they used it along with naked-eye observation, chemical analysis, and analogical reasoning in order to support their hypotheses about germination, nutrition, and growth.[21]

PLANTS AND THE AIR PUMP

The air pump was one of the most controversial inventions of the seventeenth century, because the evidence produced with it was cited by both sides in the debate over the existence of a vacuum. To determine whether a void could exist, natural philosophers placed small animals, their own arms, lighted candles, bells, and magnets into glass receivers, from which they evacuated air. After observing the asphyxiation of mice and birds, the rise of blood to the surface of an arm, the extinction of a flame, and so on, savants acknowledged that an evacuated receiver held little or no air. But they could not agree on whether or not another substance took the place of air, thereby preventing the formation of a vacuum.

It was natural that plants also be tested. Air pumps excited the experimental impulse in their early users, who compared the behavior of as many different objects as possible in their machines. But experiments on plants were not popular, for the results were never spectacular. Juice running out of a pricked fruit was less impressive than nearly suffocating a canary only to revive it by readmitting air to the evacuated chamber. Waiting several days to ascertain whether a plant would die in an evacuated bell jar was tedious by comparison with testing whether a ringing bell sounded in an airless environment. Where plants were simply part of a comparative analysis of the effects of airlessness, they were among the less interesting subjects of study.

Some savants, however, thought the air pump could shed light on botany. Sir Kenelm Digby, for one, had already suggested that air might be important to plants, and in the 1660s and 1670s Hooke tested seeds with an air pump for the Royal Society, after a request from Boyle.[22] In the Academy, Borelly once suggested testing plant distillants and earth in an evacuated receiver, and Huygens and Homberg examined the growth of plants in a vacuum. Academicians narrowed the question: instead of asking whether air was somehow important to plants, they inquired more specifically whether the presence of air was a necessary condition for the germination of seeds.

Huygens was a pioneer in the development of the air pump and was the first academician to use it. He recommended experiments with it in a 1666 memorandum to Colbert and brought his own machine to the Bibliothèque du roi when he moved in that year. In the spring of 1668 he introduced the air pump to his colleagues, starting with simple demonstrations calculated to interest his audience and drawn from his private research of 1662. He pricked the skin of an apple and placed the apple in the receiver: juice ran out of the fruit as air was removed from the jar. He made spirit of wine bubble by evacuating a receiver. Having demonstrated what the machine could do, Huygens designed an experiment that he characterized as "more important," because it would ascertain whether seeds and plants could grow in a vacuum.[23]

With this experiment, Huygens introduced the Academy to the use of the air pump in botanical research. All subsequent tests followed the same pattern: plants, seeds, branches, and soil were placed in the receiver, or bell jar, which was evacuated and removed from the pump; its contents were observed over a period of several days. Huygens evacuated the bell jar until his measuring device indicated there was no more air inside. Then he removed the bell jar from the pump and observed it for about eight days.[24]

One of the risks to botanists was that even when the initial pretext of these experiments was botanical, their ultimate interest often lay in unexpected side-effects that were inconsistent with the known properties of air. Huygens's test set the pattern in this respect as well, for he was surprised to discover a puddle of water on the floor of the bell jar, and this deflected his attention from the plants. Huygens explained the puddle by saying that the soil had exhaled some vapors that condensed on and ran down the walls of the bell jar, but he was puzzled that vapors could rise in a vacuum when feathers fell like lead. As the experiment continued, Huygens observed additional phenomena that seemed to confirm that vapors rose, and he concluded that the vapors were being converted into air.[25] As for the plants and seeds, while they did not grow or flower more, neither did they die as expected in an airless atmosphere.

Huygens later tried to re-examine how lack of air affected plants, using not an evacuated receiver but a double bottle. In May 1672 he sealed the bottle, which was one-quarter full of earth; when he brought it to a meeting in August 1675, more than three years later, academicians observed that a large amount of grass and some moss had grown in it, even though air had not entered the bottle.[26] Both of Huygens's experiments puzzled academicians, for they seemed to show that air was not necessary for the growth of plants.

The Academy dismissed such issues from its collective mind until Homberg revived them two decades later.[27] Like Huygens, Homberg perceived experiments on germination in a vacuum primarily as tests of the nature and functions of air. He performed only two such experiments with plants. The first was brief and improved on Huygens's test of 1668 by including a control, while the second was more elaborate and more carefully considered for its implications about germination.

In the first, Homberg sowed seeds in two boxes, placing one under a glass dome and the other in a receiver, which he evacuated and then placed beside the glass dome in a window with a southern exposure. On the first evening, he noticed that the earth in the vacuum had split in several places and that drops of water covered the sides of the receiver; on the base of the receiver was more water, which seemed to be the liquid used to moisten the soil at the start of the experiment. In the control box the soil was not so cracked and the glass dome was covered with less water; some water had fallen onto the stone base, but because the base had cracked and water had escaped, Homberg could not measure it. He attributed the excess liquid in both containers to evaporation and condensation, and he thought there had been less evaporation under the glass dome than in the evacuated receiver. He did not try to explain why the earth cracked, nor did he continue the experiment long enough to see whether seeds germinated in one container or the other. Homberg explained evaporation and condensation in a vacuum, which had troubled Huygens in 1668, as the effect of "ethereal matter," which forced vapor against the sides of the receiver, where it became water and ran to the bottom.[28]

Homberg's second experiment lasted about six weeks, from 1 May through 12 June 1693, and like the first had been designed with a control. He sowed the seeds of five plants—purslane, cress, lettuce, chervil, and parsley—in two small boxes. One box he placed in a receiver, the other he left in the open air. He evacuated the receiver every morning and watered each box every third day.[29]

Homberg recorded the weather and the daily appearance of the shoots, noting withering and death, the sizes of the shoots, and any continued growth. In the evacuated receiver, chervil and parsley never germinated, and purslane, cress, and lettuce began to grow much later than did their counterparts in the open air; purslane died the day after its shoot appeared in both boxes; cress in the receiver died five days after it first appeared; lettuce grew very tall in the receiver, but its leaves were small and, after an initial rapid growth, it stopped growing entirely.

Homberg decided to test whether seeds that had not germinated in the vacuum would do so when exposed to air. On 25 May he admitted air to the receiver and then closed the stopcock to prohibit air from passing into or out of the vessel. Chervil, purslane, cress, and parsley germinated, but shoots that had developed in the vacuum did not change. On 7 June Homberg removed the box from the receiver, but by 12 June all had died.[30]

Because some seeds had germinated in the evacuated receiver, Homberg decided that neither the weight nor the elasticity of air could be the principal cause of germination. But fewer seeds germinated in the receiver than in the open air, and none of the seedlings in the evacuated bell jar grew normally or survived more than a few days. Homberg therefore concluded that air was "at least an accidental cause" of germination.

Trying to explain how a vacuum could injure seeds, he argued that the seeds contained air which "dilates because of its spring rather more easily in the void, where nothing impedes it, than in the air where it is pressed on all sides." Seeds exposed to a vacuum were, therefore, damaged by the expansion of the air they contained, and "since the organs that serve to carry and to distribute nourishment are ruptured, germination cannot take place." A damaged seedling could grow but would not develop into a healthy plant.[31]

Homberg noted three additional phenomena in the evacuated receiver: the soil swelled, a fungus grew, and drops of water formed at the tops of the seedlings. He believed the soil changed because of moisture and air. Moisture penetrated and broke down small masses of earth; it filled the small spaces between clumps of soil, making the earth seem oily, soft, and slimy. Assuming that air was "mixed with the new water" used to moisten the soil, and that air expands in a vacuum, Homberg argued that its "effort" to escape accounted for the "swelling and bubbling" he had noticed in the soil.[32]

The second striking phenomenon was the appearance of what was probably the fungus *pythium* on the surface of the soil. Homberg first noticed this on the eighth day of the experiment, when he thought that "the earth enclosed in the vacuum had changed color" and seemed from certain angles to be "grayish and shining." Examining the surface of the soil with a microscope, he observed "a lot of small transparent and grayish filaments that looked like a spider's web":

> Some of these filaments were straight, others rested on the soil and were attached to small protrusions of earth; since they were crisscrossed, they formed a sort of fabric so strong and tight that the water with which we wetted the soil was held there and formed drops as big as beans, without moistening the soil.[33]

He tasted them, expecting a flavor of saltpeter because they resembled the mildew on the walls of cellars. But he noticed no taste at all and did not test the substance with any of the chemical reagents commonly used in the Academy's laboratory.

Finally, Homberg observed guttation induced by the vacuum: at the top of each seedling in the void there was always a drop of water; when it ran down the stem, a new drop formed immediately. Homberg claimed that "this water did not come out of the pores of these sprouts," but attributed it instead to vapors formed in the void by the action of ethereal matter on damp earth. He believed these droplets were the equivalent of the drops of water he had observed on the glass walls of the receiver in his previous experiment.[34] Homberg was perplexed and wondered, just as Huygens had twenty-five years earlier, how drops of water could rise in a vacuum, when light objects fell.

Homberg believed that an airy matter was present in substances such as water and seeds.[35] This led him to expect air to reappear in an evacuated, well-sealed bell jar, if it held any substances containing airy matter. But he did not classify plants as air-producers, and neither he nor Huygens thought that plants could produce the drops of water observed on leaves. Analogies with dew or with the condensation of water on a glass dome were so persuasive that Huygens and Homberg did not suspect that plants might exude water, even though contemporaries like Mariotte had shown that leaves could absorb moisture, and John Wallis had demonstrated transpiration experimentally.[36] But Homberg never drew on such notions to explain the phenomena he observed with the air pump, seeing his studies as useful to botany but not vice versa.

Some of the peculiarities of pneumatic research on plants were due to its having been borrowed: the air pump was initially developed in other countries, and those who performed experiments with it were trained in and inspired by other disciplines. The air pump never caught on as an experimental tool at the Academy, despite Huygens's enthusiasm, and such experiments as Huygens performed in Paris were often modeled on those of his English colleagues. Thus, Boyle influenced Huygens as much in his choice of experiments as in his development of the pump itself. Huygens was also familiar with Digby's discourses on vegetation, which, with their arguments that plants depend on a vital nutritious substance in the air, may well have stimulated Huygens's tests of 1668 and 1672.[37]

Claude Perrault was unusual in connecting Huygens's studies of the vacuum to his own explanations of how sap rises. He also used pneumatic observations to refute Duclos's claim that an "expulsive faculty" in branches and trunks propelled sap into the roots. Perrault pointed out that "plants that are filled with a lot of sap" simply "let it flow," but when the receiver is evacuated, so that "the compression of the air is reduced, the sap dilates and becomes less condensed than it had been."[38] He believed that this experiment, which was not one of those demonstrated by Huygens in 1668, showed that the weight of the air—and not any expulsive faculty—pushed juices from or through plants, thus causing sap to move.

Curiously, while animal physiology often influenced plant physiology, it had relatively little effect on how academicians interpreted vegetable phenomena in the vacuum, even though academicians also tested animals with air pumps. Perhaps Homberg derived his theory that a vacuum destroyed the internal organs of seeds from observing the damaged muscles and organs of animals that had died in evacuated bell jars.[39] Otherwise academicians explained the physiology of seeds with references to the passive, absorptive qualities of dead sponges. Thus, seeds were said to absorb water like sponges and to germinate by swelling, until they burst their shells and developed into new shapes; savants wanted to know whether moist seeds would inflate in a vacuum as sponges did. This question, not the example of small animals dying in evacuated receivers, motivated Homberg to study seeds.[40]

Experiments on plants in an evacuated receiver held little interest for academicians, and Huygens's and Homberg's tests mostly went unnoticed. Since air pumps were used primarily to test the weight and elasticity of air, it was always these properties that academicians cited to explain what happened to seeds and plants. In any case, vegetable behavior seemed less interesting than the other phenomena associated with botanical pneumatics. Putting plants in the air pump represents not a successful interdisciplinary exchange but isolated studies whose anomalies sidetracked academicians from answering questions about plants.

CONCLUSION

Scientific instruments did not play a major role in botanical research at the Academy. They were used only infrequently to examine plants, and very often the issues addressed were not botanical. Plants in a vacuum tested theories about the properties of air and the effects of airlessness. Instead of using microscopy to explore plant anatomy systematically, academicians sought only such observations as would address specific physiological theories.

Academicians had no program, and their research was piecemeal and lacked continuity. Patronage was irrelevant, because the microscopes, thermometers, and aerometers were inexpensive enough that academicians could have bought them with their own funds, while the air pumps were designed and owned by the academicians themselves. In these respects the Academy was no better than the private societies. Yet it was less Baconian than its rivals and its institutional continuity counteracted the centripetal individualism of physiological research.

The two principal instruments used to study plants, the microscope and the air pump, were of different value to botanists. The microscope had a more established place, because it seemed to provide clearer and more definitive evidence. Mariotte, La Hire, Tournefort, and their contemporaries knew how to describe what they saw. They could relate their observations to zoological models and could fit them into accepted patterns of thought about plants. Plant microscopists evoked two paradigms, one factual and the other logical, and such appeals to zoology and to analogical reasoning enabled botanists to explain what they saw through the lens.

With the air pump, however, savants were unsure of themselves. Experiments with plants were designed to ascertain whether plants were like animals in requiring air for their vital functions, but plants behaved inconsistently in the vacuum: some seeds sprouted while others did not, some seedlings died immediately while others survived and grew for a day or two. Unexpected and confusing phenomena—the excessive production of water, the appearance of gray filaments, the cracking and expansion of the soil—occurred in every experiment. Botanical pneumatics fell prey to disputes between conflicting theories about air and the vacuum. Academicians were cautious, therefore, in interpreting their observations of plants in the vacuum and did not build on this research.

Scientific apparatus offers decisive evidence only within the context of a paradigm that allows savants to identify and understand crucial phenomena. But in the late seventeenth century, botany was changing, and its new accretions depended on borrowed theories that sometimes failed. When academicians used new tools to study plants, their research seemed meaningful if the cross-disciplinary analogies worked, but otherwise they could not interpret their findings. Without a satisfactory theory, studies of plants simply accumulated evidence on both sides of debates about other natural phenomena, as the case of botany and the new instruments reveals.

PART IV

The Academy and the Larger Community

CHAPTER 13

Medical Motivations and Social Responsibility

Although the Academy was an elite organization whose proceedings were private and secret, it was pulled toward the outside community as a result of both obligations and opportunities. Its responsibilities were intellectual, political, and social, while its opportunities included scholarly interchange with savants who were not members.

The Academy's foremost obligation was the intellectual one of advancing knowledge about the world, and to this end academicians researched and published their findings. In these respects the institution was an extension of the individuals who composed it and would have performed these activities anyway, but the institution facilitated their work.

The Academy's foundation entailed additional responsibilities, including the political one of honoring and serving the king, and these heightened the utilitarian propensities of its program. The political liabilities created by royal patronage were discharged by the Academy with every handsome publication or discovery, but also more particularly when academicians surveyed for the water supply of Versailles, tested the chemical composition of the waters supplying royal palaces, reviewed inventions with military, agricultural, or industrial potential, planned cartographic projects of the kingdom, or focused botanical research on medical applications. The Academy met its political responsibilities by associating the king's name with its accomplishments, by serving as technical consultant to the crown, and by studying ways to improve health and industry, navigation and cartography.

By comparison, the Academy's social responsibilities were less defined and less urgent. Most savants, academicians included, mouthed the maxim that natural philosophy would improve society, but fewer mounted concerted programs to achieve specific benefits. At the Academy, moreover, the distinction between responsibilities to the king and those to the populace would have been blurred by the traditional theories of kingship, which stressed the monarch's duties on behalf of his subjects. By serving the crown, the Academy was serving the kingdom. Yet academicians' own training and inclinations led them to go beyond this general tendency and to articulate a notion of accountability not only to Louis XIV and his subjects but even to all of humankind. They saw their work as potentially useful to medicine. Furthermore, religious inclination as well as the terrible condition of French peasants focused one academician's interests on a study of medicine for the poor.

Such political and social responsibilities were to some extent imposed on the Academy. Scholarly exchange, in contrast, was basic to the scientific community of which academicians were members. The Academy, however, acted sometimes as a barrier to exchange by regulating discourse between academicians and savants outside the Academy. The Academy had its own name to protect, in addition to the reputations of its individual members, and tried to control the flow of information to its and their advantage. Creation of the Academy altered the nature of dealings within the scientific community not only in Paris but also throughout France and Europe. The Academy's attributes and prerogatives turned some outsiders into admirers or skeptics, sycophants or rivals, aspirants or failures, while others remained disinterested fellow scholars.

The Academy's external relations and influence form the subject of the three chapters in this part. The first chapter provides a specific case study, focusing on how academicians discharged some of their responsibilities by combining medical interests with botanical research. With this as illustration, the second chapter considers the scientific community and its audience, and it canvasses the resources the Academy could rely on and contribute to in Paris. Finally, the third chapter examines the Academy's place in the local and international scientific communities.

The Academy's external relations had primarily an intellectual dimension but also included socioeconomic, religious, and chauvinistic aspects. Furthermore, savants claimed to seek the disinterested dissemination of knowledge but were also competing for priority, fame, and the rewards associated with success. Thus, the scope of the subject is so vast that it can be treated only partially here, with botany providing the specific illustrations and context.

MEDICAL INTERESTS

The fields of botany and medicine were closely allied, and many botanists would have said of their discipline what Clave said of chemistry, that good health was its principal aim. The Academy rejected such a single-minded purpose and subordinated medical aims to scholarly ones. Nevertheless, medical interests remained important to academicians and decisively affected their research, partly because of their previous training and partly because of demands made by the Academy's protectors.[1]

Botanical research for medical purposes took several forms at the Academy. Academicians examined the nutritive value of plants, their uses as materia medica, and the hazards of ingesting diseased plants; they also investigated chemical medicine. Such research was important from an institutional point of view, because academicians used medical goals to shield their controversial and apparently unproductive research on plants. When asked to disband an obstinate project, academicians were able to continue some aspects of their research, albeit minimally, on the grounds that they were looking, as required, for its medical applications.

But medical goals were not an artificial construct intended only to conceal deeper, more controversial interests. On the contrary, the conviction that their activities were valuable to medicine unified a group of researchers whose approaches to botanical studies were sometimes incompatible. Academicians hoped to serve society by improving medical practice and by suggesting legislation to protect the health of the French populace. In 1689, for example, they investigated remedies for dysentery, a disorder that had afflicted Paris the year before.[2] Nowhere, however, is their sense of obligation to the public, and especially the poor, clearer than in Dodart's analysis of the cause of ergotism, as will be seen subsequently.

The continuity and strength of medical interests are explained in part by the education of the many academicians who had been trained as physicians, surgeons, or apothecaries. Seven of nineteen in 1667 had medical backgrounds. Claude Bourdelin had been apothecary to the Dauphin, had a shop for medicaments in Paris, and is said to have practiced medicine, although he was not a physician. Louis Gayant was a Paris surgeon who died serving the king's armies in that capacity. La Chambre, a graduate of Montpellier, was ordinary physician to Louis XIII and taught at the Jardin royal; Duclos was ordinary physician to Louis XIV. Nicolas Marchant studied medicine at Padua, while Jean Pecquet took his degree at Montpellier, was physician to Fouquet, and was indebted to Gayant in his anatomical research.[3]

Before the reorganization in 1699, eleven more medical practitioners were appointed. Dodart, Jean Méry, and Simon Boulduc served members of the French royal family; Langlade was later first physician to the queen of Spain.[4] Homberg studied medicine at Padua and Bologna and took his degree at Wittenberg, while Moyse Charas, a Protestant apothecary who moved to England in 1685 and took a medical degree there, had attended Huygens during a serious illness in 1670 only to be lumped together with Huygens's physicians as being timid, ignorant, and reliant on "Galenical methods & prescriptions."[5] Du Verney had ties with the medical faculty at Avignon.[6] Of all the physician-academicians admitted before 1699, only Dodart, Tournefort, and Tauvry took degrees in medicine at the Parisian faculty, although others had ties of family and friendship with that faculty. Tournefort had studied at Montpellier and Orange before coming to Paris, and Tauvry took his first medical degree at Angers. The foreign associate Domenico Guglielmini was a doctor in medicine from Bologna, and nearly all that is known of Morin de Toulon is that he also was trained as a doctor.[7]

Members' medical training or experiences flavored their contributions to meetings of the Academy.[8] Several wrote books about Galenic and chemical medicines, mineral waters, and the treatment of specific illnesses, both before and after their entry into the Academy.[9] With 29 percent of its members before 1699 trained for medical professions, the Academy had a stronger representation of such interests than the Royal Society, where 14 to 20 percent of members between 1663 and 1687 were medical practitioners.[10] The domination of the biological sciences by medical practitioners was a most important characteristic of the Company, as of the biological sciences generally in Europe at this time.[11]

Had academicians' previous training and research not been enough to sustain their interest, there was also the stimulus of official pressure. This became acute in 1686, when Louvois criticized the Academy's work in botany and chemistry because it could not cure the king. Louvois wanted academicians to emphasize the practical, especially medical, uses of plants. In particular, he wanted them to challenge controversial empirics and hidebound faculties of medicine.[12] The last concern was a traditional strategy of royal patronage, whose iconoclastic favorites included Théophraste Renaudot, the *Journal des sçavans,* and the Jardin royal.[13]

Despite official encouragement, however, the Academy did not wish to become an arbiter of medical theory and practice, although it did hope to improve medical knowledge.[14] It found earlier treatises deficient because their authors did not explain the effects of plants on humans, describe the frequency or size of doses, state for what illnesses a remedy was most appropriate, or explain when in the course of an illness to take the medicine.[15] Academicians proposed to remedy these defects by testing medicines, poisons, and antidotes on human subjects, but they were thwarted. Dodart wanted to try out antidotes on criminals condemned to death but could not obtain permission to do so; instead he suggested checking safe medicaments on humans and hazardous ones on animals.[16] Bourdelin planned in 1667 to assess all remedies listed in chemical treatises and to provide them to hospitals for experiments with patients, but he was barred from doing so by hospital guardians.[17] Instead academicians studied bloods, dissected cadavers, solicited advice from physicians, analyzed unfamiliar remedies chemically, and used as guinea pigs various animals, their own patients, and themselves.[18] The Academy also investigated chemical medicine, with Duclos a staunch advocate of potable gold and other controversial remedies.[19] Above all, members analyzed plants, selecting for study those believed to have medical or nutritional value.[20]

Although the Academy had no desire to challenge the medical faculties outright, its discussions of medical issues were empirical and thus antipathetic to the way French universities taught medicine.[21] Its work was implicitly reformist, moreover, for academicians tried to purify known medications, to produce new ones from distillants, and to publish hitherto secret cures.[22] The Academy embraced members with divergent positions,[23] and individual academicians were eclectic on medical issues, being neither strict Paracelsians nor unreformed Galenists. The Paracelsian Duclos used Galenic terminology to describe the action of drugs.[24] Dodart valued chemical analysis and sought new remedies from the distillants of plants, but his theory of digestion followed Galenic principles rather than contemporary acid-alkali theories.[25] Charas's pharmacopoeia promoted the chemical preparation of remedies from animal, vegetable, and mineral substances, especially by distillation, but allowed individual mineral cures on pragmatic grounds.[26] His empirical eclecticism is representative of the Academy's therapeutics.

Medical interests had both negative and positive effects on the Academy's research. The Academy established its scientific program to be independent of medical research. Yet its natural philosophical inquiries remained episodic partly because they were irrelevant to medicine. When the crown demanded that academicians address medical needs, this injured the broader program without yielding many practical results. Yet no academician argued that the Academy should refuse to seek medically useful information, and in one important respect that search benefited the Academy. That is, in a Company split by personal rivalries and disparate ideologies, threatened during the 1680s by ministerial intervention, and discouraged by intransigent chemical analysis, medical interests were a unifying force. When other aims and activities foundered, research on plants for medical purposes was acceptable to government, chemists, botanists, and natural philosophers—indeed to all with responsibility for the activity and development of the Academy.

ERGOTISM, ILLNESS OF THE POOR

The interplay of medical motivations and botanical research at the Academy is apparent in Dodart's study, published in 1676, of how ergot injured the health when ingested in bread. His research also clarifies how the Academy served as the focus for a network of physicians interested in the medical problems of the poor.[27]

Ergot grains grow on rye as the result of an infestation of the plant with *Claviceps purpurea*. The effect of this infection is to replace developing grain with ergot, a fungus that "contains several toxic principles," including the alkaloid lysergic acid diethylamide, or LSD.[28] When eaten after being ground with rye into flour, the fungus causes the deadly illness then known as Saint Anthony's fire and now called ergotism, which takes either a gangrenous or a hallucinogenic form.

Both ergot and the malady it causes were well known before Dodart wrote, although the connection between them was not. Descriptions of the symptoms and course of ergot poisoning were published by German authors in the 1590s and the first decade of the seventeenth century. Some writers confused ergotism with other illnesses, and some pointed out that bad food caused the attacks, but no one linked ergot grains to the illness they caused. Although some sixteenth- and seventeenth-century literature on ergot described its obstetrical uses or associated ergotism with the "honey-dew" stage of the fungus, Dodart was the first to publish the view that ergot grains caused Saint Anthony's fire and to explain why he thought so.[29]

Dodart's article and the Academy's research on ergot were stimulated by correspondence from physicians who already understood the causal relationship between ergot and Saint Anthony's fire. Four physicians—the Montpellier-trained N. Bellay, Paul Dubé, and a man named Tuillier and his son—plus the surgeon Chatton sent the Academy their observations and samples of infected rye. All of them came from the rye-growing region of France that included the Sologne, Blois, and Montargis.[30] The correspondence began when the practitioners from Blois and Montargis wrote to Perrault and Bourdelin of their suspicions that spurred rye caused gangrene; they knew no warning signs of the illness and had found medication and surgery ineffective in treating patients. In 1674, after having received several communications, the Company instructed Dodart to investigate.[31]

The Academy's informants described the sufferings of patients afflicted with Saint Anthony's fire. The illness brought on "malignant fevers accompanied by drowsiness and dreams"; this was perhaps a reference to hallucination. It dried the milk of nursing mothers and "caused gangrene in the arms, and especially in the legs, which it usually struck first." Gangrene of the limbs was preceded by "a certain numbness in the legs," and as the illness continued its painful progress, physicians observed that there was

> some swelling without inflammation, and the skin becomes cold and pallid. The gangrene begins in the center of the limb and appears in the skin only after a long time, so that it is often necessary to open the skin in order to find the gangrene inside.

Sometimes surgeons amputated the infected limb in the hope of halting the spread of the gangrene. If a limb was not amputated, it became "dry and thin, as if the skin were glued to the bones, and of a dreadful blackness, without rotting." Nonsurgical treatment included ardent spirits, volatile spirits, "orvietan," and a tisane of lupines.[32] If physicians could not agree on the course of the illness or the efficacy of various treatments, that was, according to Dodart, because the illness varied "according to time and place," which made it necessary to examine spurred rye from different areas in France.[33]

Ergotted rye had been found "nearly everywhere," but especially in "Sologne, Berry, the country around Blois," and in the Gâtinais. It was most likely to appear where the soil was light and sandy, and it was common "during wet years," and "especially when excessive heat followed a rainy spring."[34] Given these conjunctions, air, rain, and soil were the principal suspected causes of ergot. Based in Paris, the Academy could not test provincial air and rain, but Marchant grew rye in sandy soils brought to Paris from areas where ergot was common, and Bourdelin tested soils and grains.[35]

Dodart studied ergot grains and compared spurred rye with other cereals.[36] The fungus, called "ergot" in Sologne and "bled-cornu" in the

Gâtinais, appeared "black on the outside" and "rather white inside." When dried, it was harder and denser than rye grains, and Dodart found its taste not unpleasant. At the base of some ergot grains, he noticed "a substance with the taste and consistency of honey." This was the mucus, called "honey dew," which was the second or conidial stage in the development of ergot and which caused the growth of the sclerotium, or the ergot grain itself. "Infected grains" grew longer than normal grains, and Dodart observed that some were as large as thirteen or fourteen *lignes* long and two *lignes* wide. On a single blade there might be seven or eight spurs (plate 4).[37] Academicians and their contemporaries were uncertain whether ergot was the rye itself, distorted in shape and wholesomeness, or rather "foreign bodies produced among several grains of rye." Adherents of the former, incorrect view cited the resemblance of ergot to rye and the similar taste of breads made from ergot and from rye.[38]

Although it was widely doubted that the rotten rye caused the gangrenous sickness, Dodart believed that the absence of that malady except in persons who ate only rye bread, and the correlation between the appearance of ergot and the prevalence of the illness, argued in favor of ergot's being the cause. To verify this hypothesis, the Academy, like the elder Tuillier, ordered that bread made of ergot and rye be fed to animals.[39]

Dodart and his colleagues recognized that ergotism respected class lines. It was a malady of the country poor because rye bread was so important in their diet.[40] Seventeenth-century medical treatises routinely blamed mediocre food for illness among the poor.[41] Modern research has revealed just how bad that food was. In the Beauvaisis, a wheat-producing area, 75 percent of the peasants were "condemned to suffer hunger" in good years and "to starve to death" when the harvests were bad. The diet of peasants was not nutritious: it rarely included meat, milk, cheese, or fruit of good quality. Bread, gruel, and legumes formed the basis of a diet that was "both heavy and lacking in nutrition, insufficient during winter and increasingly so as spring approached."[42] The conditions in Beauvaisis resembled those in other areas of seventeenth-century France.[43]

Even during good years the peasants were chronically ill, and when times were bad, starvation and death were common. Thus, if the poor consumed rye they had grown themselves, hunger and ignorance prevented them from discarding spurred rye; sometimes hungry persons begged to be given the ergot already separated from rye, in order to make their flour go further.[44] Heavy demand for cereals, exacerbated by the army, large cities, and famine, tempted the unscrupulous to sell the ergot with rye.[45] Ignorance and circumstance led peasants to use rye infested with ergot.

Plate 4. Ergotted Rye. (Plate 111 of Bulliard, *Histoire des plantes vénéneuses et suspectes de la France*. Paris: A. J. Dugour et Durand, [1798]; photograph courtesy of Hunt Institute for Botanical Documentation, Pittsburgh.)

When poor harvests threatened starvation, the populace traditionally looked to government for relief, demanding official intervention against private hoarding and high prices. Local and royal governments accumulated stores of grain for sale when there was a dearth and attempted to prevent export of foodstuffs from a producing region whose own population required them for survival.[46] Operating within this tradition, Dodart recommended legislation and hoped local officials would prevent the use of ergot as food. The Academy would assist by studying spurred rye from every region in France, in order to correlate the variations in ergotism with differences in rye and ergot. Academicians would continue to publish their findings so that magistrates could warn the people about the danger, require that all grain be sorted, and forbid millers to grind rye mixed with ergot, "which is so easy to recognize that it is impossible to mistake it" for good rye.[47]

Dodart was probably the first to publish the connection between ergot and the gangrenous malady, and academicians and others continued his research in the eighteenth century.[48] But many medical practitioners rejected the claim that eating ergot caused Saint Anthony's fire, and ergot poisoning was neglected even in treatises that discussed malnutrition and famine.[49] Because maladies were defined in terms of symptoms rather than causes, Saint Anthony's fire was usually conflated with erysipelas, scurvy, and gangrene as a skin disease. Even Dubé explained Saint Anthony's fire simply as "a Mixture of bileous and pituitous Humours" without mentioning ingestion of ergot.[50] Dodart's important article, therefore, had only a limited effect on magistrates, medical practitioners, or the principal victims of the malady.

The Academy's medical interests and Dodart's awareness of the social discrimination of certain illnesses may suggest that the Academy was sensitive to the needs of Louis's most numerous but least privileged subjects. But academicians were isolated by birth and training from most of the populace. They were academicians because they were known personally or by reputation to those in power, and indeed many of the medical practitioners admitted to the Company had served the royal family in some capacity. Academicians analyzed meat, fish, vegetables, and fruits, but these foods mostly represented the diet of only a quarter of the population of France. The Academy's notion of social responsibility was mainly irrelevant to the needs of the poor.

The Jansenist Dodart was more interested than his colleagues in such problems: he studied medicine for the poor,[51] treated the poor free of charge, and died as a result of an illness contracted from one of his indigent patients.[52] But his sympathies did not prevent him from approving the use of prisoners as guinea pigs. Attitudes molded by social class shaped academicians' concepts of their social responsibilities. Dodart's work on ergotism represents only a modest effort by the early Academy to develop knowledge and legislation in the interests of the poor. Academicians, like their contemporaries, sought to improve the lot of the poor through ad hoc measures and took the social order as given. Thus, the Academy's posture is consistent with the entire pattern of old-regime reform, which conceived change always within the context of contemporary social and political structures.[53]

CONCLUSION

Academicians hoped their work would have practical results, and especially that it would benefit health. Such considerations influenced the Academy's natural history of plants. Dodart's article relating spurred rye to ergotism epitomizes many features of the botanical studies of the seventeenth-century Academy, from its indebtedness to outsiders and use of chemical analysis to the medical interests that influenced its research. Academicians' search for the practical, medical benefits of their work stemmed from previous training and experience and also from the urgings of the Academy's protectors. By pursuing their medical interests, academicians could fulfill institutional responsibilities, protect their theoretical research when it was threatened, and put to good use their contacts with those outside the Academy. The nature of that external community and the character of the Academy's ties to it are addressed in the following chapters.

CHAPTER 14

Scientific Paris at the End of the Century

The Academy was part of a larger community interested in science. This community comprised philosophers and experimenters, authors and debaters, travelers and collectors, instrument makers and teachers, medical and mathematical practitioners, amateurs and patrons. It was international and also local. While the Academy addressed theoretical issues of international concern, it was simultaneously part of a French community, to which it owed its foundation, its support, and most of its members. Far from monopolizing the practice of science in Paris or in France, academicians enjoyed ties to other savants with similar interests, as the Academy's study of ergot revealed.

The relations between the Academy and the larger scientific community — whether French or international — clarify the interests of both. During an era when cooperation was much vaunted within the scientific community, the formation of an elite, closed institution altered scientific intercourse. The Academy's association with outsiders throws into relief the hierarchy of the late seventeenth-century scientific community, the nature of the audience for science, some features of the Parisian scientific community, and the benefits academicians and nonacademicians derived from their contacts with one another.

THE SCIENTIFIC COMMUNITY

The early modern scientific community was stratified both socioeconomically and intellectually. Although biographical data are insufficient to assess its socioeconomic structure in detail, the intellectual hierarchy of the scientific community is somewhat more accessible. During this period most scientific savants did not make a living from research, teaching, or publication, but supported themselves instead as clergy, magistrates, physicians, or the clients of nobles and princes. Indeed, the word *scientist* had not yet been coined, and thus modern criteria of profession are often inapplicable.[1] There was, nevertheless, a clear recognition within the learned community that some of its members were more worthy than others. Some individuals gained reputations as savants, geometers, natural philosophers, anatomists, or botanists, for example, because of their accomplishments; others were known as amateurs or intelligencers, because their interests were more general and their contributions more modest.[2] There were "athletes," "talkers," and "listeners," as Le Gallois put it in the 1670s, and the athletes (that is, the vigorous experimentalists) learned little from the others, who predominated in the private learned societies of Paris.[3] Beyond the scientific community was the larger public, which variously absorbed, ignored, rejected, or was unaware of what was published by others.

The hierarchy of the scientific community depended, therefore, on the value of each member's actual contributions, and different worths were assigned to theory and raw data, with explanation more highly prized than uninterpreted information. Martin Rudwick has recently analyzed the scientific community as composed of two principal components: the elite, who determine which theories are plausible, usually preferring hypotheses put forward from their own ranks, and the amateurs, whose theories are normally rejected by the elite who may nevertheless examine the data amateurs provide. In addition, there is the interested public, whose data and theories are both suspect in the eyes of the elite, and which is regarded by both elite and amateurs primarily as having the function of audience. Individuals can move up and down the ladder, and a polymath may fit all three categories at once, but theory mostly trickles down, while some data filters up.[4]

This analysis, developed for nineteenth-century London, is suggestive for seventeenth-century Paris. But it does not convey the enthusiasm for experiment or the faint mistrust of hypothesizing characteristic of seventeenth-century savants. Thus it does not correspond wholly to the distinctions that academicians and their contemporaries made about their community. Le Gallois's athletes were experimentalists but not necessarily theorists. What mattered was that the elite be innovative, because the next rank in esteem—whose members were called amateurs, virtuosi, intelligencers, talkers or listeners—was imitative. The scientific community also included mathematical and medical practitioners who earned a living from surveying, making instruments, performing surgery, or composing medicaments, for example, and contributed in complex ways to early modern science. Finally, although the Academy had its own internal hierarchy, it constituted an elite institution vis à vis the rest of the scientific community.[5]

The seventeenth-century Academy and Royal Society were composed of these various groups in different proportions. Unlike the Academy, the Royal Society depended on its members' annual subscriptions for funds and admitted larger numbers of amateurs, so long as they could pay the price.[6] In the Academy, there were few amateurs and intelligencers—men like Thévenot, Du Hamel, and Fontenelle—but many students and practitioners, such as Niquet, Pivert, Bourdelin, and others, who did not as a rule theorize at meetings. The elite at the Academy were Cassini, Huygens, Dodart, Perrault, Duclos, Mariotte, La Hire, and others, who dominated planning and publishing and gave most of the papers at meetings. Most academicians were "athletes" in the broadest sense of the word; that is, most were serious researchers and writers, working at the edge of their respective fields.

MODEST PUBLIC INTEREST IN SCIENCE

The audience for science included the entire scientific community—elite and amateur—as well as those members of the literate public who were curious about the nature of the world. Some early modern writers recognized the importance of public interest in science. Both Bacon and Descartes, for example, emphasized the benefits that science offered to society, and Bacon thought that in exchange the public ought to supply data, while Descartes believed financial support was more efficient.[7] At the heart of the relationship between specialists and the public were mutual benefits and overlapping interests. The scientific community and the public were united by a curiosity about the universe born from the conviction that understanding it was interesting, important, and potentially useful.

During the early modern period, science became a recreation for ever larger numbers of people, who came from ever broader cross-sections of the total population. The popularity of scientific literature in the vernacular, the publication of scientific treatises for the general reader, the development of lecture-demonstrations in the eighteenth century, the changing holdings of personal libraries, and patterns of borrowing from circulating libraries all signify this trend.[8] Nevertheless, science remained the interest of a minority, and in some circles it was downright unfashionable. The audience for science during the early modern period was heterogeneous, and it remains inadequately defined.

The "battle of the dictionaries" during the 1680s highlights French attitudes towards the sciences at the end of the seventeenth century. Two dictionaries competed against each other — one prepared by the Académie française, the other by Antoine Furetière, one of its members. At stake was the nature of the language. Furetière's *Dictionnaire universel*, with its technical vocabulary from the arts and sciences, was highly regarded despite a campaign by the Académie française to suppress it. In contrast, the Académie's own dictionary was widely criticized for excessive purism. One of the issues dividing the authors was the respectability of the sciences. Purists claimed that the lowly social origins of savants and the vernacular etymologies of scientific words made natural philosophy not respectable and thus justified the exclusion of its vocabulary. Latin, and to a lesser extent Greek, had traditionally been the language of scientific savants, and when science was a bookish, scholarly preserve, Latin's technical vocabulary and international compass made it indispensable. But as the nature of scientific inquiry changed and the ranks of natural philosophers were swelled by practitioners and others lacking university degrees or knowledge of Latin, the scientific vocabulary not only expanded but even took many of its neologisms from the vernacular.

During the sixteenth century especially, the vocabulary of the French language grew because literati explored many different subjects, championed to some extent the language of the people, and learned the technical terms of various disciplines. In the seventeenth century there was a purist reaction to this expansion of the language. But the triumph of Furetière's dictionary marks the partial defeat of those sticklers who disqualified words that referred to unseemly objects and activities or that lacked Greek or Latin forebears. In the best tradition of the previous century, Furetière remarked that architects, engineers, and mathematicians spoke good French and that a dictionary must include the language of practical disciplines.[9]

Even scientific savants disagreed about the propriety of introducing harsh, technical terms into the delicate French language. The old-fashioned La Chambre, a physician renowned for his elegant prose and a member of both the Académie des sciences and the Académie française, urged physicians to conform to the highest literary standards so as to make medical literature acceptable in good society. But this would have entailed omitting such terms as *capillaire, botanique, amputation, alimenteux, impénétrabilité, effervescence, balsamique, chirurgical, anastomose, aneurisme,* and *aorte* from the language, even though many were essential for discussing the most timely scientific issues.[10] Other academicians were more receptive to innovation. Perrault and Tournefort welcomed the word *botanique,* Dodart and the Marchants coined new words for the natural history of plants, and Borelly, Blondel, and Auzout helped Furetière to master up-to-date scientific, technical, and medical vocabulary.[11]

Most scientific institutions in the late seventeenth century adopted the vernacular and published their transactions in Italian, English, or French. Although the majority of books and articles by academicians appeared in French, the publications of the Academy's two permanent secretaries reflect a transition. Du Hamel's excellent command of Latin helped earn him his post, and he wrote his history of the Academy in that traditional scholarly tongue; in contrast, Fontenelle published the history and memoirs of the Academy and his eulogies of academicians in French. By the end of the century, science was commonly discussed in the vernacular in France, and scientific ideas and words had become useful metaphors in the language.[12]

But at the very time when the vernacular began to replace Latin as the language for the sciences, literacy in France was declining. Carlo M. Cipolla has calculated that, for early modern Europe as a whole, one to three teachers per thousand persons would have been necessary to increase the proportion of those who could read and write. But "in 1672 there were in Paris 332 teachers . . . and about 480,000 people," or fewer than "seven formal elementary school teachers for every 10,000 people." This ratio was low by comparison with the late middle ages. In all of France between 1686 and 1690 only 25 percent of the persons "who contracted marriage . . . could sign their names." If the data for the Narbonne region apply generally, then "literacy among merchants and bourgeois was as high as 90 percent and more," while "among urban artisans it was about 65 percent, and among the rural population it ranged between 10 and 30 percent."[13]

Literacy alone was no guarantee of an appetite for scientific literature, which in turn led only exceptionally to study of the new theoretical sciences. The reading public preferred religion, history, the ancient classics, and French literature to the sciences and philosophy. Moreover, this was the era when the fairy tale was in vogue and when the taste for the marvelous attracted the educated to study folktales, superstitions, and prodigies. Thus, natural histories reported monsters and other curiosities, and even academicians were not immune to fashion, although they tried to reform it.[14] Popular treatises on science, some intended for the literate artisan or small shopkeeper, were practical, old-fashioned, and superstitious. Almanacs and books on medicine for the poor, or on arithmetic, astrology, or travel were aimed at popular audiences, but devotional literature enjoyed a much larger share of that market. Only 5 percent of the books in private Parisian libraries in the second half of the century were scientific.[15] Even Nicolas Blegny, the physician who compiled a book of useful addresses in Paris for 1691 and 1692, listed more music teachers (seventy-five) than physicians, and scientific practitioners and bookshops were in a minority.[16] Except for the popular or the pseudosciences, science seems to have interested only a small proportion of the French population, and the principal audience for natural philosophy, as Henri-Jean Martin has shown, was among the upper robe, the politicoeconomic elite that dominated the cultural life of Paris.

Personal libraries reveal the kinds of scientific treatises collected in the period. Books by Bacon, Galileo, and Gassendi appeared frequently, those by Rohault and La Chambre occasionally, and Malpighi's anatomy of plants rarely. Parisian readers evinced little interest in chemistry, disdained perhaps as the domain of "sooty empirics," or in medicine, the province of specialists who made a living practicing it. The most popular fields were architecture, fortification, cartography, geography, and botany. A large number of the titles represent sixteenth-century authors. Of the official publications of the Academy, only Dodart's *Mémoires des plantes* appears in the inventories Martin has analyzed. Blondel was the most widely read academician, and his treatises on architecture, fortification, and geometry had appeal beyond any works by other members of the Academy. Thus, not many Parisians kept up with technology and science, and those who did preferred military subjects and natural history.[17]

In general the audience for science was not strongly inclined to theory. It was dominated by amateurs who found scientific subjects entertaining or useful. Their relatively superficial interest is evident when their libraries are compared with those owned by the producers of scientific knowledge. In contrast to the 163 scientific books Martin identified in more than two hundred private collections, the library of Nicolas and Jean Marchant (whose work was descriptive rather than theoretical) contained more than two hundred titles on botany and medicine alone.[18] Producers of science, therefore, were its most avid consumers, and the market for scientific books was small.

SCIENTIFIC GOODS AND SERVICES IN PARIS, 1660–1700

Although the sciences were becoming more respectable, natural philosophers and their audience formed a relatively small, indeed sometimes intimate, community. Within France, Paris was a focal point of scientific activity. Paris attracted not only those who wished to make their fortunes and attain high office but also others who sought success within literary, musical, artistic, and intellectual spheres. It had become the cultural center of the kingdom, and once the Academy was established there, the allure of the city increased for scientific adepts.

The city offered its inhabitants and visitors variety within a small geographic range. Paris furnished the Academy's immediate theater of operations. In it lived the power brokers who controlled patronage, the printers and instrument makers who supplied the tools of the trade, and an audience eager for discourse and demonstrations. Academicians thus belonged to networks that wielded power, stimulated the intellect, and provided services. The Company's relationship to the community was twofold: it depended on local facilities for many of its activities, but it also affected the way science was done in Paris and how Parisians perceived scientific savants. To examine the Academy in isolation, therefore, would be to ignore the reciprocity between the institution and its environment.

But because the Academy eclipsed, both in fact and in historiographic tradition, the scientific community in which it functioned, the task of reconstructing that community is not only important but also difficult. In such a quest, biographies, notarial records, guide books, travel memoirs, correspondence, scientific literature, the minutes of the Academy and other learned institutions, and the personal papers of scholars, patrons, and their associates, would be of obvious help. Because scholarly intercourse relied mainly on frequent conversations that were rarely recorded, however, it is elusive. Here it is possible to offer only an outline of the Parisian scientific community at the end of the century—not, as would be ideal, as a commonwealth of competing ideas, each with its adherents and opponents, but, more practically, as a community of goods, services, and their consumers, which formed a network of overlapping interests that included the Academy.

When the Academy was founded, Paris was a city of nearly 500,000 inhabitants and 24,000 to 30,000 houses. The poet Paul Scarron wrote of it as a confused mass of bridges and filthy streets, of churches, palaces, prisons, houses, and shops, whose inhabitants were people of all physical and moral types. Its narrow streets, paved and unpaved, were lit at night by the reflecting candle lanterns installed in 1667. When Martin Lister visited at the end of the century, he observed that its houses were made of "hewen Stone . . . or whited over with plaister." He also remarked that very few, and then only small, signs were permitted on the streets, while statues of the king and his forebears abounded. Although Paris lacked numerous public squares, it seemed to a foreigner to be a public city, where people liked to come "together to see and be seen" and to converse out-of-doors. The English visitor noted not only many monks but also lawyers and their wives, with "their trains carried up." In the streets, coaches traveled at speeds that endangered pedestrians, including Tournefort, who died of injuries sustained in just such an accident. The air, like that of other large seventeenth-century cities, was polluted, although it was not so unhealthy as in London. Parisian institutions that figured prominently on maps and in guidebooks included hospitals, the courts of justice, the university, the Royal Garden, monasteries and churches, the *hôtels* of the wealthy, and the Academy's Observatory. The city offered products ranging from glass eyes to carpenters' nails, and services from the conservation of paintings to air disinfection.[19] As a center of culture and commodities, of architectural sights and learned gatherings, Paris inspired admiration. Locke reported ironically that Paris must "be heaven, for the French with their usuall justice extol it above althings on earth."[20]

The city (plate 5) consisted of seventeen quarters, each with a distinctive character. Three areas especially influenced the intellectual life of Paris. On the Left Bank, the university quarter with its bookshops dominated; it was flanked by the Jardin royal to the east and by monasteries, convents, *hôtels*, and hospitals to the west. On the Right Bank, the *hôtels* of the nobility and the upper robe overshadowed in magnificence and power the crowded artisans' quarters to the east. The administrative and regal center of the city—perched on the Seine, controlling the Cité, and running along the river's edge on the Right Bank—was situated in the Tuileries Gardens and Palace, the Louvre, the Palais de Justice, the Châtelet, the Hôtel de Ville, the Arsenal, and the Bastille.

Members of the scientific community could be found in most of the city's quarters, but they were often clustered around major scholarly landmarks. On the Left Bank (plate 5b), the university quarter, for example, contained not only savants like Gallois who lived in the colleges but also many printers and their shops. Some instrument makers set up their premises on rue Mazarine and rue neuve des Fossés, streets that marked the boundary between the university quarter and that of monasteries and *hôtels*. At the western and eastern limits of the city, the Hôtel des invalides and the Jardin royal housed several savants, including members of the Academy. To the south, at the end of rue du faubourg Saint Jacques, was the Observatory, where several academicians lived; between the Observatory and the university quarter was the faubourg Saint Jacques, where certain parishes, salons,

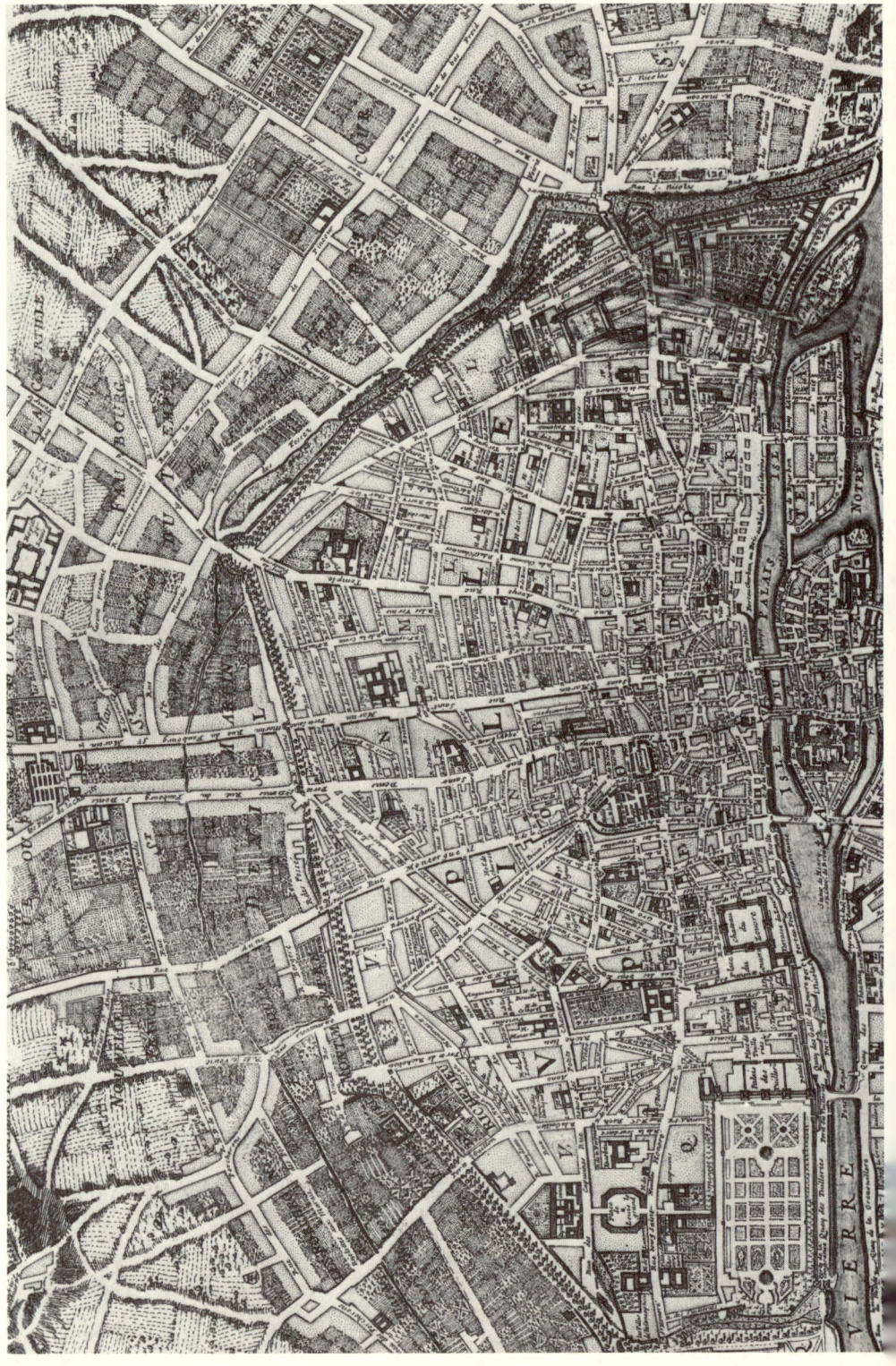

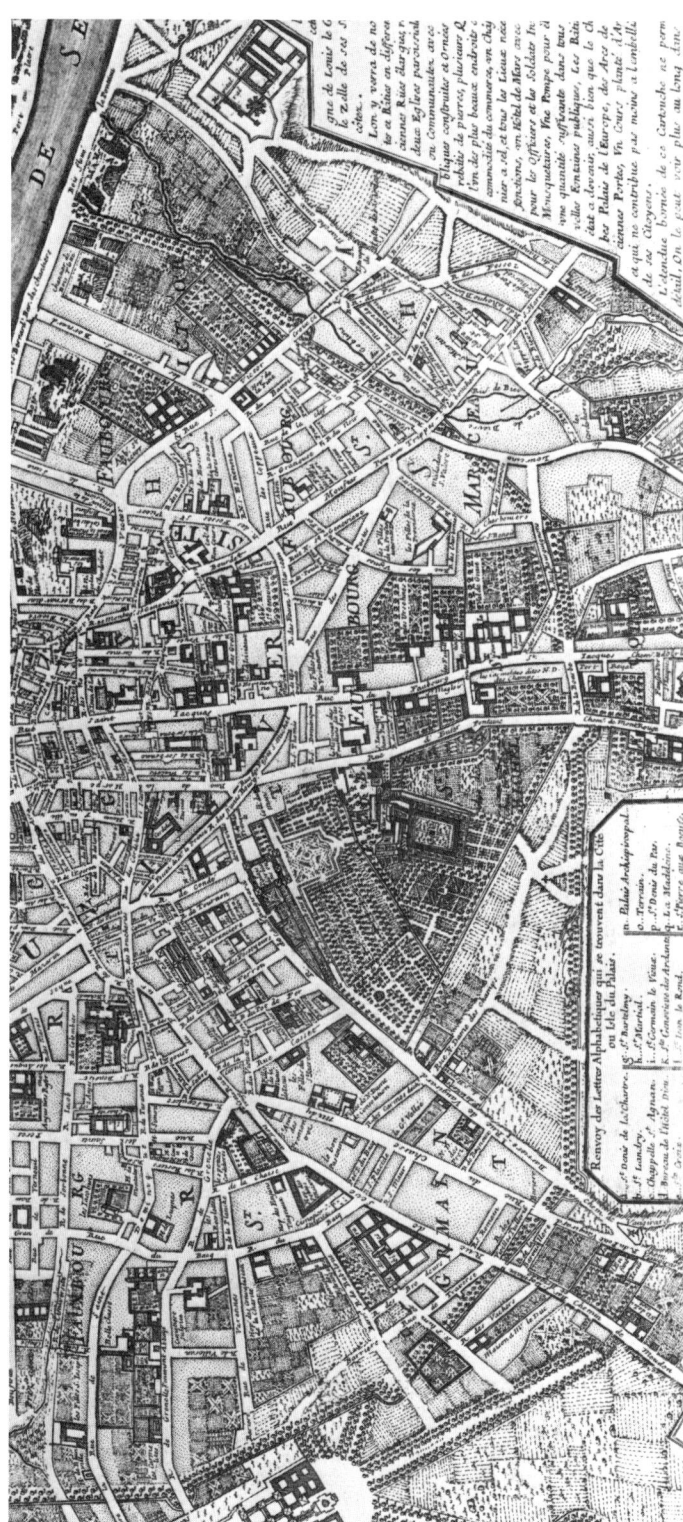

Plate 5. Paris circa 1700 (engraved by Nicolas de Fer; photograph courtesy of Bibliothèque Nationale, Paris).

and sites attracted a community of Jansenist savants, among whom academicians counted many friends. Pierre Varignon lived there with his friend and benefactor, the abbé de Saint Pierre.

On the Right Bank (plate 5a), the district around the Palais royal knew both scholars and their patrons, and the *hôtels* of the latter often housed their learned protégés. The Bibliothèque du roi was home to the Academy and a handful of savants, including academicians and the scholars responsible for the King's Library. Nearby, on rue Saint Pierre, an instrument maker had established his premises. To the east in the Marais, Dodart owned the family house in rue Sainte Croix de la Bretonnerie, not far from the *hôtel* of a tax farmer.

The administrative center of Paris formed a bridge between the Left and Right Banks, and contained several focal points for savants. Instrument makers who enjoyed royal sponsorship made and sold their wares from the Galeries du Louvre, and savants lived in the Louvre as well, as had Cassini briefly while awaiting completion of the Observatory. The rue du Harlay and the quai de l'Horloge in the Cité also attracted instrument makers and map makers. Another shop where instruments could be purchased, on the quai Peletier (now quai de Gesvres), was only a short walk from the quai de l'Horloge across the Pont au Change and along the Seine toward the Hôtel de Ville.

In the late seventeenth century, the Right Bank challenged the university quarter as the intellectual center of Paris. When the upper robe increased its patronage of learning and the arts, some savants, learned institutions, and business enterprises migrated to that district of the city. Colbert had stimulated that shift by moving the Bibliothèque du roi from the rue de La Harpe (which ran parallel to the rue Saint Jacques in the university quarter on the Left Bank), to the rue Vivienne, just north of the Palais royal where it was surrounded by the *hôtels* of Colbert himself, Louvois, and Pontchartrain. The appeal of these new sources of demand is clear, because shops offering scientific instruments and maps sprang up in the Cité and the Right Bank, despite the pull of the Observatory and the university on the Left Bank.[21]

Paris supported and encouraged scholarly effort. It was a center for the manufacture and trade of books, scientific instruments, and other objects that savants and dilettantes accumulated and used. Some of its inhabitants owned private collections that might be opened to like-minded individuals for their contemplation and admiration. The city was provided with teachers of specialized disciplines and with private societies for learned colloquy. Furthermore, it benefited from royal patronage, which singled out the city itself for attention and also situated there various institutions devoted to the theoretical and applied sciences. In short, Paris was the focus of scientific research, teaching, and discussion in France.

Publishing was essential to the learned community, and Paris was a center of the French book trade, which was in decline largely as a result of the general economic contraction. Other negative influences included royal censorship, a flourishing clandestine trade in controversial books published abroad and smuggled into France, a new expectation on the part of authors that they should profit from the sales of their books, and a preference on the part of printers for cheaper, smaller books that would reach a more numerous audience. The effects of royal censorship were obvious to Lister in 1698: he noted how few weekly gazettes were sold and how difficult and dangerous it was "to vend a Libel" in Paris.[22] The Imprimerie royale, operating outside the normal economic constraints, could specialize in prestigious works in the large formats that had become a thing of the past for most private printers but remained attractive to the crown for propagandistic reasons. In aggregate, private Parisian printers offered only a few hundred titles a year, and fewer than two hundred of those were new.[23] The book trade as a whole was in a weakened condition.

Nicolas Blegny's guide to goods and services available in Paris during the early 1690s told its users where to look for books on specific subjects. Persons interested in gardening could buy at the shops of Sercy and Barbin, in the grande salle of the Palais de Justice and on the steps of Sainte Chapelle, respectively. Several shops selling scientific books were located in the university quarter, on rue Saint Jacques. One specializing in foreign literature offered a life of Descartes and treatises on gout and surgery. Michallet, an imprimeur du roi, sold not only the *Ordonnances pour la marine* and *La connoissance des temps,* both official works, but also Lémery's *Chimie* and books on medications and the conquest of New Spain. Laurent d'Houry's shop, opposite the Saint Severin fountain, emphasized medical treatises. Cusson offered the *Journal des sçavans,* a weekly edited by the academician Gallois; it reviewed the latest books and contained articles (written in the form of letters to the editor), on the subjects of natural philosophy, history, and theology.[24]

Coignard, an imprimeur du roi who sold the *Dictionnaire* of the Académie française, also sold geography books and Vauban's treatise on fortifications. An arithmetic for engineers by La Londe, Boyle's specifics, Duncan's *Mithologie phisique,* and Cordemoy's *Discours philosophique* were available from the widow Nion on the quai de Nesle. She sold several medical books, including Blegny's treatises, and a volume said to be the works of Aesculapius, which Michallet and d'Houry also sold. Books by academicians on mathematics were available from Michallet and from Jombert, on the quai des Augustins. Savants like Étienne Baluze, a *garde* at the Bibliothèque du roi, spent their afternoons at a favorite bookshop, which offered not only items for browsing and purchase but also a place where like-minded individuals could meet and chat.[25]

Books and articles by members of the Academy constituted a major proportion of all the literature on mathematics and the sciences published in Paris between 1650 and 1700.[26] Academicians also published in related areas, such as the art of war, the fine arts, technology and scientific instruments, medicine, and the hermetic sciences. By establishing the Academy, therefore, the crown stimulated scientific publication, not only because the Imprimerie royale printed works by its members but also because the status of an academician apparently induced private printers to try new titles that otherwise they would have refused.

Like books, scientific apparatus—instruments, maps, globes, and other equipment—ranged from the state-of-the-art tools necessary for research to items more appropriate to the amateur or collector. Thus, they were produced and sold by makers of varied skills.

Maps were widely available and very popular; some reflected the Academy's cartographic advances. Two locations—the Galeries du Louvre and the quai de l'Horloge—were especially attractive for map makers at the end of the century because the best instrument makers were also situated there, but map makers had traditionally produced and sold their wares near rue Saint Jacques, the focal point of the book trade until demand extended significantly outside university circles. The cartographer Sanson, whose work had to be corrected after the Academy surveyed the northern coast of France, sold his maps out of the Galeries du Louvre. Mlle de Val sold maps and books on geography from the quai de l'Horloge, but a better product could be had further along the same quai from Nicolas de Fer, a royal geographer who based his *Atlas curieux* of 1704 on the cartographic work of the Academy and whose historical maps of Paris appeared in Delamare's *Traité de la police*.[27]

The Academy relied on the best makers to construct and repair its instruments, and other savants used their services as well. Thus, for mathematical instruments Blegny sent his readers to the elite of instrument makers: to Le Bas, who as a royal artisan had a shop in the Galeries du Louvre, and to Chapotot and Butterfield on the quai de l'Horloge.[28] The *armuriers* or gunsmiths Gosselin and Lagny, on retainer to the Academy for repairing mathematical instruments, occupied a three-story house owned by the king and located on rue St. Honoré opposite St. Roch.[29] Macard had bought up Sevin's stock and sold it from a shop on the quai de Morfundus at the sign of the astrolabe.[30]

Apparatus for measuring time was available at the Galeries du Louvre from Isaac and Jacques Thuret, father and son; on retainer to the Academy, they also sold watches and pendulum clocks to anyone who could pay the price, including perhaps owners of private observatories.[31] For meteorological instruments, such as barometers, thermometers, hydrometers, and aerometers, the work of Hubin, an enameler, was highly regarded. Hubin also offered a line of air pumps at his shop on rue Saint Denis, across from the rue aux Ours; he was a friend of Denis Papin, Huygens's one-time collaborator, and perhaps developed his air pumps through this association. Hubin supplied and repaired aerometers for the Academy's chemical laboratory, and his thermometers were later admired by Réaumur. Do, another enameler, offered simpler and cheaper meteorological instruments from his shop on rue du Harlay.[32]

André Dalesme, an inventor who became an academician in 1699 and was paid from the 1680s for assisting the Academy, had a shop on rue Saint Denis near the Queen's Fountain. At it he sold the latest gadgets, such as quill pens with special nibs, an iron sheet-metal tube suitable for burning wood without a fireplace because no smoke was produced, a "thermolampe," and a "machine à vapeur."[33] Chemists could get furnaces and crucibles at shops on the place de l'Hôtel de Cluny, on the rue Mazarine, and in the faubourg Saint Jacques.[34]

Apparatus for the new experimental sciences was itself new and experimental, and instrument makers helped savants keep up with the latest inventions or trends. Locke collected information in his travels from anyone who was willing to share it, including "Mr. Oury, a watchmaker" whose shop was on the quai Peletier and who informed Locke in 1677 that he had "given off the way of makeing watches with very great ballances in imitation of pendulums because, but several jogs in one's pocket, they are apt some times to stand still." Butterfield showed Locke a new leveling instrument that was described in the *Journal des sçavans* in 1677 and 1678.[35] When a serious experimenter at the forefront of his field needed custom-made equipment, instrument makers and lesser artisans helped design and construct it. Many of the makers recommended by Blegny worked with savants to develop new apparatus, and other, nameless artisans with the homely skills of boring tubes or blowing glass were crucial to seventeenth-century experimentalism.[36]

But many instruments were sold for entertainment rather than for research. Most mechanical timekeepers, for example, were decorative or instructive, not precision, instruments. Probably the work of Daniel Deaubonne, who was "a simple monk for more than thirty years," falls into this category. He made lenses and microscopes at the Abbey of Saint Germain des Prés and sold them "with the permission of the superiors, who saw that the money went to the poor."[37] Pouilly, on rue Dauphine, made mathematical instruments, appealing to amateurs with a curious calendar designed for their studies and claiming that he could increase the power of lodestones or that his microscopes had extraordinary magnifying power. Cylindrical mirrors were available at the shop of Amielle near Saint Hilaire and from a Theatine monk who also made magic lanterns.[38]

Natural history could also be entertaining and collectible, and menageries and gardens were fashionable showpieces of curiosity and artifice. Some curiosities provided popular spectacles and profit for entrepreneurs. Enticed into a booth advertising four exotic animals, Lister was disappointed to find only the "very ordinary" leopard and raccoon, but he enjoyed seeing in Paris a trained elephant that "bent the Joints of her Legs very nimbly in making her Salutes to the Company." Locke visited the royal menagerie half a league from Versailles, where he saw ostriches and other birds, and an elephant that ate "fifty lb of bread per diem & sixteen lbs. of wine with rice."[39] The *pépinerie,* or royal nursery located in the faubourg Saint Honoré, provided a model of horticultural ingenuity and lavishness. Regarded by visitors as "worth seeing," it maintained greenhouses and an "infirmery of sick Orange-Trees," and in four years supplied "eighteen Millions of Tulips, and other Bulbous Flowers" to Marly alone. Affluent Parisians also cultivated rare plants amid fountains and prided themselves on individual methods of pruning; Louvois's garden was "one of the neatest . . . in Paris."[40] To assist enthusiasts of gardening, Blegny listed sellers of plants, trees, fruits, and vegetables, and he informed his readers that Tournefort and Plumier taught how to grow plants, with Tournefort specializing in medicinal varieties.[41] Individual academicians were skilled gardeners, but as an institution the Academy avoided horticulture and agriculture, deferring to the expertise of Jean de La Quintinie, director of gardening at Versailles, and to others.[42]

To compensate for the impractical or ephemeral aspects of keeping live specimens, collectors could turn to the taxidermist Colson, who assisted the Academy and also offered his services to the general public from premises in the faubourg Saint Antoine opposite rue de Charonne. They could also purchase paintings of flowers and animals from Nicolas Robert, whom Gaston, duke of Orléans, and the Academy had favored; after his death de Fontenay, near the Palais royal, and Huilliot, on the rue Bourlabé, were said to offer the best renditions of these subjects.[43]

Savants and amateurs alike cultivated gardens and collected books, medals, instruments, maps, shells, skeletons, and other items, many of which found their way into Parisian *cabinets* along with paintings and objects of art. Collectors might be physicians, chevaliers, or administrators of the Hôpital général; some were women, and many were wealthy and powerful royal officials. Like Seneca's millionaire — whose slaves recited the classics from memory and thereby displayed their master's erudition — they displayed their culture by opening their *cabinets* to fellow enthusiasts.[44]

Since one of the objects of collecting was to exhibit acquisitions, the reports of learned travelers abound with descriptions of the most pleasing, famous, or useful collections. These accounts reveal eclectic tastes that joined art and science, mingling beauty, utility, and curiosity. Du Vivier's rooms at the Arsenal were filled with Chinese porcelains, books, and paintings by the masters. Pierre Michon Bourdelot, first physician to the house of Condé and sponsor of a learned academy, owned "a perfect collection of all the books of philosophy and medicine."[45] Charles César Baudelot de Dairval, author of *De l'utilité des voyages* published in 1686, collected coins, medals, and Greek marbles and also owned a two-pound stone recovered from a dissected horse. M. de Gaignières collected engravings of statesmen, oil portraits, manuscripts, playing cards, and maps. Nicolas Boucot, a keeper of records, owned paintings, statues, and nearly sixty drawers of shells, a fish given to him by "Lady Portsmouth, possibly out of King Charles's Collection," and a zebra skin.[46] Certain monasteries enjoyed renowned collections. The discalced Augustinians owned a *cabinet* of natural history, and the *cabinet des machines* made by Sébastien Truchet attracted Peter the Great and other distinguished visitors to the Carmelite monastery.[47]

The Academy maintained its own collection of natural marvels in the Jardin royal, and it owned one of Galileo's telescopes, which it had received as a gift; its *salle des machines* in the Observatory functioned as an educational collection for visitors. Individual academicians were also collectors. Huygens's apartment contained scientific instruments and works of art, Tournefort owned a large herbarium and shells, and Louis Morin had an impressive "Musaeum of Natural History ... and of comparative Anatomies," which included "a Cabinet of Shells, another of Seeds, among which were some from China: Variety of Skeletons, etc." Blondel's well-known *cabinet* contained not only mineralogical objects but also works of art by Italian and French masters.[48]

Amateurs and natural philosophers alike had broad interests, and scientific objects found a place in their collections as the odd curiosity or as a subject of more systematic interest. Proprietors who showed their collections sometimes performed an important service to serious scholars, and this was especially true of the large private libraries. The monks at the Abbey of Saint Germain des Prés, for example, were allowed to use the Bignon collection and they modified their financial policy in order to retain this right.[49] Access to the Bibliothèque du roi depended on having access to the persons responsible for it, although it was opened for public viewing two days a week. When Lister visited, he found that books in some rooms were wired to the shelves for safekeeping,[50] a wise precaution, for at least one sponsor canceled scholarly meetings after noticing thefts from his home.[51] Many libraries had substantial scientific holdings, including the Bibliothèque du roi, the collections of individual savants, and perhaps some monasteries.[52] Blegny opened his own personal library to the public; it was part of his Jardin médicinal, where he lectured on surgery, drugs, and novelties in the natural sciences.[53] Academicians had routine admission to the Bibliothèque du roi and may have been allowed to use the private libraries of Colbert and the Bignon family when their own failed them.[54]

Scientific discourse complemented research, collecting, and publication. It could take the form of private correspondence and conversations between scholarly intimates or of meetings attended by a dozen or more savants and amateurs, eager to see a new experiment performed or to debate a new idea. Jacques Rohault gave private lessons on Cartesianism and held weekly public conferences; Thévenot, Bourdelot, Denis, and others also sponsored scholarly meetings.[55] In the early 1690s Blegny directed what he called the Société royale de la médecine; the abbé La Roque held natural philosophical discussions on Thursday afternoons; while on Saturday afternoons M. le Chevalier Chassebras du Bréau held meetings on history and science, and de Fontaney sponsored mathematical sessions.[56]

Establishment of the Academy did not dismantle private scientific societies, which often included academicians. At such meetings, academicians conversed with highly regarded natural philosophers, intelligencers, nobles, and foreigners, both men and women. Particularly during the 1690s, when the Academy did not know how to fill its weekly meetings, many of its members participated in the learned group of Mathieu François Geoffroy, father of the chemist who joined the Academy in 1699: Du Verney dissected and Homberg demonstrated chemical operations; Cassini brought his planispheres, Sébastien Truchet his machines, and Joblot some lodestones. Geoffroy's meetings were convivial and stimulating, and they brought academicians together with members of the Compagnie des arts et métiers.[57]

The modest interest in natural philosophy and mathematics among the public, and its superficial and eclectic character even among aficionados, can be blamed on education. Learning about natural philosophy was difficult because schools offered little instruction in the sciences or mathematics. Most advocates of such teaching saw these subjects as having a fairly limited utilitarian function. Jansenist pedagogues recommended teaching mathematics and physics (along with history, geography, religion, and the classics) in order "to develop the child's judgment,"[58] and Jesuit colleges offered a thorough grounding in mathematics and its applications in gnomonics and fortifications.[59] Parisian elementary schools for the bourgeois offered religion and commercial arithmetic, while noble youths learned riding, dance, military exercises, and mathematics.[60] Private lessons were available, and teachers of applied mathematics, such as surveying, sometimes advertised in *Journaux d'avis* or at Bureaux d'adresse.[61] Paris boasted at least fifty masters of arts who taught Latin, Greek, philosophy, and mathematics; others offered geography, foreign languages, fortifications, surveying, and architecture. Such teaching emphasized the basic skills necessary for performing military, medical, or mercantile operations. It was the collèges and university that introduced students to the new natural philosophy, primarily by trying to refute it.[62]

For instruction in the controversial ideas of the new science and mathematics, a student did best to teach himself or herself, work with individual savants, and attend meetings of private scientific societies. This was how many academicians had developed their knowledge and skills. In Paris savants like Lémery, Rohault, and Ozanam, some of them autodidacts themselves, offered instruction in their specialties.[63] Although the Academy was not designed as a teaching institution, some of its members gave lessons at the Observatory, with Rolle for example specializing in algebra.[64] The crown sponsored tuition in pure mathematics and the applied sciences. The Collège royal, where Varignon, La Hire, and Gallois were professors, was known for its teaching of mathematics.[65] Blondel and La Hire were professors in the Royal Academy of Architecture, which was like a craft organization in that its purpose was to teach the principles of architecture.[66] The Jardin royal tried to improve medical practice and reached a large audience with public lectures on botany, chemistry, and anatomy; its garden was "open also to Walk in, to all People of Note," and once replanted to illustrate Tournefort's classification, it educated in theoretical as well as medical botany.[67] The concentration of patronage and scientific savants, goods, and services, therefore, made Paris a center of education in science and mathematics, and especially of informal tuition in the new science.

Publication was important to the scientific world, but so were the personal contacts among scholars, fellow collectors, artisans, teachers, and students. All sought opportunities to study, converse, circulate manuscripts, or display some novel curiosity, and academicians participated along with everyone else. Paris encouraged a healthy mingling among the elite, amateurs, and the public, and by attracting provincials and foreigners, it broadened the contacts available to everyone. Just as it was a necessary stop on the scholarly grand tour of Europe, Paris was also an attractive destination for the ambitious savant who hoped to make a name for himself. The Academy and its members were important in both respects.

CONCLUSION

Scientific Paris abounded in shops and institutions, meetings and lectures. Science was to some extent a sociable discipline, and Parisians with an inclination toward natural philosophy took advantage of the varied opportunities the city offered. The Academy was an important participant in Parisian scientific circles. It contributed to a slight topographic shift of the scholarly community from the Left to Right Bank. Its transactions with instrument makers encouraged the trade, and its improved cartographic techniques were adopted by map makers. Its Observatory was open to visitors who could learn from the equipment and maps displayed there. Academicians participated in other scientific gatherings. They also collected books, natural curiosities, and instruments, and they educated the public through their publications, in private lessons, and by teaching at institutions sponsored by the crown. Scholarly interchange with persons outside the Academy remained important to academicians, who sought contacts in Paris and abroad. Members of the Academy were not cloistered, but avidly sought contacts. The use they made of these connections, however, was complicated by personal competitiveness among savants and by institutional interference.

CHAPTER 15

Academicians and the Larger Scientific Community

Because scientific issues went beyond linguistic and geopolitical boundaries, early modern savants, like their modern counterparts, maintained far-flung contacts. The Baconian and Cartesian programs advocated sharing research and information, and savants found it advantageous to do so. As a result, they sponsored and attended learned meetings, and contributed and subscribed to periodical literature. They also engaged in voluminous correspondence and traveled to visit well-informed individuals, famous collections, or learned academies.

Academicians were no exception. They, too, valued the idea of cooperation, which they tried to put into practice among themselves and with those outside the Academy. In seeking out like-minded persons for intellectual stimulation, they were not dependent on the local community, and in fact their Parisian ties were often less important to their work than more distant ones. The surviving correspondence, published and unpublished, of savants suggests how important were such exchanges of information and opinion and how wide was the geographical reach of seventeenth-century scholarly networks.[1]

As an institution, the Academy favored collaborative research, correspondence, and publication as means of sharing data. As individuals, academicians participated in every kind of scientific activity, locally and worldwide. But the Academy demanded a particular allegiance from its members, one that limited their freedom to exchange scientific information. In that respect the Company differed from most other contemporary scientific societies. It offered privileges—such as pensions, facilities for experiment and observation, and subsidized printing—that its members could not get elsewhere, and in exchange, it closed its meetings to those who were not members and required academicians to keep its business confidential. Membership in the Academy conferred privilege and status, but at a price. In exchange for privilege, the Company exacted a pledge of secrecy and tried to censor what reached the public.

PRIVATE CONTACTS BETWEEN ACADEMICIANS AND OTHER SAVANTS

Although by and large savants would have agreed with Mariotte that progress in natural philosophy depended on cooperation, there were many obstacles to their doing so.[2] The impediments to effective scholarly collaboration derived from not only personal traits and socioeconomic circumstance but also allegiances to religious, intellectual, or political views.

An overriding motivation for scholarly research was the quest for personal *gloire*. Savants like Du Hamel accepted this as a part of human nature, while Mariotte and others deplored it as a sign of "bad faith" and a cause of stagnation in scientific inquiry.[3] But other individual traits also affected learned collaboration, sometimes negatively. Personal inclination made some natural philosophers gregarious but others reclusive: Lister and Locke pursued every lead, using their travels to visit anyone with a promising reputation or collection; in contrast, the Jansenist botanist Louis Morin took as his motto, "visitors honor me, those who stay away give me pleasure."[4] Some savants had incompatible personalities, and still others would not collaborate with each other because of previous clashes.

Differences in social status were important barriers to learned discourse: on the one hand, inferior social standing disqualified some persons from attending scholarly meetings, while on the other, social superiority made some savants reluctant to dirty their hands in the practical labors necessary to perform certain experiments and closed their minds to the contributions made by artisans.[5] Financial limitations hampered scholars, preventing them from building their libraries, purchasing equipment and supplies, or traveling to educate themselves; Ray was grateful to his friend, the wealthy amateur Francis Willughby, who made it possible for Ray to travel by paying his way as a learned companion on Willughby's grand tour of the Continent.[6] Ignorance of Latin and foreign vernaculars circumscribed a savant's field of inquiry, and neophytes simply did not know where to go or whom to seek out; to some extent such problems resulted from social and economic circumstances that limited educational opportunities.

Religious, political, and intellectual influences were also decisive. Religious discord—whether between Protestant and Catholic, between Jesuit and Jansenist, or between rival religious orders—made some savants antagonistic toward one another and could affect how scholars formulated scientific questions or hypotheses. Political instability also impeded scholarly intercourse in several ways. During the reign of Louis XIV, French, Dutch, and English savants struggled to avoid embargoes on the exchange of correspondence and books; Lister's pleasure in visiting Paris during 1698 was heightened because France had been inaccessible to English scholars during the recent wars. Some savants were more seriously affected if they used their scholarly pursuits as a cover for clandestine activities on behalf of one side or the other, allowed their scholarly inclinations to be swayed by chauvinistic feelings, or lost their lives in battle.[7] Ultimately, intellectual sectarianism also divided scholarly ranks and obstructed reasoned discourse; Mariotte was distressed by the blind devotion to ill-chosen scientific camps that he believed was prevalent, and he hoped that a sound logic could overcome this problem.[8]

Within the limits imposed by such hindrances, however, academicians maintained a lively exchange of data and opinions both within France and internationally. Because their ties were intellectual, political, religious, social, and occupational, the stimuli to investigate scientific questions often came from unexpected quarters. Mariotte, for example, was influenced by Jean Baptiste Lantin to investigate plant physiology, and his published works mention correspondents in distant corners of Europe.[9] Duclos influenced the chemist Le Febvre and corresponded with Paul Ferry of Metz about books and religious concerns.[10] Thévenot urged Mabillon to arrange a correspondence with Malpighi for the benefit of the Academy.[11] Dodart was active among Jansenists as physician and confidant and, at Antoine Arnauld's urging, represented Jansenist interests to the king.[12] Tournefort stayed in touch with a wide circle, including not only Plumier and Fagon in Paris but also correspondents throughout Europe.[13] Dodart, Bourdelin, and Duclos associated profitably with Vallant, the physician to Mme de Sablé and the duchesse de Guise; the four colleagues shared gossip about medicine, foreign customs, and scientific findings.[14]

International contacts were prized by all savants, not least because, as Fontenelle later pointed out, different lands yielded different opportunities to savants.[15] Travelers sought meetings with their foreign counterparts, and acquaintance begun in person might survive at a distance through correspondence, with both parties conveying information about other savants.[16] Many savants studied abroad, often formally at university, as did Nicolas Marchant at Padua. Homberg had traveled extensively to study and earn a living in Europe before settling in Paris, and his ties with German and Swedish savants—he learned from Kunckel how to make phosphorus and studied mining in Sweden—influenced his contributions to the Academy.[17] Savants who could not travel depended on correspondents to keep them informed about the latest news and to send observations and natural curiosities.

Anglo-French contacts have been investigated more thoroughly than other contacts because of the comparisons to be made between the Academy and the Royal Society. But ties between French and English scientists were common well before the two scientific societies were established. Nicolas Marchant and Robert Morison, for example, had been colleagues in Blois at the garden of the duke of Orléans, and in later years too had similar ideas about how to study plants.[18] The private French societies that preceded the Academy included many members with English connections, and the Royal Society hoped to establish a regular correspondence with Montmor about the activities of his academy.[19] There was considerable curiosity about the Royal Society's membership, purposes, and activities, and its experimental program excited admiration.[20]

This tradition continued after the Academy was established, to the benefit of both institutions and their members. The rival societies measured their own success by each other's accomplishments and envied one another's advantages: the seventeenth-century Academy had no Oldenburg, no *Philosophical Transactions,* but when Fontenelle began publishing the annual *Histoire et mémoires* at the beginning of the eighteenth century, Lister worried that the Royal Society would "sink apace" unless it followed suit.[21] Huygens, Du Hamel, Blondel, Charas, Roemer, and Thévenot traveled, studied, or worked in England; among the academicians appointed before 1699, Auzout, Cassini, Huygens, Fontenelle, Lagny, and Varignon were or became Fellows of the Royal Society.[22] After his travels in England, Tournefort corresponded with several English scientists whom he had met, including Lister and Edward Lhwyd.[23] Dodart sent inscribed copies of the 1679 edition of the *Mémoires des plantes* to Morison, Grew, and Locke.[24]

All the travel was not in one direction, and English savants valued their contacts with academicians and other savants. When Locke visited France during the 1670s, he became acquainted with dozens of scientists, including the Protestant Charas (who was not yet an academician and with whom he stayed briefly), Picard, Auzout, Cassini, Roemer, and Thévenot.[25] He visited the Observatoire and the Bibliothèque du roi, and in April 1679 he saw the garden that future academician Louis Morin maintained at the Hospital of the Incurables in the faubourg Saint Germain, admiring "amongst other things there Thlaspi semper virens semper florens," which he found "extraordinary" because it always flowered.[26] News of academicians reached England, through letters and hearsay, and in the 1670s Oldenburg kept the Royal Society informed about Mariotte's work and urged Fellows to imitate his studies of winds.[27]

Correspondence reveals a substantial thread of Anglo-French botanical cooperation: seeds, plants, and books were exchanged and friendships initiated,[28] and Perrault claimed that the correspondence between La Quintinie and the English would fill three printed volumes.[29] Henri Justel offered in 1682 to establish regular communication between academicians and English scientists,[30] and in 1684 he tried to arrange an exchange of plants between Jean Marchant and the bishop of London.[31] At least some scholarly exchanges were motivated by the hope among English savants of obtaining patronage from Louis XIV, but others seem disinterested.[32]

Huygens's correspondence reveals the range of acquaintances and the merits of scholarly exchange for a seventeenth-century academician. His familiarity with English and Dutch scientists affected several disciplines at the Company, including botany. Huygens visited London in 1661, 1663, and 1689, going to meetings at Gresham College, being admitted to the Royal Society in 1663, and attending its meetings then and in 1689. He followed with interest the activities of his English colleagues and reciprocated with advice, details of his own research, and information about continental savants. Robert Moray, Henry Oldenburg, and other English correspondents supplied him with publications such as Hooke's *Micrographia* and Digby's *Vegetation,* and sent word of Grew's and Malpighi's botanical studies. Huygens kept in touch with Boyle through intermediaries; Boyle's books were sent to him regularly on publication, and Huygens reciprocated with his *Horologium oscillatorium*.[33] As an intermediary between the French and the Dutch, Huygens performed the valuable services of introducing Hartsoeker to the Academy and of translating Leeuwenhoek's correspondence for the Academy.[34]

Informal contacts between outsiders and academicians as individuals enriched their personal and intellectual lives. They also influenced botanical research at the Academy in five specific ways. First, they stimulated Huygens to develop the air pump and to experiment with the effects of airlessness on plants. Second, they led Huygens to develop the spherical microscope, with which he and other academicians viewed animalcules and pollen. Third, Mariotte examined the composition and nutrition of plants in response to questioning from Lantin, who was therefore partly responsible for Mariotte's and Perrault's debates on the subject in 1668. Fourth, the Marchants obtained many varieties of rare plants from foreigners who sent seeds and dried samples. Finally, correspondents drew the attention of Perrault, Bourdelin, and Dodart to the correlation between ingesting ergot and suffering from Saint Anthony's fire; this also affected Dodart's investigation of medicine for the poor. Without the influence of Boyle, Leeuwenhoek, Hartsoeker, Lantin, the Marchants' correspondents, and several French doctors, botanical research at the Academy would have been much impoverished during the seventeenth century.

INSTITUTIONAL REGULATION OF CONTACTS

But private contacts between individual savants do not tell the whole story, not only because the Academy sometimes tried to regulate such connections but also because it entered as an institution into official relations with individuals and societies. In looking at the ties between the Academy and nonacademicians, two distinctions are important. The first has to do with conflicts between institutional and individual interests, the second with conflicts between principle and practice.

Academicians were both private scholars and members of the Academy, and their individual activities sometimes conflicted with institutional policy. Before joining the Academy, their contacts with other savants had been limited only by opportunity, personal caution, or taste. But as academicians they were expected to circumscribe their external exchanges if these were made on behalf of the Academy or if information conveyed about an individual's ideas might betray the activities of the institution as a whole.

Viewed from this perspective, the actual contacts between academicians and outsiders reflect varying degrees of conflict between individual and institutional interests. Least threatening to the institution were the private lessons that many, perhaps most, academicians gave, or the medical services performed by about a quarter of them. More worrying were Cassini's and Homberg's demonstrations of experiments and apparatus at private conferences, because these might reveal secrets and because participation in such meetings diluted the two academicians' allegiance to the Academy at a time when it was in trouble. In more direct challenges, many academicians defied the policies of the institution. Huygens, for example, corresponded about the very questions debated in meetings of the Company, and Duclos published a book against the Academy's wishes.

At both the individual and the institutional level, however, theory and practice were at odds. Savants espoused cooperation, sought widespread contacts with other scientists, and valued exchanges of ideas and information, so long as such exchanges did not endanger their own priority of invention, discovery, or explanation. Scholarly exchanges among individuals were therefore sometimes circumspect.

In seeking to establish formal relations between itself and other scientific societies in France and abroad, the Academy institutionalized the inclinations expressed, the compunctions felt, and the restraints exercised by individuals. At the same time it deliberately held aloof from certain scholarly habits of the time. Colbert organized reciprocal relations between the Academy and its provincial counterpart in Caen,[35] and the Academy and the Royal Society established irregular means of communication.[36] But the Academy also changed the traditional rules of the scholarly community: mathematicians, for example, often issued challenges in the form of difficult problems, depositing their own solutions with trusted third parties; but when one academician solved such a problem, he refused to forward his solution to the bookseller named as referee by the challenger, claiming instead that the Academy was the supreme referee.[37]

The Academy set itself apart from and above the rest of the scholarly community in still another respect. In contrast to the scholarly ideal of cooperation and exchange of information, and unlike the scientific academies organized by Montmor, Rohault, or Bourdelot that mostly welcomed the interested public to their meetings, the Academy established a rule, recorded in the minutes of 15 January 1667, that "the business of the Academy should be kept secret and... communicated to outsiders only with the approval of the Company."[38] Twenty-one years later another prohibition was announced: academicians were forbidden to publish without the permission of the Company, such approval to be granted only after examination of the manuscript in question.[39]

There were several reasons for the Academy to adopt the rules about secrecy and publication. "Fear of public satire" may well have influenced academicians.[40] Henri Justel thought so and wrote to Oldenburg in December 1667 that the Academy was keeping its activities secret "because there are those here who seek to make fun of it."[41] Such a fear would have been reasonable in light of the ridicule endured by the Royal Society, ranging from Dr. South's oration in Oxford's Sheldonian Theatre in 1669 and Henry Stubbe's mockery, to plays by Thomas Shadwell and Aphra

Behn.⁴² On the French stage as well, in the works of Molière and others, philosophers and physicians were figures of fun.⁴³ Huygens, melancholic and jealous of his own reputation, believed outsiders were envious of the Academy and especially of him.⁴⁴

Dread of mockery certainly motivated the judgment in 1677 that Duclos should not publish his book on the principles of natural mixts. His manuscript was read by a committee composed of Du Hamel, Blondel, Mariotte, and Perrault, who voted against publication on the grounds that Duclos's views would offend "some delicate Philosophers, who cannot suffer what seems to them to look like Platonism." They claimed further that Duclos's views were no longer novel, since they had been expressed by Athanasius Kircher in 1667.⁴⁵ After the committee refused permission to print the work, Duclos abided by the decision for three years while he revised the book,⁴⁶ and then sent it to Amsterdam for publication; he was posthumously vindicated when the Academy, better established and less fearful for its reputation, included the treatise in an eighteenth-century edition of its collected works.

The Academy behaved in this fashion not only to protect itself from gibes but also because academicians wanted priority and fame for themselves. There were two principal threats to such ambitions: academicians feared one another and they feared outsiders. In 1686 they took measures against the misappropriation of material by immediate colleagues, requiring that the Academy examine any manuscript a member wished to publish and reserving the right to set the record straight about works already published.⁴⁷ Cassini's personal power and anger with La Hire lay behind this particular ruling, but there had been other disputes among academicians, notably Mariotte and Huygens, and the problem continued to aggravate members of the Academy.⁴⁸

As for outsiders, academicians worried about simultaneous discovery and preemptive publication. Even Du Hamel, who was generally enthusiastic about exchanging discoveries with foreign academies, justified secrecy in order to forestall success by others. Research was motivated, he said, by the hope of gaining fame through priority.⁴⁹ This view was no less common in the seventeenth than in the twentieth century, despite protestations in both eras about cooperation. Thus, the Italian mathematician and physiologist, G. A. Borelli, a member of the Accademia del Cimento, wrote to Prince Leopold along similar lines: he wanted to know what Montmor's Academy was doing, but he feared that the French would

> make themselves the authors and discoverers of the inventions and speculations of our masters, and of those that we ourselves have found. This fear makes me

go slowly in beginning this correspondence with those gentlemen of the Parisian academy, since in writing, one cannot do less than communicate something or other, and I fear that this may give those foreign minds an opportunity to rediscover the things; I am speaking of the causes, not the experiments.[50]

Fontenelle later described the basis for such concern:

for persons familiar with a particular field, sometimes only a word is necessary to make them understand all the nicety of an invention, and perhaps then they will carry the matter further than the original authors. That is what Galileo did with respect to telescopes.[51]

The aim of academician and Academy alike was to balance reticence and publication, keeping in mind that absolute secrecy was impossible for a group whose members boasted numerous ties with other circles and enjoyed discussing their activities.[52]

If individual members were jealous of their reputations, so was the Academy of its own. When secrecy backfired—as when Fagon credited Boccone for a plant that the Academy had already engraved—academicians raged about a conspiracy against the institution.[53] Books produced by a team of academicians posed a special problem: should the preface identify each contributor or should it simply say that the volume represented the work of the Academy? In 1676 the Academy adopted the former practice for its *Mémoires des plantes,* but eleven years later it preferred the latter for its *Mémoires des animaux.*[54] The opposite difficulty surfaced when La Hire prepared a treatise on magnetic variation and a new compass. He was reluctant to publish under his own name, but the Academy did not want the book to be an official treatise, because unlike the Royal Society it did not disassociate itself from the views of members.[55] La Chapelle wrote to Huygens about the matter, explaining that

this topic, which is one of the most sensitive in Natural Philosophy, is subject to many contradictions. This means that the uncertainty of all these hypotheses, while provoking controversy and settling nothing, will simply incite others to make new discoveries.[56]

When an academician sought recognition by risking the reputation of the Academy rather than his own, he encountered official resistance, especially when his inconclusive findings might enable others to surpass his work.

The security of the kingdom was a further reason for delaying publication. As père Léonard explained it, academicians were subject to royal censorship, just like any other writer, and not until 1705 did they gain the right to publish the works of the Academy without first obtaining an "approbation."[57] Thus, after Blondel had written his *Art de jeter des bombes* in 1675, "Louis XIV forbade its publication at that time lest his enemies profit by it."[58] Delay of course opened savants to the risk of seeing their ideas published by others first or of discovering, as the Florentine Accademia had with its *Saggi,* that their treatises were no longer timely.

The greatest obstacle to publication was the crown. Academicians were readier to publish than were their ministerial protectors. The latter could undermine projects by a remonstrance or refusal of funds, and they often delayed or thwarted publication by turning down requests to publish or by halting printing once it had begun. Thus, when Louis XIV visited the Academy in December 1681, academicians presented him with a list of works ready for publication, but few ever appeared. During the 1680s La Hire planned to edit selected papers of academicians for an official publication, "especially since I do not see that anyone is presently disposed to having our registers printed as we would have hoped." He obtained permission from Louvois to publish on this reduced scale only after making "several requests." But in the late 1680s and early 1690s, Louvois prevented the Imprimerie royale from completing the printing of several works.[59]

Academicians were eager to discuss and publish their work. While they did not want to help rivals solve problems before them, they aspired to praise for their accomplishments and fretted lest once timely writings become disappointing relics. The Academy's official protectors were somewhat more reluctant to publish, however, because they were solicitous of the reputation of the king and the Academy, because they favored some projects at the expense of others, or because the treasury could not afford it. Publication was taken seriously as a means of enhancing the name of sponsor, Academy, and academician, and it was planned with ambition and vigilance.

The Academy resolved conflicts between its objectives and its fears pragmatically. During the 1660s it released news of its dissections anonymously in the *Journal des sçavans.* During the 1670s, the Imprimerie royale printed several books for the Academy on mathematics, botany, and anatomy, and academicians like Huygens signed the articles they published in the *Journal des sçavans.* After 1678 treatises by individual academicians appeared regularly, mostly via private publishers (some of them also *imprimeurs du roi*) rather than the Imprimerie royale. Only one collective work—the observations of the Jesuits in the Far East edited by Thomas Gouye—came out during the 1680s, even though the Imprimerie royale had begun printing at least two other such volumes and La Hire was editing the treatises of deceased colleagues for publication.[60] In the 1690s official works reappeared and a new format—monthly articles printed by the Imprimerie royale—was tested. Throughout the entire three decades, academicians wrote articles and treatises, asked the crown for permission to publish at the Imprimerie royale, submitted their manuscripts for vetting by colleagues, and then sought ways of evading the negative recommendations of protectors or committees.

Taken as a whole, corporate and individual publications earned prestige for the Company. They were reviewed in both the *Journal des sçavans* and the *Philosophical Transactions,* which also printed articles by academicians.[61] Such publicity and other methods ensured extensive dissemination of the Academy's work. Thus, news of the chemical laboratory reached Sweden,[62] and academicians' views on germination and the classification of plants influenced Swedish botanists.[63] Tournefort's publications circulated most widely; his system of classification was adopted by botanists in England and France, and his posthumous *Relation d'un voyage au Levant,* published in 1717, became one of the most popular books in eighteenth-century Paris. The practical and theoretical writings of Dodart, Mariotte, Perrault, Lémery, Charas, and others also attracted attention in the eighteenth century, when many were reprinted.[64]

The Academy espoused collective endeavor, communal publication, free exchange of information, and progress in knowledge. These ideals were tempered by the hope of individual renown and the desire to develop a strong reputation for the Academy. Insofar as academicians' publications were well received, the Academy's program was successful.

THE CHARACTER AND BENEFITS OF CONTACTS

Complete corporate exclusiveness was impossible and undesirable, and controlling publication in order to enhance the institutional reputation was only a small aspect of the Academy's relations with outsiders. Academicians depended on outsiders for information, inspiration, and practical assistance, and they reported weekly what they had gleaned from collaborative work, discussion, or reading. The Academy had an exclusive character but many external interests. As a rule, it controlled its contacts with outsiders so as to preserve its own advantage whenever possible.

As a body, the Academy interacted with nonmembers in several ways: it received unsolicited communications, employed assistants, admitted visitors to meetings, solicited information, and kept abreast of the literature. The content of these official exchanges, the stature of the persons contacted, the flow of raw data to the Academy and of publications from it, the motivations for contacts, and the way such relations influenced it reveal a Company at once dependent on and isolated from contemporary savants. Since only a fraction of the information about contacts between the Academy and the rest of the scientific community survives, however, conclusions cannot be pushed too far, especially when they depend on negative evidence.

Unsolicited communications between academicians and those outside the Company were plentiful. Nonacademicians approached the Academy personally and through letters or intermediaries. They offered information, inventions, observations, and experiments. Some dedicated books to the Academy or its members.[65] These scientists and amateurs of uneven capabilities were not reimbursed, and their motives varied. Some hoped to become academicians,[66] while others sought official approval of an invention or technique,[67] and still others were disinterested and simply wished to help the research of academicians.[68] Sometimes these outsiders believed that the Academy had taken advantage of them by stealing their ideas.[69]

Unsolicited letters were primarily disinterested and had the least important content of the many forms of intelligence received; they contained reports of medical remedies or, more often, observations of eclipses, and accounts of curious phenomena and experiments.[70] Occasionally correspondents sent data directly pertinent to the Academy's work on plants.[71] Materials submitted unasked announced friendship or support; many were self-interested ploys. These unsolicited communications were also part of ordinary scientific discourse and reflect genuine public interest in scientific novelties and natural curiosities. But they were often of little value to academicians and do not bespeak real collaboration between the Academy and the larger community.

Academicians hired and trained scientific practitioners whose special skills—from dissection, taxidermy, distillation, and illustration to surveying and astronomical observation—were required for certain projects. They also commissioned scientific instruments and models of new inventions and machines from instrument makers, artisans, or others.[72] The natural philosopher Nicolas Hartsoeker established his career as a result of his close association with the Academy during a twelve-year sojourn in Paris. He made lenses for the Observatory and supervised the production of scientific apparatus for the Jesuits who represented church, king, and Academy in the Far East; in 1699 he became a foreign associate of the Academy.[73] The instrument maker Michael Butterfield made the expensive silver planisphere which on one side showed the Tychonic and Copernican systems, and on the other the stars in the latitude of Paris. At the end of the century, when he was a royal engineer and had lived in Paris so long that he wrote English awkwardly, Butterfield passed information and books between academicians and Lister.[74]

When academicians' assistants were scientists or instrument makers of independent reputation, the prestige of being an academician may well have been blurred for the student or other lesser members. Academicians Niquet and Couplet, for example, fell into "an inferior category and... were there only to listen and to carry out whatever was decided by the Company, and especially to make the observations it needed."[75] Compared with David du Vivier, who collaborated in the preparation of academicians' maps and became royal geographer in 1680, their opportunities for advancement or even to contribute to the Academy's work may well have been limited.[76] Couplet's name figures along with Dalesme's among those who were paid for constructing models and instruments; Claude Perrault and his brother thought of Couplet as merely "the usher of the Company and then . . . caretaker of the Observatory."[77] Some assistants got paid more for their work than did some academicians. The distinction between lesser academician and hired assistant was further weakened when assistants like Chazelles[78] and Dalesme became academicians.

The contributions of these hired assistants were more important than most of the unsolicited communications, since the assistants actually participated in the research of the Academy. But these were usually not instances of cooperation between intellectual equals, but rather were associations of teacher and pupil, master and assistant. Practitioners supplied specific skills and the extra hands necessary for carrying out certain tasks, and they were not expected to display the theoretical insights or breadth of interest that characterized academicians. Thus, as was the case with the suppliers of data, the ensuing collaboration was more often between unequals than between peers.

Occasionally outsiders were invited to attend a meeting, usually to read a paper or to demonstrate an invention. The privilege was extended to very few, for academicians and their protectors understood that their effectiveness depended on freedom from intrusion.[79] Cassini told Francis Vernon that no one "of what quality soever" who was "not of their own body" was admitted to their conferences,[80] but the rule was relaxed for those who could contribute something scientific.[81] Some visitors eventually became academicians. Homberg, appointed academician in 1691, had demonstrated experiments with an air pump and made phosphorus for the

Academy in 1683 and 1687.[82] Papin became a corresponding member in 1699; he had visited the Academy when he was working with Huygens on the air pump.[83] But because secrecy was a rule of the Academy, it closed its doors to nearly everyone.

The common feature of all these contacts with outsiders—via unsolicited communications, hired assistants, and visitors—is that academicians received useful information, aid, and demonstrations. This trait is further epitomized in the information submitted in response to requests. Academicians convinced the public to share information with them. Fontenelle reported that by 1686 the Academy had

> adopted some Correspondents who learned from it how to question Nature correctly, and to study things with the eyes of Philosophers; very often the Academy has been enriched by Foreigners who have hastened to share the rarities and curiosities that Nature has sown in their province.[84]

Among the fields that profited most from this kind of support was astronomy. Many of Cassini's informants began their association with the Academy through unsolicited communications. As a result of his encouragement letters flowed to him from the provinces whenever there was an eclipse; the data were recorded in the minutes and published in the eighteenth-century annual *Histoire et mémoires*.[85]

Dodart exhorted his readers to convey botanical observations, experiments, and criticisms to the Academy. He hoped "to stimulate the Public to cooperate" with the Academy "to the advancement of" its natural history project. For the sake of the public good, Dodart appealed to all persons who understood botany and the chemical analysis of plants "to communicate their thoughts" to the Company; in return he promised that future publications on plants would acknowledge by name those whose ideas had been helpful, even when academicians had to be content with anonymity behind the corporate name. He also signaled academicians' intentions to send "Memoirs to the medical doctors with whom we have dealings and to give an account to the public of what they inform us." Dodart followed this practice in writing about the *remède des pauvres* and ergot. Indeed, his article on ergot published what a few other physicians already knew, for the Academy mainly tested and verified the hypotheses of its informants, acting as a clearing-house of information about the disease and its cause.[86]

Other botanists in the Academy obtained information and samples from travelers. Thus, before departing for China the Jesuits visited the Academy in order "to learn what matters of natural history the Academy would like them to correspond about," especially with respect to plants.[87] As in the case of ergot, accounts of the Jesuits' scientific observations were published by the Academy,[88] with the justification that this was its own work, since the authors "wrote it in concert with the Academy, and in accordance with the instructions they had received."[89] Academicians and their protectors put requests to French diplomats,[90] and letters describing the flora of Smyrna,[91] drawings of plants, animals, and other objects seen in the Straits of Magellan,[92] and a paper on ginseng[93] were among the prizes obtained from travelers.

Nicolas and Jean Marchant, with the help of Huygens, Justel, and others, fostered contacts around the world in order to obtain rare seeds and cuttings. Flora came from Portugal, the Americas, Italy, the south of France, Holland, and England; the Academy's suppliers included Vespasian Robin (of the Jardin royal), Pierre Magnol (professor and director of the botanical gardens at Montpellier), the academician Jean Richer (who brought plants from Cayenne), and Bishop Compton of London.[94] Such success seemed to justify Nicolas Marchant's confidence that nonacademicians would communicate information, advice, and materials in response to the requests published in the *Mémoires des plantes*.[95]

Academicians pointedly sought help in two ways from outsiders: they paid for skilled assistants and they solicited information. In both cases, the Academy absorbed the contributions of outsiders into its own research and publications. The royal institution transformed raw data into hypotheses, verified theories, and took part of the credit for the labors and observations of outsiders.

Finally, since an important part of scientific research consists in keeping abreast of current literature, the Academy corporately reviewed recent books. Fontenelle later explained:

> It was one of the Academy's occupations, and not the least useful, to examine books that appeared on subjects it had undertaken, especially those which merited particular attention because of the reputation of their Authors. Whether we adopted their views or surveyed their errors, we always profited.[96]

Journals were especially important for keeping abreast of the latest books, inventions, or ideas. Indeed, for a short period the Academy arranged for extracts from the *Philosophical Transactions* to be translated so that it could stay current with English developments.[97] Huygens valued the *Philosophical Transactions* and the books his English acquaintants sent, and he followed the *Journal des sçavans* so intently that his colleagues kept it from him when he was ill, to prevent him from overworking.[98]

The natural philosopher whose books were most thoroughly reviewed in meetings was Robert Boyle. Duclos reviewed his *Certain Physiological Essays* during thirteen meetings,[99] and Mariotte read his own translation of the hypothesis on acids and alkalis.[100] The Academy examined portions of Boyle's other works[101] and repeated some of the experiments.[102] Academicians challenged some of Boyle's views. Duclos, for example, criticized Boyle's corpuscularianism, and the permanent secretaries defended their colleague on the grounds that Duclos had "a more chemical cast of mind" than Boyle,[103] perceiving corpuscularianism as a philosophical, not a chemical, explanation. Academicians also developed color indicators for use in their chemical analyses of plants independently of Boyle, whose influence on academic chemistry was negligible.

Despite the availability of journals, important new botanical literature was often neglected, especially by the academicians who studied the natural philosophy of plants. The Marchants had an extensive botanical library, and drafts and notes for the natural history of plants refer to early modern writers, including Ray and Morison.[104] But Grew and Ray were never mentioned in the minutes of meetings before 1699, and only the anatomist Du Verney discussed Malpighi's theories about plants.[105] Before Tournefort's books and articles of the 1690s, it was rare for an academician to analyze contemporary botanical works.[106] This neglect is especially surprising because at the very time (1676) when Dodart expressed bewilderment about classifying plants, Ray's *Tables* and Morison's *Praeludia botanica* were available.[107] But Dodart never referred to them, even though Marchant knew Morison personally and Dodart was later to send both Morison and Grew copies of the *Mémoires des plantes*.

English and French botany developed independently during the latter third of the seventeenth century. Investigations of sycamores and the flow of sap which Ray, Willughby, and others published during the 1660s and 1670s in the *Philosophical Transactions* seem not to have influenced Jean Marchant's similar studies during the 1680s. Experiments described in the *Philosophical Transactions* as proving the circulation of sap only rarely resembled those cited by Mariotte and Perrault.[108] Mariotte's and Perrault's studies of germination differ from Grew's report of the germination of white beans and squash, and Grew's terminology was never adopted by French botanists.[109] When La Hire, Charas, and Tournefort considered whether kermes and cochineal were seeds or insects, they never alluded to the evidence Lister had published during the 1670s in the *Philosophical Transactions*, although they did rely on Plumier's opinion.[110] Perrault and La Hire both stated that there were valves in the vessels of plants—a notion Grew attempted to disprove, as the review of his book in the *Philosophical*

Transactions pointed out—and neither referred to Grew.[111] The inevitable conclusion is that most academicians who studied plants did not read the *Philosophical Transactions*. Linguistic ignorance does not explain this neglect, for the Academy could commission translations, and academicians even disregarded the Latin treatises of Malpighi and the French translations of Grew.[112] Their isolation had both positive and negative consequences: although academicians before Tournefort did not benefit from the accomplishments of their botanical contemporaries, they also escaped from their errors and thus did not follow Malpighi and Grew in claiming to have seen air vessels in plants.[113]

In the case of the Academy's botanical research, outside influence was least felt through contemporary literature, visitors to meetings, or hired assistants. More important to both the natural history of plants and the study of medical botany were informal communications, and especially those generated through the personal contacts of individual academicians rather than by the formal agency of the Academy. While few academicians show familiarity with contemporary theoretical treatises about botany, they combed older literature for information about individual plants and were inspired to explore plant physiology by accomplishments, both recent and contemporary, in other disciplines. Finally, although the works of Boyle were frequently debated, chemical studies of botany owed more to continental than to English chemistry.

Where academicians failed to acknowledge outside influence, it may have been due to a habit of mind that preferred to cite experiments and observations rather than literature. Like Thomas Sprat who denounced Samuel Sorbière for imputing to the Royal Society the intention of developing a library, or Oldenburg who criticized the early French societies for discoursing rather than experimenting,[114] academicians emphasized experiment and observation above all. Yet many scientists were avid readers of current literature, and their correspondence summarizes the latest books or requests items for their personal libraries. Academicians owned sizable collections. They also had access to the scientific holdings of other libraries, and they referred to the books of their predecessors.[115] Nevertheless, although their botanical research was in the vanguard of late seventeenth-century efforts, it remained curiously aloof from contemporary influences until Tournefort entered the Academy.

CONCLUSION

Both the individual and the institutional dynamics of research limited academicians' ties with the larger scientific community. At the individual level, scientific discourse was sometimes a struggle in which the contestants tried to get as much information as possible from each other at least cost, with the lesser figures taking the greater risks. Savants wanted the acclaim that came from the opportune delivery and convincing proof of a well-formed hypothesis; they wanted to publish their discoveries under their own names when the time was ripe. The wish to make a name for oneself was an important limit on cooperation and on the exchange of information during the seventeenth century. Besides the striving for personal *gloire,* there were other limitations on cooperation among individual savants, which derived from circumstances, personality, and systems of belief.

The formation of the Academy of Sciences altered the balance of scientific discourse, because the Academy erected an institutional barrier between academician and nonacademician and formalized the relations among savants. It also introduced new elements into the quest for recognition by savants: the Company's own name had to be protected and enhanced, and its prestige augmented that of its correspondents and members. The Academy's ministerial protectors sought renown for the institution and its royal patron, and they expected the Company to gain respect through its publications. For these reasons, the Academy limited cooperation and exchange of information between academicians and outsiders.

Establishment of the Academy also affected the structure of the scientific community. Because the Academy enjoyed advantages of funding, prestige, and power, it could establish itself as an arbiter of acceptable theory and correct data. Into a scientific community consisting—according to Rudwick's criteria—of theory-formulating elite and data-collecting amateurs, the crown injected the Academy of Sciences, an elite institution that arrogated to itself the right to judge what scientific knowledge was admissible.[116] Not all academicians were among the scientific elite themselves, but that did not matter. The Academy's reputation was enhanced by the inclusion of highly regarded savants among its members. But its standing was higher than that of its members collectively, partly because royal patronage enabled it to set itself above the rest of the scientific community.

The Academy's status as an elite institution is clarified by its official contacts with outsiders. The Academy controlled the dissemination of news about its activities, emphasizing the subjects on which it would publish treatises and inviting communications from the public. It tried to keep the ideas of its members secret until their hypotheses were ready for publication, and it treated most nonacademicians as amateurs, collecting from them data that members applied to their own projects. When the

Academy reviewed the publications of outsiders, as in the case of Boyle's chemical treatises, it was critical and preferred the work of its own members. It used ties with other royal establishments, such as the Jardin royal or the *pépinerie,* to support its own research, and in obtaining local services and goods it kept the Parisian practitioners in a subordinate position.

The result was that the Academy's external ties had a mixed effect on its own research and on the larger scientific community. For the Academy's own botanical research, at least, the personal connections of academicians were more effective stimuli to their research than were the associations generated by the institution. Institutional interchanges about plants were plentiful and had a wide geographical reach; they were best at attracting data. Personal interchanges could obtain data, though not necessarily from so many lands, but were also likely to stimulate new ideas. By conducting their research on plants with so little regard to contemporary work, academicians struck out independently but unsystematically.

The larger scientific community was both animated and discouraged by the Academy. The institution became a clearing-house of information but was an obstacle to royal patronage of other individuals or groups. It undertook projects too vast for any one individual but did not publish sufficiently. It attracted many who sought membership but it admitted few candidates. By restricting the flow of information and using contacts to its own advantage, the Academy deprecated the larger community. But by raising the standard of work in some fields, by training personnel in several disciplines, and by airing the difficulties of solving certain scientific problems and by solving others, the Academy improved the practice of science, both theoretical and applied, in the larger scientific community. The Academy's scholarly accomplishments and connection to the crown gave it the special position academicians claimed as its right. Thus, royal patronage of the Academy had manifold effects on both science and the scientific community.

PART V
The Effects of Patronage

CHAPTER 16

The Academy as an Instrument of the Crown

The Academy of Sciences was a royal foundation with overlapping functions, some of which were scholarly, others political and social. As a learned institution, it was responsible for reforming the sciences and for putting medicine and technology on a sound theoretical footing. Meeting these scholarly obligations enabled the Academy to perform its political and social duties, that is, to honor the king with its accomplishments and to advise the crown about technology and public health.

Colbert founded the Academy as an instrument of reform, and the king was to preside over its successes. Troubled by the uncertainty of scientific knowledge, yet impressed by its promise, Colbert and his advisers hoped that patronage would speed discovery. They expected the Academy to renovate natural philosophy. Its accomplishments would redound to the king's glory and to the benefit of the practical arts. The Academy thus owed its genesis to both the French crown and the scientific revolution.

Its mismatched parentage produced results that were on the whole positive. The beneficiaries included scientific inquiry and knowledge, the institution and its members, the larger scientific community, and the patron himself. But the nature of patronage affects the work supported, and the motives of the French crown in launching and protecting the Academy were mixed. Thus, the program and accomplishments of the Academy reflect not only the cultural and intellectual milieu in which academicians operated but also the motives—both high-minded and intensely self-interested—of its protectors.

The Academy addressed matters of substantive knowledge and issues of reasoning and method. Its program was comprehensive: no aspect of the sciences as then conceived was neglected. Its comparative anatomy of animals, proposed treatise on mechanics, and ill-fated natural history of plants all exemplify the Academy's efforts to reform science, with each project intended to make a clean sweep of existing work.

Academicians planned to reestablish science on the double footing of induction and deduction. They were dedicated to observation and experiment, but that did not exhaust their aims. For Mariotte, La Hire, and others, the idea was to establish inductively the axioms of natural history and natural philosophy. From these they would deduce further consequences. Thus, academicians were haunted by a Euclidean-Cartesian deductive model that promised to systematize their disciplines.[1] Tournefort's *Élémens de botanique,* intended to establish the foundations of botany by classifying plants, paid implicit homage to Euclid in its title.

Theory was never isolated from practice at the Academy, many of whose projects had a special appeal to the state. Whether at the instruction of a ministerial protector or at the initiative of an academician, utilitarian concerns informed much of the Academy's work. The institution examined medicine, technology (especially military), inventions, hydraulics, and cartography, and it accomplished several tasks. Some of these were hit-or-miss inquiries with a narrow focus. But academicians and their protectors had broader utilitarian goals. They believed that scientific inquiry should establish a theoretical basis for the practical arts. This ideal—when combined with the savant's plan of discovering general principles experimentally and with the bureaucrat's hope of a general reform—offered a powerful incentive for patronage on a generous scale.

The Academy's own corporate status, that of a "moral person" responsible for protecting and nurturing the sciences,[2] burgeoned at the expense of outsiders and of the normal rules of conduct in the scientific community. Despite its claim to be an arbiter of scientific knowledge, however, the Academy was circumspect in adjudicating theoretical disputes, often taking an agnostic position and calling for more data before a theory could be adopted or rejected. As an institution it sometimes ignored the most provocative and influential hypotheses of its time.

Careful observation and experiment, proper reasoning, cautious theorizing, and elite status—these were the methods of the Academy. But for its reforms to take effect they had to be publicized. This the academicians accomplished formally and informally, by publishing, corresponding, and debating. The Academy also taught assistants who used the new skills in various practical occupations; thus, it assisted Colbert's general reform by training future military engineers and cartographers, many of whom were employed by the crown.

The Academy was an instrument of royal propaganda. Louis's academies were the domestic counterpart of foreign conquest, especially when foreign savants became luminaries in them. They were at once symbols of his beneficence and instruments in the royal program of censorship. Scientific treatises by academicians increased the number of acceptable publications and were identified as the product of the *Royal* Academy of Sciences. The crown preempted scientific patronage by founding a better academy and funding it generously, by requiring academicians to attend meetings from which outsiders were excluded, and by taking over and improving projects begun under other auspices, such as the botanical illustrations initiated by Gaston d'Orléans. Louis XIV made his Academy the envy of the learned world. It was an institution whose exclusiveness, research program, and accomplishments emphasized its royal connections.

The Academy of Sciences, along with the other royal academies, was an apologist for a specifically French cultural flowering. Its deliberations and nearly all its publications were in the vernacular. Indeed its members forced foreign associates to converse and write in French. Academicians expanded the technical vocabulary of the language, keeping it healthy and bringing science into the mainstream of French written culture. The Academy helped the French language eclipse Latin as the tongue of French savants.[3]

The Academy offered its ministerial protectors a sphere of relative autonomy, because the king knew and cared little about natural philosophy. Colbert, Louvois, and Pontchartrain designed the Academy. It was they who approved its regulations and oversaw its research program, who appointed academicians and determined their pensions. The ministers could authorize or withhold funding. They encouraged or discouraged publishing, and they also arranged privileges and special associations for academicians. They even made it a haven for some Jansenists. Their supervision of the Academy belies the image of a monarch who prided himself on his personal rule over every aspect of his realm.

Royal foundation of the Academy affected the sciences in several ways. The very establishment of a scientific institution — whose corporate identity was symbolized by the Observatory (a building designed and constructed for the Academy's sole use) and expressed in regulations about secrecy, publication, and attendance — dignified and protected the sciences. By pensioning academicians and encouraging their work, the crown helped transform scientific research into a profession in its own right. More specifically, the Academy offered status, material reward, and incentive to the intelligentsia of the third estate, giving prestige to a scholarly life that was all too often practiced in a nether world of dependency, favoritism, and poverty. By connecting pensions with scientific research in general and by making membership in the Academy a lifetime appointment (although no sinecure), the crown changed its patronage of science from a reward for piecework into a permanent annual stipend for serious scholarship. The Academy also provided a standard of success by which other savants could measure their careers: its pensions and subventions were powerful incentives, and many aspired to a place in the Academy of Sciences.

The crown made it possible for academicians to undertake research that would otherwise have been impossible. A program on the scale of the Academy's was, as Bacon had put it, "a Thing of very great size," which could not "be executed without great labour and expense; requiring as it does many people to help, and being . . . a kind of royal work."[4] By funding the Academy, the crown encouraged projects beyond the capacity of individuals or private scientific societies. The well-funded official projects also provided a protective shadow in which academicians carried out their smaller-scale individual studies. Moreover, institutional inertia protected inconclusive inquiries, and against ministerial intervention academicians could sometimes continue recalcitrant research on the grounds that it might have utilitarian consequences.

The Academy and its protectors, however, also endangered scientific inquiry. Some highly qualified savants were excluded from membership. Corporate esprit sometimes inhibited individual initiatives. The interests of a few powerful figures controlled planning, funding, publication, or the interpretation of research findings, sometimes to the detriment of talented colleagues. Another danger lay in the precarious balance of theoretical and utilitarian interests in the Academy's research program. Here academicians and their protectors sometimes clashed. Through funding and persuasion, ministers pressured academicians to stress the practical side of their theoretical research. When utilitarian pressures tipped the balance, the consequences could be disastrous, as the case of botany shows.

Botany was being transformed in the late seventeenth century, and it proved fragile under utilitarian prodding. From ancient times plant study had been mainly a branch of medicine, although some savants had also searched for family resemblances that would lead to a system of classification, and others had studied plant reproduction and related topics.[5] In the late seventeenth century, scholars began to examine plants afresh because of analogical and experimental influences. Botany was henceforth not principally ancillary to medicine but became an independent branch of natural history and natural philosophy. Three main features of the scientific revolution brought this about: the premium placed on experiment and observation, the development of new apparatus such as the air pump and the microscope, and the interdisciplinary relations of the sciences that led savants to apply techniques and concepts from one field to another. Savants who thought of plants as resembling animals applied novel theories and experimental procedures to plants. Academicians redefined even the traditional natural history of plants by searching for the basic chemical constituents of plants.

Botany as practiced at the Academy looked to other biological sciences for a model. It vacillated between chemical and mechanical modes of explanation and relied extensively on analogy. It was an exciting but susceptible discipline, threatened by personal rivalry between competing editors, by the tensions inherent in collaborative research, by the sporadic character of individual research, by the loss of manuscripts, and by the appearance of rival publications. But ministerial intervention on the side of utilitarian interests decisively injured it.

Louvois's interference in 1686 is important because it reveals just how delicate was the balance of royal patronage and scholarly initiative at the Academy. When Louis was dangerously ill Louvois sought a scapegoat, and as a result he scolded the Academy for wasting its time on curious research instead of improving medicine. Louvois was impatient with the Academy's still incomplete natural history of plants which, in trying to unite traditional and innovative elements, failed. Yet academicians persisted with the research because their goals were so important. In an ironic twist the minister probably never understood, the very aspect of botanical research that was singled out for blame—the chemical analysis of plants—was the sole survivor of the harangue of January 1686, because it promised medical results. For the six years from 1686 until the end of 1691, scarcely any theoretical botanical research was done at the Academy. Only when a new minister appointed Tournefort and Homberg to the Academy and refrained from telling academicians how to conduct their studies, could theoretical botanical research recommence.

In fields such as animal anatomy and physiology, astronomy, or mathematics, the Academy enjoyed greater success and ministerial interference was not so deleterious. But for botany, which was only beginning to emerge as a discipline in its own right, interventionist patronage that demanded immediate and practical results could be disastrous.

As an institution with political and scholarly functions, with practical responsibilities and theoretical interests, the Academy of Sciences was both privileged and vulnerable. The very favor it received from the crown was double-edged. Louis and his ministers were no disinterested protectors, nor did they expect merely to bask in any reflected intellectual glory. Above all, they wanted immediate and tangible benefits from the institution they supported. The return on their investment took many forms, ranging from scientific and technical to cultural and propagandistic.

When they heeded good advice, protectors could work harmoniously with academicians. But when practical and political interests prompted protectors to interfere with theoretical research—when the crown tried to direct not only research goals but even research methods—the Academy was threatened. Inconsistent and narrowly conceived patronage damaged the Academy more than it did the royal image. Patronage at its best, however, provided opportunity, protection, and encouragement to scientific savants. Ultimately it integrated scientific endeavor into the larger political, cultural, intellectual, and social structures of the time.

Abbreviations Used in the Appendix, Notes, and Bibliography

Complete citations for all references in the notes can be found in the bibliography.

AdS	Paris, Académie des Sciences, Archives.
AdS, Reg.	Paris, Académie des Sciences, Archives, Registre des procès-verbaux des séances.
AN	Paris, Archives Nationales.
Annales	*Annales. Économies. Sociétés. Civilisations.*
Arch. anc. rég.	*Archives de l'ancien régime.*
ARdS	Académie royale des sciences.
BA	Bibliothèque de l'Arsenal.
BdR	Bibliothèque du roi.
BMHN	Paris, Bibliothèque du Muséum d'Histoire Naturelle.
BN	Paris, Bibliothèque Nationale.
Bodleian	Oxford, Bodleian Library.
BU	*Biographie universelle.*
CdB	*Les comptes des bâtiments du roi sous le règne de Louis XIV,* ed. Guiffrey.
CRA	*Comptes rendus de l'Académie des sciences.*
d.	denier or deniers.
DBF	*Dictionnaire de biographie française,* ed. Balteau.
DSB	*Dictionary of Scientific Biography,* ed. Gillispie.
Est.	Establishments.
Estampes	*Estampes pour servir à l'histoire des plantes,* Bosse et al.
Histoire	Fontenelle, *Histoire de l'Académie royale des sciences.*
Histoire . . . (date)	Académie royale des sciences, *Histoire . . . , avec les mémoires . . . (1699–1790);* see the section "Histoire."
Historia	Du Hamel, *Regiae Academiae Scientiarum Historia.*

227

IB	Institut de France; Académie des Sciences, *Index biographique des membres et correspondants de l'Académie des Sciences.*
JdS	*Journal des sçavans* or *savants.*
JR	Jardin royal.
lv.	livre or livres.
Mémoires	Vols. 3–11 of Académie royale des sciences, *Histoire et mémoires . . . depuis 1666 jusqu'à 1699.*
Mémoires . . . (date)	Académie royale des sciences, *Histoire . . . , avec les mémoires . . . (1699–1790);* see the section "Mémoires."
NBU	*Nouvelle biographie universelle,* ed. Hoefer.
Obs.	Observatoire.
OED	*The Oxford English Dictionary,* ed. Murray et al.
Phil. Trans.	*Philosophical Transactions.*
Rés.	Réserve.
RHS	*Revue d'histoire des sciences.*
s.	sol or sous.

Note: In giving titles and quotations from seventeenth-century materials I have reproduced accents and spelling as they are found in the source.

APPENDIX
The Record of Expenditure, 1666–1699*

* Note: Due to information obtained after Stroup, *Royal Funding* was published, there are minor discrepancies between data in tables 1, 7, 13, 14, 15, 16, and 17, and the figures in the earlier analysis of the Academy's finances.

TABLE 1. Pensions Paid to Academicians and Their Assistants

a. Pensions Paid by Colbert, 1666-1683[a]

	1666	1667	1668	1669	1670	1671	1672	1673	1674	1675
	lv.	lv.	lv.	lv.	lv.	lv.	lv.	lv.	lv.	lv.
i. Academicians										
Carcavi, 1666-1684	1,500	2,000	2,000	2,000	2,000	2,000	2,000	2,000	2,000	2,000
Huygens, 1666-1682	5,000	6,000	6,000	6,000	6,000	6,000	6,000	6,000	6,000	6,000
Roberval, 1666-1675	1,500	1,500	1,500	1,500	1,500	1,500	1,500	1,500	1,500	
Frenicle, 1666-1675	1,200	1,200	1,200	1,200	1,200	1,200	1,200	1,200		
Auzout, 1666-1668	1,500	1,500								
Picard, 1666-1682	1,200	1,200	1,200	1,500	1,500	1,500	1,500	1,500	1,500	1,500
Buot, 1666-1680	1,200	1,200	1,200	1,200	1,200	1,200	1,200	1,200	1,200	1,200
Du Hamel, 1666-1706		1,500							1,500	1,500
La Chambre, 1666-1671	2,000	2,000	2,000	2,000						
Perrault, 1666-1688	1,500	1,500	2,000	2,000	2,000	2,000	2,000	2,000	2,000	2,000
Duclos, 1666-1685	2,000	2,000	2,000	2,000	2,000	2,000	2,000	2,000	2,000	2,000
Bourdelin, 1666-1699	1,500	1,500	1,500	1,500	1,500	1,500	1,500	1,500	1,500	1,500
Pecquet, 1666-1674		1,200	2,400	1,200	1,200	1,200	1,200	1,200		
Gayant, 1666-1673	1,200	1,200	1,200	1,200	1,200	1,200	1,200			
Marchant, N., 1666-1678	1,200	1,200	1,200	1,200	1,500	1,500	1,500	1,500	1,500	1,500
Niquet, 1666-1675	800	800	1,400	1,000	1,000	1,000	1,000	1,000	1,000	
Couplet, C. A., 1666-1722			600	600	600	800	800	800	800	800

Continued on next page

TABLE 1. CONTINUED

a. Pensions Paid by Colbert (continued)

	1666	1667	1668	1669	1670	1671	1672	1673	1674	1675
	lv.	lv.	lv.	lv.	lv.	lv.	lv.	lv.	lv.	lv.
i. Academicians (continued)										
Richer, 1666-1674	800	1,000	1,000	1,000	1,000	2,000	1,000	1,000		
Pivert, 1666-1672		600		600	800	800				
La Voye, 1666-1677		800		600						
Mariotte, 1668-1684			1,500	1,500	1,500	1,500	1,500	1,500	1,500	1,500
Gallois, 1668-1707			1,500	1,500						
Blondel, 1669-1686						2,000	1,500	1,500	1,500	1,500
Cassini I, 1669-1712			3,000	6,750	9,000	9,000	10,500	9,000	9,000	9,000
Borelly, 1670-1689						1,200	1,200	1,200	1,200	1,200
Dodart, 1671-1707					900	1,125	1,500	1,500	1,500	1,500
Roemer, 1672-1681							1,000	1,000	1,000	1,000
Du Verney, 1674-1730									750	2,250
La Hire, P. de, 1678-1718										
Marchant, J., 1678-1738										
Sédileau, 1681-1693										
Pothenot, 1682-1696?										
Subtotal, i	24,100	29,900	34,400	38,050	37,600	42,225	42,800	40,100	38,950	37,950

Continued on next page

TABLE 1. CONTINUED

a. Pensions Paid by Colbert (continued)

	1676	1677	1678	1679	1680			1681	1682	1683	Subtotal, 1666-1683		
	lv.	lv.	lv.	lv.	lv.	s.	d.	lv.	lv.	lv.	lv.	s.	d.
i. *Academicians (continued)*													
Carcavi	2,000	2,000	2,000	2,000	2,000.	0.	0	2,000	2,000		33,500.	0.	0
Huygens	3,000		6,000	3,000	6,000.	0.	0	4,500			81,500.	0.	0
Roberval											13,500.	0.	0
Frenicle											9,600.	0.	0
Auzout											3,000.	0.	0
Picard	1,500	1,500	1,500	1,500	1,500.	0.	0	1,500			23,100.	0.	0
Buot	1,200	1,500	2,000	2,000							18,700.	0.	0
Du Hamel	1,500	1,500	1,500	1,500	1,500.	0.	0	1,500	1,500		15,000.	0.	0
La Chambre											8,000.	0.	0
Perrault	2,000	2,000	2,000	2,000	2,000.	0.	0	2,000	2,000		33,000.	0.	0
Duclos	2,000	2,000	2,000	2,000	2,000.	0.	0	2,000	2,000		34,000.	0.	0
Bourdelin	1,500	1,500	1,500	1,500	1,500.	0.	0	1,500	1,500		25,500.	0.	0
Pecquet											9,600.	0.	0
Gayant											8,400.	0.	0
Marchant	1,500	1,500									16,800.	0.	0
Niquet											9,000.	0.	0
Couplet, C. A.	800	800									7,400.	0.	0

Continued on next page

TABLE 1. CONTINUED

a. Pensions Paid by Colbert (continued)

	1676	1677	1678	1679	1680			1681	1682	1683	Subtotal, 1666-1683		
	lv.	lv.	lv.	lv.	lv.	s.	d.	lv.	lv.	lv.	lv.	s.	d.
i. Academicians (continued)													
Richer											8,800.	0.	0
Pivert											2,800.	0.	0
La Voye											1,400.	0.	0
Mariotte	1,500	1,500	1,500	1,500	1,500.	0.	0	1,500	1,500		22,500.	0.	0
Gallois											3,000.	0.	0
Blondel	1,500	1,500	1,500	1,500	1,500.	0.	0	1,500	1,500		18,500.	0.	0
Cassini I	9,000	9,000	9,000	9,000	9,000.	0.	0	9,000	9,000	6,750	135,000.	0.	0
Borelly	1,200	1,200	1,200	1,200	1,200.	0.	0	1,200	1,500		15,600.	0.	0
Dodart	1,500	1,500	1,500	1,500	1,500.	0.	0	1,500	1,500		17,625.	0.	0
Roemer	1,000	1,600	1,400	1,600	2,600.	0.	0				12,200.	0.	0
Du Verney	1,500	1,500	1,500	1,500	3,500.	0.	0	1,500	1,500		15,500.	0.	0
La Hire, P. de			1,500	1,500	1,500.	0.	0	1,500	1,500		7,500.	0.	0
Marchant					1,200.	0.	0			1,200	2,400.	0.	0
Sédileau								400	1,000		1,400.	0.	0
Pothenot									700		700.	0.	0
Subtotal, i	34,200	32,100	37,600	34,800	40,000.	0.	0	33,100	28,700	7,950	614,525.	0.	0

Continued on next page

TABLE 1. CONTINUED

a. Pensions Paid by Colbert (continued)

	1666	1667	1668	1669	1670	1671	1672	1673	1674	1675
	lv.	lv.	lv.	lv.	lv.	lv.	lv.	lv.	lv.	lv.
ii. Assistants										
Laury	800									
Du Vivier			1,500						1,000	
Pa[s]quin[e]				600	600	600	600	600	600	600
[à/de] Kemps				350	600	600	600	1,200		
Dupuy/Dupuis					500	500	500	600		
[Le] Vavasseur						250	600	600		
Dissection Assistant										
Truchet										
Dalesme										
Guenier/Guerrier				400						
Subtotal, ii	800		1,500	1,350	2,150	1,950	2,300	3,000	1,600	600
Total, i and ii	24,900	29,900	35,900	39,400	39,750	44,175	45,100	43,100	40,550	38,550

Continued on next page

TABLE 1. CONTINUED

a. Pensions Paid by Colbert (continued)

	1676	1677	1678	1679	1680			1681	1682	1683	Subtotal, 1666-1683		
	lv.	lv.	lv.	lv.	lv.	s.	d.	lv.	lv.	lv.	lv.	s.	d.
ii. Assistants (continued)													
Laury											800.	0.	0
Du Vivier		1,000	1,000	2,000	3,333.	6.	8	2,000			12,283.	6.	8
Pa[s]quin[e]	600	600									5,400.	0.	0
[à/de] Kemps											3,350.	0.	0
Dupuy/Dupuis											2,100.	0.	0
[Le] Vavasseur											1,450.	0.	0
Dissection Assistant			600	600				600	600		2,400.	0.	0
Truchet										400	400.	0.	0
Dalesme										600	600.	0.	0
Guenier/Guerrier											400.	0.	0
Subtotal, ii	600	1,600	1,600	2,600	3,333.	6.	8	2,600	600	1,000	29,183.	6.	8
Total, i and ii	34,800	33,700	39,200	37,400	43,333.	6.	8	35,700	29,300	8,950	643,708.	6.	8

b. Pensions Paid by Louvois, 1684-1690[b]

TABLE 1. CONTINUED

	1684	1685	1686	1687	1688	1689	1690			Subtotal, 1684-1690		
	lv.	lv.	lv.	lv.	lv.	lv.	lv.	s.	d.	lv.	s.	d.
i. Academicians												
Du Hamel	1,500	1,500	1,500	1,500	1,500	1,500	500.	0.	0	9,500.	0.	0
Perrault	2,000	2,000	2,000	2,000	2,000					10,000.	0.	0
Duclos	2,000									2,000.	0.	0
Bourdelin	1,500	1,500	1,500	1,500	1,500	1,500	500.	0.	0	9,500.	0.	0
Couplet	500	500	500	500	500	500	166.	13.	4	3,166.	13.	4
Gallois		3,000	1,500	1,500	1,500	1,500	500.	0.	0	9,500.	0.	0
Blondel	1,500	1,500								3,000.	0.	0
Cassini I	4,500	9,000	9,000	9,000	9,000	9,000	3,000.	0.	0	52,500.	0.	0
Borelly	1,500	1,500	2,000	2,000	2,000	2,000				11,000.	0.	0
Dodart	1,500	1,500	1,500	1,500	1,500	1,500	500.	0.	0	9,500.	0.	0
Du Verney	1,500	1,500	1,500	1,500	1,500	1,500	500.	0.	0	9,500.	0.	0
La Hire, P. de	1,500	1,500	1,500	1,500	1,500	1,500	500.	0.	0	9,500.	0.	0
Marchant, J.	1,200	1,200	1,200	1,200	1,200	1,200	400.	0.	0	7,600.	0.	0
Lannion, 1679-1686	1,500	1,500								3,000.	0.	0
Sédileau	500	500	500	500	500	500	166.	13.	4	3,166.	13.	4
Pothenot	400	400	400	400	400	400	133.	6.	8	2,533.	6.	8
Le Febvre, 1682-1702			300	300	300	300	100.	0.	0	1,300.	0.	0

Continued on next page

TABLE 1. CONTINUED

b. Pensions Paid by Louvois (continued)

	1684	1685	1686	1687	1688	1689	1690			Subtotal, 1684-1690		
	lv.	lv.	lv.	lv.	lv.	lv.	lv.	s.	d.	lv.	s.	d.
i. Academicians (continued)												
Méry, 1684-1722		600	600	600	600	600	200.	0.	0	3,200.	0.	0
Rolle, 1685-1719			400	400	400	400	133.	6.	8	1,733.	6.	8
Cusset, 1685-1691?			300	300	300	300	100.	0.	0	1,300.	0.	0
Subtotal, i	23,100	29,200	26,200	26,200	26,200	24,200	7,400.	0.	0	162,500.	0.	0
ii. Assistants												
Dissection Assistant	600	600	600	600	600	600	200.	0.	0	3,800.	0.	0
Dalesme	600	600	600	600	600	600	200.	0.	0	3,800.	0.	0
Chastillon			400	400	400	400	133.	6.	8	1,733.	6.	8
Subtotal, ii	1,200	1,200	1,600	1,600	1,600	1,600	533.	6.	8	9,333.	6.	8
Total, i and ii	24,300	30,400	27,800	27,800	27,800	25,800	7,933.	6.	8	171,833.	6.	8

TABLE 1. CONTINUED

c. Pensions Paid or Budgeted by Pontchartrain, 1691-1699c

	1691			1692	1692	1694	1694	1695	1696
	lv.	s.	d.	lv.	lv.	lv.	lv.	lv.	lv.
i. Academicians									
Cassini I	6,000.	0.	0	9,000	9,000	9,000*	9,000	9,000*	9,000*
Bourdelin	1,000.	0.	0	1,500	1,500	1,500*	1,500*	1,500*	1,500*
Dodart	1,000.	0.	0	1,500	1,500	1,500	1,500	1,500	1,500*
Du Hamel	1,000.	0.	0	1,500	1,500	1,500*	1,500*	1,500*	1,500*
Du Verney	1,000.	0.	0	1,500	1,500	1,500	1,500	1,500	1,500
Gallois	1,000.	0.	0	1,500	1,500	1,500	1,500	1,500	1,500*
La Hire	1,000.	0.	0	1,500	1,500	1,500	1,500	1,500	1,500*
Marchant	800.	0.	0	1,200	1,200	1,200*	1,200	1,200*	1,200*
Méry	400.	0.	0	600	600	600*	600*	600*	600*
Couplet	333.	6.	8	500	500	500*	500*	500*	500*
Sédileau	333.	6.	8	500	500	500*			
Pothenot	266. 13.		4	400	400	400*	400*	400*	400*
Rolle	266. 13.		4	400	400	400*	400*	400*	400*
Cusset	200.	0.	0						
Le Febvre	200.	0.	0	400*	300	400*	400*	400*	400*
Varignon, 1688-1722				300	300	300*	800*	800*	800*
Tournefort, 1691-1708						1,500*	1,500	1,500*	1,500*

Continued on next page

TABLE 1. CONTINUED

c. Pensions Paid or Budgeted by Pontchartrain (continued)

	1691			1692	1693	1694	1695	1696
	lv.	s.	d.	lv.	lv.	lv.	lv.	lv.
i. Academicians (continued)								
Homberg, 1691-1715						1,500	1,500	1,500*
Boulduc, 1694-1729								
Fontenelle, 1697-1757								
Des Billettes, 1699-1720								
Jaugeon, 1699-1724								
Dalesme, 1699-1727								
Truchet, 1699-1726								
Subtotal, i	14,800.	0.	0	22,300	22,200	25,300	25,300	25,300
ii. Assistants								
Dissection Assistant	400.	0.	0	600	600	600	600	600
Dalesme	400.	0.	0	600	600	600	600	600
Chastillon	266.	13.	4	400	400	400*	400	400
Subtotal, ii	1,066.	13.	4	1,600	1,600	1,600	1,600	1,600
Total, i and ii	15,866.	13.	4	23,900	23,800	26,900	26,900	26,900

Continued on next page

TABLE 1. CONTINUED

c. Pensions Paid or Budgeted by Pontchartrain (continued)

	1697	1697	1698	1699	Subtotal, 1691-1699		
	lv.	lv.	lv.	lv.	lv.	s.	d.
i. Academicians (continued)							
Cassini I	9,000*	9,000*	9,000*	9,000*	96,000.	0.	0
Bourdelin	1,500*	1,500*	1,500*	1,500*	16,000.	0.	0
Dodart	1,500*	1,500*	1,500*	1,500*	16,000.	0.	0
Du Hamel	1,500*	1,500*	1,500*	1,500*	16,000.	0.	0
Du Verney	1,500*	1,500*	1,500*	1,500*	16,000.	0.	0
Gallois	1,500*	1,500*	1,500*	1,500*	16,000.	0.	0
La Hire	1,500*	1,500*	1,500*	1,500*	16,000.	0.	0
Marchant	1,200*	1,200*	1,200*	1,200*	12,800.	0.	0
Méry	600*	600*	600*	600*	6,400.	0.	0
Couplet	800*	500*	1,000*	950*	6,583.	6.	8
Sédileau					1,833.	6.	8
Pothenot					2,666.	13.	4
Rolle	400*	400*	400*	400*	4,266.	13.	4
Cusset					200.	0.	0
Le Febvre	400*	400*	400*	400*	4,100.	0.	0
Varignon	800*	800*	800*	800*	6,500.	0.	0
Tournefort	1,500*	1,500*	1,500*	1,500*	12,000.	0.	0

Continued on next page

TABLE 1. CONTINUED

c. Pensions Paid or Budgeted by Pontchartrain (continued)

	1697	1697	1698	1699	Subtotal, 1691-1699		
	lv.	lv.	lv.	lv.	lv.	s.	d.
i. Academicians (continued)							
Homberg	1,500*	1,500*	1,500*	1,500*	12,000.	0.	0
Boulduc	400*	400*	400*	400*	1,600.	0.	0
Fontenelle			1,500*	1,500*	3,000.	0.	0
Des Billettes				1,000*	1,000.	0.	0
Jaugeon				1,000*	1,000.	0.	0
Dalesme				600*	600.	0.	0
Truchet				1,000	1,000.	0.	0
Subtotal, i	25,600	25,300	27,300	30,850	269,550.	0.	0
ii. Assistants							
Dissection Assistant	600*	600*	600*	600*	6,400.	0.	0
Dalesme	600*	600*	600*		5,800.	0.	0
Chastillon	400*	400*	400*	400*	4,266.	13.	4
Subtotal, i and ii	1,600	1,600	1,600	1,000	16,466.	13.	4
Total, i and ii	27,200	26,900	28,900	31,850	286,016.	13.	4

Continued on next page

TABLE 1. CONTINUED

d. Total Paid or Budgeted for Pensions, 1666-1699

	Total, 1666-1699		
	lv.	s.	d.
Colbert	643,708.	6.	8
Louvois	171,833.	6.	8
Pontchartrain	286,016.	13.	4
Total	1,101,558.	6.	8

TABLE 2. THE OBSERVATORY

Fiscal Year Paid	Sources	(a) Construction[a]			(b) Porter & Maintenance[b]			(c) Marly Tower			(d) Total Observatory		
		lv.	s.	d.	lv.	s.	d.	lv.	s.	d.	lv.	s.	d.
	CdB, unless otherwise noted												
1666													
1667	1: 207, 212, 213, 216, 172	57,758.	4.	0									
1668	1: 274, 276, 278, 281, 291	99,744.	3.	4									
1669	1: 317, 376, 387, 388	135,913.	6.	0									
1670	1: 401, 413, 460, 471	139,534.	9.	0									
1671	1: 504-5, 567-68, 569	118,901.	11.	0									
1672	1: 600-601, 652	61,389.	9.	5									
1673	1: 686-87, 715-16, 717	20,846.	12.	2									
1674	1: 747, 748, 764	11,084.	8.	0									
1675	1: 818, 819, 859	13,288.	13.	0									
1676	1: 889	12,684.	15.	6									
1677	1: 947, 948	25,739.	7.	0									
1678	1: 1025, 1026	1,930.	14.	0									
1679	1: 1125, 1126	3,945.	9.	0									

Continued on next page

TABLE 2. CONTINUED

Fiscal Year Paid	Sources	(a) Construction[a]			(b) Porter & Maintenance[b]			(c) Marly Tower			(d) Total Observatory		
		lv.	s.	d.	lv.	s.	d.	lv.	s.	d.	lv.	s.	d.
	CdB, unless otherwise noted												
1680	1: 1242, 1243	4,769.	1.	6									
1681	2: 84	793.	0.	0									
1682	2: 127, 162, 219	3,423.	10.	0									
1683	2: 267, 312, 313, 356	1,957.	11.	0									
	Subtotal	713,704.	3.	11							713,704.	3.	11
1683													
1684													
1685	2: 775							4,200.	0.	0			
1686	2: 978, 979, 981							4,054.	18.	4[c]			
1687	2: 1167, 1189, 1204				657.	4.	6	1,134.	10.	11			
1688	3: 107, 124, 125, 193				1,012.	3.	4	340.	0.	0			
1689	3: 360				180.	0.	0						
1690	3: 425, 441				480.	0.	0						
1691	3: 568, 584; AN O¹2173: 153r				276.	10.	0						
	Subtotal				2,605.	17.	10	9,729.	9.	3	12,335.	7.	1

Continued on next page

TABLE 2. CONTINUED

Fiscal Year Paid	Sources	(a) Construction[a]			(b) Porter & Maintenance[b]			(c) Marly Tower			(d) Total Observatory		
		lv.	s.	d.	lv.	s.	d.	lv.	s.	d.	lv.	s.	d.
	CdB, unless otherwise noted												
1691	3: 568, 584				142.	10.	0						
1692	3: 787				230.	0.	0						
1693	3: 855, 920				304.	15.	0						
1694	3: 992, 1061				273.	14.	0						
1695	3: 1118, 1122, 1192				246.	15.	0						
1696	4: 45, 125, 273				227.	3.	0						
1697	4: 191-92, 272, 273, 277				549.	2.	0						
1698	4: 333, 334, 340, 416, 421				514.	9.	2						
1699	4: 482, 557, 558, 561				384.	10.	0						
	Subtotal				2,872.	18.	2				2,872.	18.	2
	Total	713,704.	3.	11	5,478.	16.	0	9,729.	9.	3	728,912.	9.	2

TABLE 3. SCIENTIFIC INSTRUMENTS

Fiscal Year Paid	Sources	(a) Mathematical & Astronomical			(b) Surveying & Rainfall Measurement		
	CdB, unless otherwise noted	lv.	s.	d.	lv.	s.	d.
1666							
1667	1: 230, 231, 232	3,500.	2.	0			
1668	1: 271	1,790.	0.	0	90.	0.	0
1669	1: 314, 365, 384, 388	738.	0.	0			
1670	1: 447, 448, 463, 470, 473, 475	1,535.	10.	0			
1671	1: 502-3, 552, 576	2,689.	0.	0			
1672	1: 646, 647, 659	2,500.	10.	0			
1673	1: 712, 723	6,590.	0.	0			
1674	1: 780, 781, 791	4,762.	10.	0			
1675	1: 853, 854, 863	2,059.	0.	0			
1676	1: 924	1,292.	15.	0			
1677	1: 948, 990, 1002	875.	0.	0			
1678	1: 1025, 1026, 1084, 1098, 1111	1,140.	0.	0	583.	5.	0
1679	1: 1126, 1203-4, 1217, 1231	3,503.	0.	0			
1680	1: 1243, 1342, 1343, 1354, 1367	4,866.	0.	0	60.	0.	0
1681	2: 93-94, 101, 119	2,124.	10.	0			
1682	2: 206, 253	2,500.	0.	0			
1683	2: 313, 377	1,175.	0.	0			
	Subtotal	43,640.	17.	0	733.	5.	0
1683	2: 349, 391				319.	0.	0
1684	2: 568						
1685	2: 583, 662, 666, 691, 730, 761, 780	962.	0.	0	1,098.	0.	0
1686	2: 915, 968, 1002, 1003, 1009	1,460.	0.	0			
1687	2: 1167, 1187, 1204, 1296-97	445.	15.	0[b]			
1688	3: 124, 125, 219	165.	0.	0	195.	0.	0[c]
1689	3: 376						
1690	3: 441, 503	100.	0.	0			
1691							
	Subtotal	3,132.	15.	0	1,612.	0.	0
1691	3: 650						
1692	3: 798						
1693	3: 934-35						
1694	3: 1076						
1695							
1696	AN G^{7}899, pd. July 1696	420.	0.	0			
1697							
1698	AN G^{7}902, pd. Aug. 1698	370.	7.	0[d]			
1699	4: 484						
	Subtotal	790.	7.	0			
	Total	47,563.	19.	0	2,345.	5.	0

Appendix 247

TABLE 3. CONTINUED

Fiscal Year Paid	(c) Other			(d) Retainers			(e) Jesuit Missions to Far East			(f) Total Scientific Instruments		
	lv.	s.	d.	lv.	s.	d.	lv.	s.	d.	lv.	s.	d.
1666												
1667												
1668												
1669	7,000.	0.	0[a]	300.	0.	0						
1670				200.	0.	0						
1671				200.	0.	0						
1672	286.	0.	0	500.	0.	0						
1673				500.	0.	0						
1674				500.	0.	0						
1675				500.	0.	0						
1676												
1677				500.	0.	0						
1678				500.	0.	0						
1679				500.	0.	0						
1680				500.	0.	0						
1681				500.	0.	0						
1682				500.	0.	0						
1683												
	7,286.	0.	0	5,700.	0.	0				57,360.	2.	0
1683				500.	0.	0						
1684				500.	0.	0						
1685	9,000.	0.	0[a]	500.	0.	0	3,495.	0.	0			
1686	1,389.	10.	0[a]	500.	0.	0						
1687				500.	0.	0	660.	0.	0			
1688				500.	0.	0	2,152.	10.	0			
1689				500.	0.	0						
1690				500.	0.	0						
1691												
	10,389.	10.	0	4,000.	0.	0	6,307.	10.	0	25,441.	15.	0
1691				500.	0.	0						
1692				500.	0.	0						
1693				500.	0.	0						
1694				300.	0.	0						
1695												
1696												
1697												
1698												
1699	110.	0.	0[a]									
	110.	0.	0	1,800.	0.	0				2,700.	7.	0
	17,785.	10.	0	11,500.	0.	0	6,307.	10.	0	85,502.	4.	0

TABLE 4. EXPEDITIONS AND SPECIAL PROJECTS

Fiscal Year Paid	Sources	(a) Mapping the Généralité de Paris			(b) Establishing the Meridian & Measuring the Earth		
	CdB	lv.	s.	d.	lv.	s.	d.
1666							
1667							
1668	1: 270, 278	700.	0.	0	1,000.	0.	0
1669	1: 361, 379	3,900.	0.	0			
1670	1: 471, 476, 480	4,000.	0.	0			
1671	1: 547, 549-50, 554	5,400.	0.	0			
1672	1: 646, 647	2,765.	0.	0	400.	0.	0
1673	1: 712				300.	0.	0
1674	1: 780, 781				1,800.	0.	0
1675							
1676	1: 925	1,000.	0.	0			
1677	1: 990	1,000.	0.	0			
1678	1: 1067, 1084, 1085, 1091, 1107, 1111	1,000.	0.	0			
1679	1: 1181, 1203, 1209						
1680	1: 1313, 1315, 1343						
1681							
1682	2: 236						
1683	2: 377-78				9,190.	0.	0
	Subtotal	19,765.	0.	0	12,690.	0.	0
1683	2: 378				802.	0.	0
1684	2: 496, 497, 564				800.	0.	0
1685	2: 759				600.	0.	0
1686	2: 1009, 1077				330.	0.	0
1687							
1688	3: 114						
1689							
1690							
1691							
	Subtotal				2,532.	0.	0
1691							
1692							
1693							
1694							
1695							
1696							
1697							
1698							
1699							
	Subtotal						
	Total	19,765.	0.	0	11,722.	0.	0

TABLE 4. CONTINUED

Fiscal Year Paid	(c) Surveying for Water-Supply of Versailles			(d) Other			(e) Total Expeditions & Special Projects		
	lv.	s.	d.	lv.	s.	d.	lv.	s.	d.
1666									
1667									
1668									
1669				850.	0.	0			
1670				4,000.	0.	0			
1671				5,500.	0.	0			
1672				2,800.	0.	0			
1673				1,000.	0.	0			
1674				1,200.	0.	0			
1675									
1676									
1677									
1678	4,316.	0.	0	300.	0.	0			
1679	1,317.	3.	0	800.	0.	0			
1680	1,798.	15.	0						
1681									
1682	200.	0.	0	2,300.	0.	0			
1683									
	7,631.	18.	0	18,750.	0.	0	58,836.	18.	0
1683									
1684	1,300.	0.	0						
1685									
1686	2,800.	0.	0						
1687									
1688	600.	0.	0						
1689									
1690									
1691									
	4,700.	0.	0				7,232.	0.	0
1691									
1692									
1693									
1694									
1695									
1696									
1697									
1698									
1699									
	12,331.	18.	0	18,750.	0.	0	66,068.	18.	0

TABLE 5. THE COLLECTION OF MODELS

Fiscal Year Paid	Sources	(a) Sums Paid			(b) Type of Machine, Apparatus, Repair
	CdB, unless otherwise noted	lv.	s.	d.	
1666					
1667					
1668	1: 270, 271	1,098.	1.	0	
1669	1: 383, 384	2,238.	16.	0	Hydraulic machines, lanterns, pumps
1670	1: 447-48, 471, 473	3,528.	15.	0	Machines for cutting & threshing wheat
1671	1: 502, 503, 545	1,660.	10.	0	Machines for grinding wheat
1672	1: 646	1,410.	7.	6	Fire extinguisher
1673	1: 712	66.	0.	0	Hydraulic machine
1674	1: 780	140.	10.	0[a]	
1675	1: 853	259.	16.	0	
1676	1: 924	304.	0.	0	Repair to machine for raising water
1677					
1678					
1679	1: 1232	320.	5.	0	Repairs to models at Bibliothèque du roi
1680					
1681					
1682					
1683					
	Subtotal	11,027.	0.	6	
1683					
1684					
1685	2: 755	150.	0.	0	Balance
1686					
1687					
1688					
1689					
1690					
1691					
	Subtotal	150.	0.	0	
1691					
1692					
1693					
1694					
1695					
1696					
1697					
1698	4: 341, AN G⁷902, pd. Aug. 1698	1,279.	17.	0	Machines, tools, Samaritaine machine & pump
1699					
	Subtotal	1,279.	17.	0	
	Total	12,456.	17.	6	

Appendix 251

TABLE 5. CONTINUED

Fiscal Year Paid	(c) Supplier or Maker	(d) Total Collection of Models		
		lv.	s.	d.
1666				
1667				
1668				
1669	Danglebert, Gosselin, Gayon, Potel, Buirette			
1670	Gaultier, Coulombier & Gosselin, Buirette			
1671	Cosson			
1672	Niquet, Langenach			
1673	Hense			
1674				
1675				
1676	Potel[b]			
1677				
1678				
1679	Colson			
1680				
1681				
1682				
1683				
		11,027.	0.	6
1683				
1684				
1685	Vincent Le Roy[c]			
1686				
1687				
1688				
1689				
1690				
1691				
		150.	0.	0
1691				
1692				
1693				
1694				
1695				
1696				
1697				
1698	Couplet, Obry			
1699				
		1,279.	17.	0
		12,456.	17.	6

TABLE 6. NATURAL PHILOSOPHICAL RESEARCH

Fiscal Year Paid	Sources	(a) Anatomy		
	CdB, unless otherwise noted	lv.	s.	d.
1666				
1667				
1668	1: 270	336.	7.	0
1669				
1670	1: 448	200.	0.	0
1671				
1672				
1673				
1674				
1675				
1676				
1677				
1678				
1679	1: 1232	40.	0.	0
1680	1: 1343	378.	18.	0
1681	2: 101, 102	1,311.	4.	6
1682				
1683	2: 311	498.	17.	0
	Subtotal	2,765.	6.	6
1683				
1684	2: 536, 537	388.	13.	0
1685	BN Arch. anc. rég. 1: 24r, 35r, 40r	2.	11.	0
1686	2: 1009; BN Arch. anc. rég. 1: 66v, 67r, 73v, 74v	346.	10.	0
1687	2: 1204; BN Arch. anc. rég. 1: 75v	75.	7.	0
1688	3: 120, 121, 125; BN Arch. anc. rég. 1: 77r, 78r, 80r, 81r, 83r-v; AN O^{1}2124	386.	10.	0
1689	3: 293, 302-3; BN Arch. anc. rég. 1: 84v, 85r, 87r, 90v	10.	5.	0
1690	3: 438, 441; AN O^{1}2124	316.	10.	0
1691	3: 584			
	Subtotal	1,526.	6.	0
1691				
1692	3: 730; BN Arch. anc. rég. 2: 1r; BN MS. Clairambault 814: 643-45	4.	9.	0
1693				
1694				
1695	AN G^{7}898, pd. Mar. 1695	159.	15.	0
1696				
1697	AN G^{7}901, pd. May 1697, Jan. 1698	396.	9.	6
1698				
1699	AN G^{7}904, pd. Sept. 1699 & Jan. 1700 to M. des Granges			
	Subtotal	560.	13.	6
	Total	4,852.	6.	0

Appendix 253

TABLE 6. CONTINUED

Fiscal Year Paid	(b) Petit jardin			(c) Botany			(d) Mineralogy			(e) Total Natural Philosophy		
	lv.	s.	d.	lv.	s.	d.	lv.	s.	d.	lv.	s.	d.
1666												
1667												
1668												
1669												
1670												
1671												
1672												
1673												
1674												
1675												
1676												
1677												
1678												
1679												
1680												
1681												
1682												
1683												
										2,765.	6.	6
1683												
1684												
1685												
1686												
1687	33.	5.	0									
1688	151.	10.	0									
1689	392.	4.	0									
1690	89.	0.	0									
1691	63.	4.	0									
	729.	3.	0							2,255.	9.	0
1691												
1692	105.	10.	0									
1693												
1694												
1695												
1696												
1697				118.	2.	0	600.	0.	0			
1698												
1699				363.	10.	0						
	105.	10.	0	481.	12.	0	600.	0.	0	1,747.	15.	6
	834.	13.	0	481.	12.	0	600.	0.	0	6,768.	11.	0

TABLE 7. THE LABORATORY

Fiscal Year	(a) Bourdelin's Expenses[a]			(b) Borelly's Known Reimbursements[b]			(c) Homberg's Known Reimbursements[c]			(d) Total Laboratory		
	lv.	s.	d.	lv.	s.	d.	lv.	s.	d.	lv.	s.	d.
1666												
1667	1,223.	0.	6									
1668	715.	11.	0									
1669												
1670	1,521.	12.	6									
1671	769.	0.	0									
1672	364.	7.	6									
1673	1,000.	16.	6									
1674	1,732.	9.	6									
1675	1,543.	9.	6									
1676	2,991.	19.	0									
1677	1,864.	13.	0									
1678	2,071.	0.	0									
1679	1,981.	0.	0									
1680	2,268.	2.	0									
1681	1,745.	15.	0									
1682	1,666.	7.	0									
1683	1,431.	5.	0									
Subtotal	24,890.	8.	0							24,890.	8.	0
1683												
1684	725.	4.	0									
1685	709.	16.	0									
1686	647.	7.	0									
1687	278.	2.	6	1,394.	2.	0						
1688	409.	15.	0	1,404.	10.	0						
1689	318.	3.	0									
1690	240.	7.	0									
1691												
Subtotal	3,328.	14.	6	2,798.	12.	0				6,127.	6.	6
1691	160.	0.	0									
1692	203.	8.	6									
1693	133.	10.	0				523.	4.	0			
1694	147.	8.	0				513.	10.	0			
1695	55.	10.	0				1,122.	3.	0			
1696	168.	7.	0				1,322.	6.	0			
1697	165.	10.	0				1,536.	14.	0			
1698	191.	2.	0				1,172.	12.	0			
1699							608.	10.	0			
Subtotal	1,224.	15.	6				6,798.	19.	0	8,023.	14.	6
Total	29,443.	18.	0	2,798.	12.	0	6,798.	19.	0	39,041.	9.	0

Appendix   255

TABLE 8. THE ENGRAVINGS AND DRAWINGS FOR NATURAL HISTORY

Fiscal Year Paid	Sources	(a) Plants			(b) Animals		
	CdB, unless otherwise noted	lv.	s.	d.	lv.	s.	d.
1666							
1667							
1668	1: 271, 281, 285	1,365.	0.	0	300.	0.	0
1669	1: 359, 364				364.	0.	0
1670	1: 469	2,340.	0.	0			
1671	1: 543, 544	2,838.	0.	0			
1672	1: 641, 643	2,057.	0.	0			
1673							
1674	1: 806	1,164.	0.	0			
1675	1: 874, 875	1,174.	0.	0	560.	0.	0
1676	1: 927, 928	1,130.	0.	0	910.	0.	0
1677	1: 993	1,620.	0.	0			
1678	1: 1087	2,264.	0.	0			
1679	1: 1206, 1208	1,080.	0.	0	104.	0.	0
1680	1: 1345	540.	0.	0			
1681							
1682	2: 210	540.	0.	0			
1683							
	Subtotal	18,112.	0.	0	2,238.	0.	0
1683							
1684	2: 541	1,080.	0.	0	150.	0.	0
1685	2: 779-80, 782, 784, 785	200.	0.	0	2,119.	15.	0
1686	2: 1015	540.	0.	0			
1687	2: 1192, 1200	1,227.	0.	0			
1688	3: 109-10	1,662.	0.	0	88.	0.	0
1689	3: 296; BN Arch. anc. rég. 1: 87v-88r	562.	6.	0	88.	0.	0
1690	3: 433	726.	0.	0			
1691	3: 575	528.	0.	0			
	Subtotal	6,525.	6.	0	2,445.	15.	0
1691							
1692	AN G⁷894, pd. May, July 1692	300.	0.	0	319.	0.	0
1693							
1694							
1695	AN G⁷898, pd. Oct. 1695	246.	10.	0			
1696							
1697							
1698							
1699							
	Subtotal	546.	10.	0	319.	0.	0
	Total	25,183.	16.	0	5,002.	15.	0

TABLE 8. CONTINUED

Fiscal Year Paid	(c) Plants & Animals			(d) Total Engravings & Drawings, Natural History		
	lv.	s.	d.	lv.	s.	d.
1666						
1667						
1668						
1669						
1670	1,105.	0.	0			
1671						
1672						
1673						
1674						
1675						
1676						
1677						
1678						
1679						
1680						
1681						
1682						
1683						
	1,105.	0.	0	21,455.	0.	0
1683						
1684						
1685	1,483.	0.	0			
1686	154.	0.	0			
1687						
1688						
1689						
1690						
1691						
	1,637.	0.	0	10,608.	1.	0
1691						
1692						
1693						
1694						
1695						
1696						
1697						
1698						
1699						
				865.	10.	0
	2,742.	0.	0	32,928.	11.	0

Appendix 257

TABLE 9. THE ENGRAVINGS AND DRAWINGS FOR THE MATHEMATICAL SCIENCES

Fiscal Year Paid	Sources	(a) Map of Moon			(b) Planets & Constellations			(c) Astronomical Treatise		
	CdB	lv.	s.	d.	lv.	s.	d.	lv.	s.	d.
1666										
1667										
1668										
1669										
1670										
1671	1: 543				120.	0.	0	53.	0.	0
1672	1: 642				75.	0.	0			
1673	1: 712	450.	0.	0						
1674	1: 780, 805	720.	0.	0						
1675	1: 874	540.	0.	0						
1676	1: 928									
1677	1: 990, 994	270.	0.	0	270.	0.	0			
1678	1: 1089	1,080.	0.	0						
1679	1: 1208, 1211	810.	0.	0						
1680										
1681										
1682										
1683										
	Subtotal	3,870.	0.	0	465.	0.	0	53.	0.	0
1683										
1684										
1685										
1686										
1687										
1688										
1689										
1690										
1691										
	Subtotal									
1691										
1692										
1693										
1694										
1695										
1696										
1697										
1698										
1699										
	Subtotal									
	Total	3,870.	0.	0	465.	0.	0	53.	0.	0

TABLE 9. CONTINUED

	(d) Geometry & Mathematics			(e) Other or Unspecified			(f) Total Engravings & Drawings Mathematical Sciences		
	lv.	s.	d.	lv.	s.	d.	lv.	s.	d.
1666									
1667									
1668									
1669									
1670									
1671	55.	0.	0						
1672				95.	0.	0			
1673									
1674	1,100.	0.	0	540.	0.	0			
1675				540.	0.	0			
1676				1,080.	0.	0			
1677				540.	0.	0			
1678									
1679				270.	0.	0			
1680									
1681									
1682									
1683									
	1,155.	0.	0	3,065.	0.	0	8,608.	0.	0
1683									
1684									
1685									
1686									
1687									
1688									
1689									
1690									
1691									
1691									
1692									
1693									
1694									
1695									
1696									
1697									
1698									
1699									
	1,155.	0.	0	3,065.	0.	0	8,608.	0.	0

Appendix 259

TABLE 10. THE ENGRAVINGS AND DRAWINGS FOR PRACTICAL PROJECTS
AND UNKNOWN SUBJECTS

Fiscal Year Paid	Sources	(a) Map of the Généralité de Paris[a]			(b) Perrault's Vitruvius[b]			(c) Subject Unknown[c]		
	CdB, unless otherwise noted	lv.	s.	d.	lv.	s.	d.	lv.	s.	d.
1666										
1667										
1668	1: 281				300.	0.	0			
1669	1: 358, 362				1,230.	0.	0			
1670	1: 407, 468, 473, 474				3,300.	0.	0			
1671	1: 543, 544, 545	300.	0.	0	2,443.	0.	0			
1672	1: 642				2,626.	0.	0			
1673	1: 709	300.	0.	0	952.	0.	0			
1674	1: 806	300.	0.	0						
1675	1: 874							864.	0.	0
1676	1: 927							690.	0.	0
1677	1: 993									
1678	1: 1088	200.	0.	0						
1679	1: 1207	300.	0.	0						
1680	1: 1346							320.	0.	0
1681										
1682										
1683										
	Subtotal	1,400.	0.	0	10,851.	0.	0	1,874.	0.	0
1683										
1684										
1685										
1686										
1687										
1688										
1689										
1690										
1691										
	Subtotal									
1691										
1692										
1693										
1694										
1695										
1696	AN G[7]899, pd. Oct. 1696									
1697										
1698										
1699	AN G[7]903, 904, July, Dec. 1699[f]									
	Subtotal									
	Total	1,400.	0.	0	10,851.	0.	0	1,874.	0.	0

TABLE 10. CONTINUED

Fiscal Year Paid	(d) Arts, Crafts, Maps[d]			(e) Surveying[e]			(f) Total Table 10a-e			(g) Total Tables 8-10[g]		
	lv.	s.	d.	lv.	s.	d.	lv.	s.	d.	lv.	s.	d.
1666												
1667												
1668												
1669												
1670												
1671												
1672												
1673												
1674												
1675												
1676												
1677				75.	0.	0						
1678												
1679												
1680												
1681												
1682												
1683												
				75.	0.	0	14,200.	0.	0	44,263.	0.	0
1683												
1684												
1685												
1686												
1687												
1688												
1689												
1690												
1691												
										10,608.	1.	0
1691												
1692												
1693												
1694												
1695												
1696	300.	0.	0									
1697												
1698												
1699												
	3,097.	0.	0									
	3,397.	0.	0				3,397.	0.	0	4,262.	10.	0
	3,397.	0.	0	75.	0.	0	17,597.	0.	0	59,133.	11.	0

TABLE 11. SMALL EXPENSES

Fiscal Year Paid	Sources	(a) Academy Only		
	CdB, unless otherwise noted	lv.	s.	d.
1666	1: 151	2,500.	0.	0
1667	1: 231	739.	1.	4
1668	1: 270-71	1,909.	18.	0
1669	1: 358	500.	0.	0
1670				
1671	1: 503, 505, 552	191.	12.	7
1672	1: 600, 646	193.	5.	6
1673	1: 686			
1674	1: 747, 780, 803-4	4,414.	19.	0
1675	1: 853	622.	10.	0
1676	1: 924			
1677	1: 948, 989	1,786.	14.	6
1678	1: 1084			
1679	1: 1231			
1680	1: 1242			
1681	2: 101			
1682	2: 219, 236			
1683				
	Subtotal	12,858.	0.	11
1683				
1684	2: 538			
1685	2: 780; BN Arch. anc. rég. 1: 14r, 21r, 26v-27r	338.	2.	6
1686	2: 1009; BN Arch. anc. rég. 1: 70v	427.	18.	6
1687	2: 1204	65.	18.	6
1688	3: 124	157.	1.	0
1689	3: 304			
1690	3: 441			
1691	3: 584	180.	9.	0
	Subtotal	1,169.	9.	6
1691				
1692	3: 723; AN G⁷894, pd. May, Nov. 1692	302.	15.	6
1693	AN G⁷895, pd. Jan. 1693	892.	0.	0
1694				
1695	AN G⁷898, pd. Mar. 1695			
1696				
1697	AN G⁷901, pd. Mar., May 1697, Jan. 1698	150.	0.	0
1698	AN G⁷902, pd. June 1698, Jan. 1699	576.	11.	0
1699	AN G⁷904, pd. Mar., May, July, Oct., Dec. 1699, Jan. 1700			
	Subtotal	1,921.	6.	6
	Total	15,948.	16.	11

TABLE 11. CONTINUED

Fiscal Year Paid	(b) Observatory Only			(c) Academy & Observatory			(d) Total Small Expenses		
	lv.	s.	d.	lv.	s.	d.	lv.	s.	d.
1666							2,500.	0.	0
1667							739.	1.	4
1668							1,909.	18.	0
1669							500.	0.	0
1670									
1671	826.	6.	0[a]				1,017.	18.	7
1672	794.	0.	0				987.	5.	6
1673	425.	19.	0				425.	19.	0
1674	435.	5.	0				4,850.	4.	0
1675				848.	19.	0	1,471.	9.	0
1676				1,323.	6.	0	1,323.	6.	0
1677	400.	0.	0				2,186.	14.	6
1678				1,722.	10.	0	1,722.	10.	0
1679				668.	13.	0	668.	13.	0
1680				287.	10.	0	287.	10.	0
1681				633.	4.	6	633.	4.	6
1682	52.	0.	0[b]	956.	2.	6	1,008.	2.	6
1683									
	2,933.	10.	0	6,440.	5.	0	22,231.	15.	11
1683									
1684				163.	0.	0	163.	0.	0
1685							338.	2.	6
1686							427.	18.	6
1687	167.	13.	3				233.	11.	9
1688				433.	19.	0	591.	0.	0
1689	370.	9.	0				370.	9.	0
1690				65.	17.	6	65.	17.	6
1691							180.	9.	0
	538.	2.	3	662.	16.	6	2,370.	8.	3
1691									
1692	220.	13.	0				523.	8.	6
1693							892.	0.	0
1694									
1695				104.	0.	0	104.	0.	0
1696									
1697				549.	9.	0	699.	9.	0
1698							576.	11.	0
1699									
				13,150.	0.	0	13,150.	0.	0
	220.	13.	0	13,803.	9.	0	15,945.	8.	6
	3,692.	5.	3	20,906.	10.	6	40,547.	12.	8

TABLE 12. THE SHARED EXPENSES OF THE ACADEMY AND OTHER ROYAL ESTABLISHMENTS

Fiscal Year Paid	Sources	(a) Rent BdR lv. s. d.	(b) Maintenance BdR lv. s. d.	(c) Small Expenses BdR lv. s. d.	(d) Small Expenses & Carpentry BdR & ARdS lv. s. d.
	CdB, unless otherwise noted				
1666	1: 151, 209		803. 10. 0		7,000. 0. 0
1667	1: 221, 230, 231		2,949. 15. 0		60. 7. 0
1668	1: 233, 270, 271, 275, 277, 278, 280	3,000. 0. 0	1,090. 11. 0		
1669	1: 367, 383-84	3,000. 0. 0	5,600. 0. 0		10,000. 0. 0
1670	1: 403, 448, 471, 475, 476, 478	7,500. 0. 0	4,834. 0. 0	7,000. 0. 0	
1671	1: 502, 503, 563	3,000. 0. 0	979. 1. 8		3,962. 9. 0
1672	1: 642, 645, 646, 647	3,000. 0. 0	2,679. 13. 0	6,481. 3. 6	
1673	1: 683, 711, 712, 713	3,000. 0. 0	100. 0. 0		4,319. 5. 6
1674	1: 740, 779-80, 782	3,000. 0. 0	1,065. 3. 0		2,212.12. 8
1675	1: 853, 854, 855	3,000. 0. 0	1,543. 5. 0	11,762. 18. 0[c]	4,814. 5. 6
1676	1: 924, 925	3,000. 0. 0			5,232. 0. 0
1677	1: 989, 990, 993, 994, 995	3,000. 0. 0	656. 15. 0		3,792.17. 0
1678	1: 1081, 1082, 1084, 1088	3,000. 0. 0	360. 0. 0	2,000. 0. 0	
1679	1: 1203, 1209	3,000. 0. 0			3,873.14. 6
1680	1: 1336, 1343, 1347	6,000. 0. 0			4,326.15. 0
1681	2: 102, 108	5,000. 0. 0	44. 0. 0		6,241. 7. 0
1682	2: 222, 206-7, 214	5,000. 0. 0			
1683	2: 307, 314, 377		400. 0. 0		5,324.12. 6
	Subtotal	56,500. 0. 0	23,105. 13. 8	27,244. 1. 6	61,160. 5. 8

Continued on next page

TABLE 12. CONTINUED

Fiscal Year Paid	Sources	(a) Rent BdR			(b) Maintenance BdR			(c) Small Expenses BdR			(d) Small Expenses & Carpentry BdR & ARdS		
		lv.	s.	d.	lv.	s.	d.	lv.	s.	d.	lv.	s.	d.
	CdB, unless otherwise noted												
1683													
1684	2: 499, 500, 537, 538, 542, 563	5,000.	0.	0	157.	4.	0	2,324.	8.	0			
1685	2: 755, 778, 779, 787	5,000.	0.	0	190.	5.	9	2,653.	13.	9			
1686	2: 984, 1008, 1009, 1016	5,000.	0.	0	613.	4.	6	14,464.	15.	7d			
1687	2: 1162, 1167, 1169, 1190, 1199, 1211	5,000.	0.	0	42.	1.	7	7,079.	8.	9			
1688	3: 120, 121-22, 132	5,000.	0.	0	124.	15.	9	2,288.	5.	7f			
1689	3: 303, 309	5,000.	0.	0				5,022.	18.	10			
1690	3: 427, 429, 430, 438-39, 442	5,000.	0.	0				1,943.	11.	6			
1691	3: 566, 567, 569, 582, 584, 585	5,000.	0.	0	300.	0.	0	1,998.	0.	6			
								902.	14.	0			
	Subtotal	40,000.	0.	0	1,427.	11.	7	38,677.	16.	6			
1691	3: 568, 570-71, 582							156.	15.	6			
1692	3: 660, 698-99, 709, 714, 718, 719, 720; AN G⁷894, pd. Dec. 1692	5,000.	0.	0				1,919.	3.	5			
1693	3: 806, 842, 850, 851, 852, 863, 867	5,000.	0.	0									
1694	3: 945, 949, 954, 982-83, 988-89, 990, 997, 1001, 1005, 1008	10,000.	0.	0									
1695	3: 1089, 1091, 1119, 1120, 1137, 1138, 1146, 1148, 1202	5,000.	0.	0									
1696	4: 46-47, 54, 135												
1697	4: 139, 186-87, 198, 201-2, 271	5,000.	0.	0									
1698	4: 286-87, 335, 336, 344, 348, 430	5,000.	0.	0									
1699	4: 431, 476-77, 492, 569	5,000.	0.0										
	Subtotal	40,000.	0.	0				2,075.	18.	11			
	Total	136,500.	0.	0	24,533.	5.	3	67,997.	16.	11	61,160.	5.	8

Continued on next page

TABLE 12. CONTINUED (CATEGORIES e-i)

Fiscal Year Paid	(e) Garden BdR			(f) Engraving & Printing BdR			(g) Watercolors of Plants & Animals, by Robert & Joubert, BdR[a]			(h) Maintenance, etc. Obs., BdR, & Other Est.[b]			(i) Maintenance, etc. ARdS, BdR, JR, & Other Est.[b]			(j) Total Shared Expenses		
	lv.	s.	d.	lv.	s.	d.	lv.	s.	d.	lv.	s.	d.	lv.	s.	d.	lv.	s.	d.
1666																		
1667																		
1668							3,550.	0.	0	600.	0.	0						
1669																		
1670	167.15.	0		2,000.	0.	0												
1671	251.15.	0		2,389.	4.	0				365.	0.	0						
1672				1,182.	4.	0												
1673				1,214.	0.	0												
1674				1,000.	0.	0												
1675				2,098.	4.	0												
1676				10,500.	0.	0												
1677				3,289.	0.	0				332.	18.	0						
1678				3,769.	2.	0												
1679				2,017.	0.	0												
1680													140.	0.	0			
1681																		
1682	261.17.	6											2,500.	0.	0			
1683													1,711.	0.	0			
Subtotal	681. 7.	6		29,459.	0.	0	3,550.	0.	0	1,297.	18.	0	4,351.	0.	0	207,349.	6.	4

Continued on next page

TABLE 12. CONTINUED (CATEGORIES e-i)

Fiscal Year Paid	(e) Garden BdR			(f) Engraving & Printing BdR			(g) Watercolors of Plants & Animals, by Robert & Joubert, BdR[a]			(h) Maintenance, etc. Obs., BdR, & Other Est.[b]			(i) Maintenance, etc. ARdS, BdR, JR, & Other Est.[b]			(j) Total Shared Expenses		
	lv.	s.	d.	lv.	s.	d.	lv.	s.	d.	lv.	s.	d.	lv.	s.	d.	lv.	s.	d.
1683																		
1684				1,139.	14.	0[e]							155.	10.	0			
1685										158.	12.	6						
1686				9,250.	0.	0[g]												
1687				150.	0.	0	450.	0.	0				1,193.	5.	10			
1688							600.	0.	0				1,346.	0.	0			
1689													57.	17.	0			
1690							600.	0.	0	600.	0.	0	1,480.	18.	5			
1691							600.	0.	0	750.	0.	0	200.	0.	0			
				10,539.	14.	0	2,250.	0.	0	1,508.	12.	6	4,433.	11.	3	98,837.	5.	10
1691							300.	0.	0	140.	0.	0	100.	0.	0			
1692							1,400.	0.	0	4,076.	19.	2	396.	17.	10			
1693							750.	0.	0	2,658.	0.	8	547.	2.	7			
1694							250.	0.	0	6,959.	1.	8	3,627.	0.	6			
1695										3,903.	19.	2	383.	14.	8			
1696							600.	0.	0	713.	18.	4	482.	1.	8			
1697							600.	0.	0	784.	7.	8	550.	12.	1			
1698							600.	0.	0	737.	11.	3	200.	0.	0			
1699							7,350.	0.	0	700.	0.	0						
Subtotal							11,850.	0.	0	20,673.	17.	11	6,287.	9.	4	80,887.	6.	2
Total	681.	7.	6	39,998.	14.	0	17,650.	0.	0	23,480.	8.	5	15,072.	0.	7	387,073.	18.	4

TABLE 13. SUMMARY OF EXPENDITURE ON THE ACADEMY, 1666-1699

Table No.	Category of Expenditure	(a) Colbert			(b) Louvois			(c) Pontchartrain			(d) Total		
		lv.	s.	d.	lv.	s.	d.	lv.	s.	d.	lv.	s.	d.
	i. Direct Expenditure												
1	Pensions	643,708.	6.	8	171,833.	6.	8	286,016.	13.	4	1,101,558.	6.	8
2	Observatory	713,704.	3.	11	12,335.	7.	1	2,872.	18.	2	728,912.	9.	2
	Research												
3	Scientific Instruments	57,360.	2.	0	25,441.	15.	0	2,700.	7.	0	85,502.	4.	0
4	Expeditions & Projects	58,836.	18.	0	7,232.	0.	0				66,068.	18.	0
5	Models	11,027.	0.	6	150.	0.	0	1,279.	17.	0	12,456.	17.	6
6	Natural Philosophy	2,765.	6.	6	2,255.	9.	0	1,747.	15.	6	6,768.	11.	0
7	Laboratory	24,890.	8.	0	6,127.	6.	6	8,023.	14.	6	39,041.	9.	0
	Engravings & Drawings												
8	Natural History	21,455.	0.	0	10,608.	1.	0	865.	10.	0	32,928.	11.	0
9	Mathematical Sciences	8,608.	0.	0					0.	0	8,608.	0.	0
10	Practical & Unknown	14,200.	0.	0				3,397.	0.	0	17,597.	0.	0
11	Small Expenses	22,231.	15.	11	2,370.	8.	3	15,945.	8.	6	40,547.	12.	8
	Subtotal, Research	221,374.	10.	11	54,184.	19.	9	33,959.	12.	6	309,519.	3.	2
	Subtotal, Obs. & Research	935,078.	14.	10	66,520.	6.	10	36,832.	10.	8	1,038,431.	12.	4

Continued on next page

TABLE 13. CONTINUED

Table No.	Category of Expenditure	(a) Colbert			(b) Louvois			(c) Pontchartrain			(d) Total		
		lv.	s.	d.	lv.	s.	d.	lv.	s.	d.	lv.	s.	d.
	Summary												
	Pensions	643,708.	6.	8	171,833.	6.	8	286,016.	13.	4	1,101,558.	6.	8
	Observatory	713,704.	3.	11	12,335.	7.	1	2,872.	18.	2	728,912.	9.	2
	Research	221,374.	10.	11	54,184.	19.	9	33,959.	12.	9	309,519.	3.	2
	Total, Direct Expenditure	1,578,787.	1.	6	238,353.	13.	6	322,849.	4.	6	2,139,989.	19.	0
	ii. Indirect Expenditure												
12	Shared Expenses	207,349.	6.	4	98,837.	5.	10	80,887.	6.	2	387,073.	18.	4

TABLE 14. HOW COLBERT ALLOCATED DIRECT EXPENDITURE ON THE ACADEMY, 1666-1683

Table & Column	Category of Expense	Total Spent				The Category of Expense as % of Total Spent On			Average Spent Yearly
						Pensions, Obs., & Research	Obs. & Research	Research	
		lv.	s.	d.		%	%	%	lv.
1a	Pensions	643,708.	6.	8		41.0			35,762
2	Observatory	713,704.	3.	11		45.0	76.0		39,650
	Research								
3	Scientific Instruments	57,360.	2.	0		3.6	6.1	26.0	3,187
4	Expeditions & Special Projects	58,836.	18.	0		3.7	6.3	27.0	3,269
5	Collection of Models	11,027.	0.	6		.7	1.2	5.0	612
6	Natural Philosophy	2,765.	6.	6		.2	.3	1.0	154
7	Laboratory	24,890.	8.	0		1.6	2.7	11.0	1,383
8-10	Engravings & Drawings	44,263.	0.	0		2.8	5.0	20.0	2,459
11	Small Expenses	22,231.	15.	11		1.4	2.4	10.0	1,235
	Subtotal, Research	221,374.	10.	11		14.0	24.0	100.0	12,299
	Subtotal, Obs. & Research	935,078.	14.	10			100.0		51,949
	Total, Direct Expenditure	1,578,787.	1.	6		100.0			87,711

Continued on next page

TABLE 14. CONTINUED

Table & Column	Category of Expense	Total Spent			The Category of Expense as % of Total Spent On			Average Spent Yearly
					Pensions, Obs., & Research	Obs. & Research	Research	
		lv.	s.	d.	%	%	%	lv.
	The Research Program							
	Mathematical Sciences							
3a, d	Scientific Instruments	49,340.	17.	0				
4b, d	Expeditions	31,440.	0.	0				
9	Engravings	8,608.	0.	0				
11b	Small Expenses—Obs.	2,933.	10.	0				
	Subtotal	92,322.	7.	0	6.0	10.0	42.0	5,130
	Natural Philosophy							
6a	Anatomical Research	2,765.	6.	6				
7	Laboratory	24,890.	8.	0				
8	Engravings	21,455.	0.	0				
3c	Burning Mirror	7,000.	0.	0				
	Subtotal	56,110.	14.	6	4.0	6.0	25.0	3,117
	Practical Projects							
3b, 4c, 10e	Survey of Versailles	8,440.	3.	0				
5	Models	11,027.	0.	6				
10b	Engravings—Vitruvius	10,851.	0.	0				
4a, 10a	Map of *Généralité de Paris*	21,165.	0.	0				
	Subtotal	51,483.	3.	6	3.0	6.0	23.0	2,860

Continued on next page

TABLE 14. CONTINUED

Table & Column	Category of Expense	Total Spent			The Category of Expense as % of Total Spent On			Average Spent Yearly
					Pensions, Obs., & Research	Obs. & Research	Research	
		lv.	s.	d.	%	%	%	lv.
	Undetermined							
11a, c	Small Expenses	19,298.	5.	11				
10c	Engravings	1,874.	0.	0				
3c	Other Scientific Instruments	286.	0.	0				
	Subtotal	21,458.	5.	11	1.0	2.0	10.0	1,192
	Total, Research Program	221,374.	10.	11	14.0	24.0	100.0	12,299

TABLE 15. HOW LOUVOIS ALLOCATED DIRECT EXPENDITURE ON THE ACADEMY, 1683-1691

Table & Column	Category of Expense	Total Spent			The Category of Expense as % of Total Spent On			Average Spent Yearly
					Pensions, Obs., & Research	Obs. & Research	Research	
		lv.	s.	d.	%	%	%	lv.
1b	Pensions	171,833.	6.	8	72.1			19,093
2	Observatory	12,335.	7.	1	5.2	18.5		1,371
	Research							
3	Scientific Instruments	25,441.	15.	0	10.6	38.2	47.0	2,827
4	Expeditions & Special Projects	7,232.	0.	0	3.0	11.0	13.3	804
5	Collection of Models	150.	0.	0	.1	.2	.3	17
6	Natural Philosophy	2,255.	9.	0	1.0	3.4	4.2	251
7	Laboratory	6,127.	6.	6	2.6	9.2	11.3	681
8-10	Engravings & Drawings	10,608.	1.	0	4.4	16.0	19.6	1,179
11	Small Expenses	2,370.	8.	3	1.0	3.5	4.3	263
	Subtotal, Research	54,184.	19.	9	22.7	81.5	100.0	6,020
	Subtotal, Obs. & Research	66,520.	6.	10		100.0		7,391
	Total, Direct Expenditure	238,353.	13.	6	100.0			26,484

Continued on next page

TABLE 15. CONTINUED

Table & Column	Category of Expense	Total Spent			The Category of Expense as % of Total Spent On			Average Spent Yearly
					Pensions, Obs., & Research	Obs. & Research	Research	
		lv.	s.	d.	%	%	%	lv.
	The Research Program							
	Mathematical Sciences							
3a, d, e	Scientific Instruments	13,440.	5.	0				
4b	Expeditions	2,532.	0.	0				
11b	Small Expenses—Obs.	538.	2.	3				
	Subtotal	16,510.	7.	3	7.0	24.8	30.4	1,834
	Natural Philosophy							
6a	Anatomical Research	1,526.	6.	0				
7	Laboratory	6,127.	6.	6				
8	Engravings	10,608.	1.	0				
3c	Burning Mirror	10,389.	10.	0				
6b	*Petit jardin*	729.	3.	0				
	Subtotal	29,380.	6.	6	12.3	44.2	54.2	3,264
	Practical Projects							
3b, 4c	Survey of Versailles	6,117.	0.	0				
3b	Measuring Rainfall	195.	0.	0				
5	Models	150.	0.	0				
	Subtotal	6,462.	0.	0	2.7	9.7	12.0	718

Continued on next page

TABLE 15. CONTINUED

Table & Column	Category of Expense	Total Spent			The Category of Expense as % of Total Spent On			Average Spent Yearly
					Pensions, Obs., & Research	Obs. & Research	Research	
		lv.	s.	d.	%	%	%	lv.
11a, c	Undetermined Small Expenses	1,832.	6.	0	—	—	—	
	Subtotal	1,832.	6.	0	.7	2.8	3.4	204
	Total, Research Program	54,184.	19.	9	22.7	81.5	100.0	6,020

TABLE 16. HOW PONTCHARTRAIN ALLOCATED DIRECT EXPENDITURE ON THE ACADEMY, 1691-1699

Table & Column	Category of Expense	Total Spent or Budgeted			The Category of Expense as % of Total Spent On			Yearly Average
					Pensions, Obs., & Research	Obs. & Research	Research	
		lv.	s.	d.	%	%	%	lv.
1c	Pensions	286,016.	13.	4	88.6			31,780
2	Observatory Research	2,872.	18.	2	.9	7.8		319
3	Scientific Instruments	2,700.	7.	0	.8	7.3	8.0	300
4	Expeditions & Special Projects							
5	Collection of Models	1,279.	17.	0	.4	3.5	3.8	142
6	Natural Philosophy	1,747.	15.	6	.5	4.7	5.1	194
7	Laboratory	8,023.	14.	6	2.5	21.8	23.6	891
8-10	Engravings & Drawings	4,262.	10.	0	1.3	11.6	12.5	474
11	Small Expenses	15,945.	8.	6	5.0	43.3	47.0	1,772
	Subtotal, Research	33,959.	12.	6	10.5	100.0		3,773
	Subtotal, Obs. & Research	36,832.	10.	8		100.0		4,092
	Total, Direct Expenditure	322,849.	4.	0	100.0			35,872

Continued on next page

TABLE 16. CONTINUED

Table & Column	Category of Expense	Total Spent or Budgeted			The Category of Expense as % of Total Spent On			Yearly Average
					Pensions, Obs., & Research	Obs. & Research	Research	
		lv.	s.	d.	%	%	%	lv.
	The Research Program							
	Mathematical Sciences							
3a, d	Scientific Instruments	2,590.	7.	0				
11b	Small Expenses—Obs.	220.	13.	0				
	Subtotal	2,811.	0.	0	.9	7.6	8.3	312
	Natural Philosophy							
6a	Anatomical Research	560.	13.	6				
7	Laboratory	8,023.	14.	6				
8	Engravings	865.	10.	0				
3c	Burning Mirror	110.	0.	0				
6b, c	Botanical Research	587.	2.	0				
6d	Mineralogy	600.	0.	0				
	Subtotal	10,747.	0.	0	3.3	29.2	31.6	1,194
	Practical Projects							
5	Models	1,279.	17.	0				
10d	Engravings	3,397.	0.	0				
	Subtotal	4,676.	17.	0	1.4	12.7	13.8	520

Continued on next page

TABLE 16. CONTINUED

Table & Column	Category of Expense	Total Spent or Budgeted			The Category of Expense as % of Total Spent On			Yearly Average
					Pensions, Obs., & Research	Obs. & Research	Research	
		lv.	s.	d.	%	%	%	lv.
11a, c	Undetermined Small Expenses	15,724.	15.	6				
	Subtotal	15,724.	15.	6	4.9	42.7	46.3	1,747
	Total, Research Program	33,959.	12.	6	10.5	92.2	100.0	3,773

TABLE 17. COLBERT, LOUVOIS, AND PONTCHARTRAIN COMPARED: DIRECT EXPENDITURE ON THE ACADEMY, 1666-1699

Category of Expense	Total Spent or Budgeted			The Category of Expense as % of Total Spent or Budgeted On		Yearly Average	
				Category, 1666-1699	Total, 1666-1699	Each Minister	1666-1699
	lv.	s.	d.	%	%	lv.	lv.
Pensions							
Colbert	643,708.	6.	8	58.0	30.1	35,762	
Louvois	171,833.	6.	8	16.0	8.0	19,093	
Pontchartrain	286,016.	13.	4	26.0	13.4	31,780	
Subtotal	1,101,558.	6.	8	100.0	51.5		32,399
Observatory							
Colbert	713,704.	3.	11	98.0	33.4	39,650	
Louvois	12,335.	7.	1	1.6	.6	1,371	
Pontchartrain	2,872.	18.	2	.4	.1	319	
Subtotal	728,912.	9.	2	100.0	34.1		21,439
Research							
Colbert	221,374.	10.	11	71.5	10.3	12,299	
Louvois	54,184.	19.	9	17.5	2.5	6,020	
Pontchartrain	33,959.	12.	6	11.0	1.6	3,773	
Subtotal	309,519.	3.	2	100.0	14.4		9,104
Summary							
Colbert	1,578,787.	1.	6		74.0	87,711	
Louvois	238,353.	13.	6		11.02	26,484	
Pontchartrain	322,849.	4.	0		15.0	35,872	
Total	2,139,989.	19.	0		100.0		62,940

Notes to Appendix

TABLE 1

Payments are listed under the fiscal year in which they were made. Academicians' names are followed by their dates of membership in the Academy. I am grateful to Michael S. Mahoney for sharing information about pensions paid under Colbert and Louvois.

[a] Sources: *CdB,* 1: 158, 161–63, 226–28, 278, 283, 299–300, 377–79, 448, 449–51, 476, 564–66, 646, 647–50, 712–15, 780, 782–83, 855–56, 925–27, 990, 992–93, 1084, 1085–87, 1093, 1204–1206, 1209, 1211–12, 1342, 1344–45, 1348, 1349–50; 2: 101–103, 104, 236–38, 244, 377, 379. Colbert, *Lettres,* 5: 470–98, lists only pensions to *gens de lettres,* but academicians and their assistants were often paid under other rubrics, so that it often has less information than the *CdB*. When one source provides information omitted from the other, I have followed the former. Table 1a includes all sums listed in these two documents as pensions, supplements to pensions, moving expenses, or wages. Colbert, *Lettres,* sometimes confuses the names of Pivert, Picard, Carcavi, and Cassini. The wages of Bourdelin's *garçon du laboratoire* are included in table 7. Payments to Loir, who assisted academicians in extending the meridian and mapping the *généralité* of Paris, are included in table 4. The 2,000 lv. paid to Du Vivier in 1679 was actually for him and other, unnamed persons "qui servent avec luy" in mapping the kingdom. In 1673 Pecquet's heirs collected his pension.

[b] Sources: *CdB,* 2: 539–40, 782–83, 1011, 1012–13, 1208–10; 3: 125–26, 305–307, 440; AN O¹ 656. On Gallois's pension, see Stroup, *Royal Funding,* 91n. c. Louvois paid academicians in fiscal year 1684 for work done in 1683, in 1685 for 1683 and 1684, in 1686 for 1685, and so on; in 1690 academicians received one-third of their pensions for 1689. In 1688 Perrault's heirs received his pension; in 1689 Borelly's widow collected two-thirds of his pension. Thévenot was pensioned for his work at the Bibliothèque du roi, not at the Academy: *CdB,*

2: 541. Lannion's pension was said to be for work in belles-lettres, but he was not a member of the Académie des inscriptions or the Académie française, and he was listed with members of the Academy of Sciences.

 c Sources: BN MS. Clairambault 566: 247, 251; AN G[7] 893–94, 897–903, 973, 986–87, 992; *CdB,* 4: 268, 411, 427, 566. Asterisks indicate that no record of payment exists, but that the amount listed is consistent with the *estat* and with practice in other years. Many pensions due in fiscal years 1694, 1695, and 1696 became *rentes.* Payments in fiscal year 1691 were for work performed in 1689, in 1692 for 1690 and 1691, in 1694 for 1692 and 1693, in 1695 for 1694, in 1696 for 1695, in 1697 for 1696 and 1697, in 1698 for 1697 and 1698, in 1699 for 1698 and 1699. For details, see Stroup, *Royal Funding,* app. C, and tables 1 and 2.

TABLE 2

 a Wolf, *Observatoire,* 14–16, gives the cost of the Observatory from 1667 through 1683 as 713,954 lv. 15 s. 11 d.; my total for Colbert's expenditure on the Observatory is 713,704 lv. 3 s. 11 d. The following points explain this difference of 250 lv. 12 s. between our totals:

 (i) Wolf included two payments of 620 lv. to Danglebert for a wooden model of the Observatory (in 1667 and 1668); there was only one (*CdB,* 1: 276).

 (ii) Wolf omitted 8 s. in a payment for a furnace (*CdB,* 1: 213).

 (iii) Wolf omitted payment of 600 lv. to Sainte-Marie for guarding the construction site (*CdB,* 1: 376).

 (iv) Wolf omitted 6 s. from his total for 1669, although it appears in his subtotals.

 (v) Wolf's total for 1671 omits 200 lv. included in the subtotals.

 (vi) Wolf includes in the costs of construction 431 lv. paid to Furet in 1683 for metalwork and scientific instruments. I count that as payment for scientific instruments.

 (vii) Addition of Wolf's yearly totals yields the sum of 713,954 lv. 9 s. 11 d., not 713,954 lv. 15 s. 11 d., as printed.

I follow Wolf (*Observatoire,* 8) in accepting that only one payment of 6,604 lv. was made for purchase of the site, although two payments are listed in *CdB,* 1: 153, 216. Both Wolf and I base our calculations on the *CdB,* but the architect Robert de Cotte, a *surintendant des bâtiments du roy,* calculated in the eighteenth century that from 1667 to 1683 the Observatory cost 725,174 lv. 4 s. 8 d.: BN MS. fr. 7801: 54r–55r. His sources are unknown.

 b Some repairs are included with small expenses. See table 11b, 167 lv. 13 s. 3 d. (1687), 433 lv. 19 s. (1688), and 370 lv. 9 s. (1689).

 c As Wolf, *Observatoire,* 167–68, states, the sum paid to Claude Tricot, mason, in 1686 must have been 2,748 lv. 17 s. 5 d., not 2,748 lv. 17 s. 4 d. as written in *CdB,* since this is *parfait payement* of a total of 3,548 lv. 17 s. 5 d.

TABLE 3

 a Marks sums, totaling 17,499 lv. 10 s., paid for burning mirrors. On the mirror purchased in 1669, see Oldenburg, *Correspondence,* 6: 294–95; on that purchased in 1686, see *Mercure galant* (Oct. 1685): 252–66, and AdS, Reg., 11: 129v, 131v–32r (26 May, 3 June 1685).

[b] Part of 195 lv. 15 s. paid to Raguin was also for work on the Marly tower.

[c] Apparatus for measuring rainfall.

[d] This sum covered repairs to machines in the Observatory and a pendulum clock.

[e] From 1695 through 1699, Thuret's responsibilities were for all pendulum clocks in all royal buildings, in Paris and Versailles; his retainer of 300 lv. is listed therefore in table 12h.

TABLE 4

The cost of apparatus for these expeditions is listed in table 3. For additional costs of mapping the *généralité de Paris,* see table 1a, ii (Du Vivier, Dupuy), and table 10a.

Col. d includes De La Voye's aborted voyage to Madagascar, the trip of Meurisse and Richer to Cayenne, Picard's voyage to Denmark, Roemer's to England, and Deglos's to Gorée and St. Thomé.

TABLE 5

Some reimbursements for small expenses (table 11) included payments for the collection of models: see *CdB*, 3: 124 (1688), and AN G[7] 902 (1698).

[a] Niquet's expenses.

[b] Almost all of Potel's work for the royal buildings was with hydraulic equipment. He first worked on this model in 1669.

[c] This may have been for the *salle des machines* at the Louvre; *CdB*, 2: 499, lists a payment in 1684 to Le Roy for work there.

TABLE 6

Additional expenses for anatomical research in 1693 are included in the sum of 892 lv. divided by Colson, Homberg, and Chastillon (table 11a). Du Hamel was probably reimbursed during fiscal year 1688 for 327 lv., spent for the Academy's anatomical research from April 1686 through February 1688 (AN O[1] 2124), but no corresponding entry exists in *CdB*. From 1693 the cost of the *petit jardin* (ranging from 100 lv. to 171 lv.) was combined in the *CdB* with payments for the terrace and amphitheater of the Jardin royal (table 12i).

TABLE 7

Col. a lists expenses, cols. b and c list reimbursements, and col. d summarizes the known costs of the laboratory. Given the inadequacy of the record, we do not know when the crown reimbursed Bourdelin; thus this table, unlike the others, is not a record of what the crown spent in each fiscal year. The record of expenditure is incomplete: Bourdelin's recorded expenses are much greater than his reimbursements in the *CdB,* and no record survives of Duclos's, Borelly's, or Homberg's expenses. Some may have been charged to the Library (table 12c–d).

ᵃ Based on BN MSS. n. a. fr. 5147 and 5149. For 1678 and 1681, when the sources disagree, I have followed BN MS. n. a. fr. 5149, which records the actual *quittances* submitted by Bourdelin. The *CdB* records only 7,032 lv. 4 s. paid to Bourdelin and his suppliers under Colbert (1: 231, 271, 503–504, 647, 781), and 3,341 lv. 18 s. 6 d. paid under Louvois (2: 378, 779, 1009, 1204; 3: 438, 584). AN G⁷ 894–95, 987–99, and 901–903 show reimbursements totaling 678 lv. 17 s.; see Stroup, *Royal Funding,* table 5.

ᵇ *CdB,* 3: 120, 305.

ᶜ From AN G⁷; see Stroup, *Royal Funding,* table 5, for details.

TABLE 8

The artists were Bosse, Chastillon, Le Clerc, and Robert.

Some entries in *CdB* for 1688 and 1689 combine payments for the Academy's engravings and other engravings. I have calculated the Academy's share by dividing proportionally or by applying the normal price scale. Chastillon, for example, earned 300 lv. for engraving a conquered city, 90 lv. for engraving a plant, and 88 lv. for revising a plate of animals. Unidentified plates by Chastillon that cost 90 lv. in 1675, 1676, and 1677 are assumed to be plants.

Lister heard that the plates for Tournefort's *Élémens* cost the crown 12,000 lv.; since no confirmation of that figure has been found in any of the sources searched to date, it is omitted here.

TABLE 9

The artists were Jean Patigny and Gilles Jodelet, sieur de La Boissière. Patigny illustrated planets in 1671 and 1672, constellations in 1677. The payment of 1,080 lv. in 1678 for the map of the moon was also for illustrations of the satellites of Jupiter. Le Clerc drew the "mer des humeurs" and the phases of an eclipse of the moon on 14 and 15 Sept. 1671 (Wolf, *Observatoire,* 168), but payment for this does not appear in the *CdB.*

TABLE 10

ᵃ This map, engraved by [De] La Pointe, required nine plates. The final payment is described as "parfait payement de 1300 livres"; that total seems to be mistaken.

ᵇ The engravers for Perrault's Vitruvius were Gérard Edelinck, Gantrel, Grignon, Le Clerc, Le Pautre, Patigny, Pitau, Scotin, Tournier, and Van der Banc; Papillon prepared wood cuts. The cost of Le Clerc's work for this book is not included because his engravings are not identified as such in the *CdB.*

ᶜ The engravers were Le Clerc (1675, 1676, 1680) and Chastillon (1678). Chastillon's portion of 892 lv. paid in 1693 to him, Homberg, and Colson (table 11a) also fits this category.

ᵈ The artists were Chastillon and Simonneau.

Notes to Appendix

e Engraved by Chastillon.
f See Stroup, *Royal Funding,* 146 n. g.
g This total excludes the costs of the engravings for Tournefort's *Élémens de botanique;* see table 8, n. a.

TABLE 11

a Although reimbursements to Couplet total 826 lv. 6 s., the reference in *CdB* to "parfait payement de 321 lv. 7 s. 6 d." implies that he received 926 lv. 6 s.: Wolf, *Observatoire,* 42.
b Spent in preparation for the king's visit.

TABLE 12

a A payment of 952 lv. (*CdB,* 1: 218) in 1667 to Robert may also belong here.
b Lists payments for work on several royal establishments including the Academy, Observatory, Bibliothèque du roi, or, in certain contexts, the Jardin royal.
c For books, flowers, and animals purchased in the Levant.
d Paid to Thévenot for books, maps, etc.
e For printing plates for the *Histoire des animaux* and for an inventory of books involving Chazelles, Pothenot, and Gallois.
f Includes the cost of two copies of a catalogue of the garden in Leiden.
g Includes payment for printing part of the Academy's "Voyages et Observations astronomiques."

TABLE 16

Because Pontchartrain treated the pensions of several academicians as *rentes,* the actual sums he paid are unknown, although the amount budgeted was a matter of record. The figures reported here for pensions, therefore, represent the total amount budgeted.

TABLE 17

Because Pontchartrain treated the pensions of several academicians as *rentes,* the actual sums he paid are unknown, although the amount budgeted was a matter of record. The figures reported here for pensions under Pontchartrain, therefore, represent the total amount budgeted.

Notes

1: PORTRAIT OF AN INSTITUTION

1. Watson, "Early Days," 557–61; see also Bénézit, *Dictionnaire,* "Le Clerc."
2. Watson, "Early Days," 566–70; Méjanès, "Le Cabinet du roi et la Collection des planches gravées"; Porcher, "La création du Cabinet des Planches gravées." Le Clerc also worked with Cassini on astronomical drawings in September 1671: Wolf, *Observatoire,* 168.
3. Wolf, *Observatoire;* Cassini, *Anecdotes;* Stroup, "Louis XIV." On propaganda during the reign, see Klaits, *Printed Propaganda.*
4. See, for example, the standards of admission to Rohault's learned society: Clair, *Rohault,* 26–27.

2: MEMBERS AND PROTECTORS

1. For an eighteenth-century summary of the secretary's responsibilities, see BA MS. 7464: 48–52.
2. Bourdelin referred to Gallois as Colbert's representative: BN MS. n. a. fr. 5147: 108v (20 Nov. 1682).
3. AdS, Reg., 9: 5r, 93v–95r, 100v, 101v–3r; 11: 162v (27 Feb. 1686); Huygens, *Oeuvres,* 9: 250–52, 260–61, 400, 421 (2 Dec. 1679, 18 Jan., 31 May 1681); Saunders, *Decline and Reform,* 120–23.
4. *Histoire . . . 1722,* 128; AdS, dossier "Pierre Couplet"; Wolf, *Observatoire,* 42–44, 94–95; Ch. Perrault, *Mémoires,* 47; Cassini, *Anecdotes,* 290, 304, 309. Other academicians, including Bourdelin, Carcavi, Du Hamel, and Thévenot, shared the responsibility; on Bourdelin, see BN MS. n. a. fr. 5147: 101r (3 Oct. 1681).
5. During the 1660s and 1670s, meetings lasted from 3 to 7 or 8 o'clock:

Oldenburg, *Correspondence,* 6: 143. During the 1690s, they were shorter, starting at 2 and ending at 5 o'clock: BN MS. Clairambault 566: 252v.

6. Biographical information about academicians in aggregate is based on *IB, NBU,* Nicéron, *Hommes illustres, DSB,* and the eulogies by Fontenelle and Condorcet; other sources are indicated in the notes.

7. Cipolla, *Literacy,* 61. Recruitment for the Academy resembled that for the Maurist order, whose monks came mainly from dioceses north of the Loire river: Ultee, *Abbey,* 46.

8. Very few academicians came from outside France. Only seven of sixty-two, or 11 percent, were foreigners, whereas during the eighteenth century foreigners accounted for 36 percent of the members. In the seventeenth century, however, foreign academicians had a better chance of receiving pensions; their eighteenth-century counterparts were usually corresponding members: McClellan, "Académie," 554.

9. See, for example, the biography of Gilles Ménage, in his *Dictionnaire étymologique.* Some of the academicians born in Paris came from families that had only recently moved to the capital. Claude Perrault's lawyer father, for example, came from Tours: Condorcet, *Oeuvres,* 2: 43.

10. Mousnier, *Paris au XVIIe siècle;* Martin, *Livre;* Fisher, "Development of London."

11. McClellan, "Académie," 554–55; in the eighteenth century most academicians from the provinces were corresponding members.

12. See Condorcet, *Oeuvres,* 2, and Fontenelle, *Éloges,* for biographies of academicians; Brockliss, *French Higher Education;* Huppert, *Bourgeois Gentilshommes;* Martin, *Livre;* Mousnier, *Institutions,* 1: 112–210, 236–74.

13. Lister, *Journey,* 97; Hirschfield, *Académie,* chap. 2; Whitmore, *The Order of Minims,* 190, 240. On Colbert's hostility to the Jesuits, see Dainville, *Géographie des humanistes,* 434; Colbert, *Lettres,* 5: 513–14. Duclos, however, was a Paracelsian.

14. Condorcet, *Oeuvres,* 2: 1–4, 5–15, 33–39.

15. Huygens, *Oeuvres,* 9: 378, 264, 204, 129.

16. Information about academicians is taken from *CdB, Histoire, Historia, IB, DSB,* and eulogies by Fontenelle and Condorcet. Other sources are indicated in the notes on individuals.

17. On Perrault, see Ch. Perrault, *Mémoires,* and *Hommes illustres,* 1: 67–68; Nicéron, *Hommes illustres,* 33: 258–68; Condorcet, *Oeuvres,* 2: 43–54; Hallays, *Les Perrault;* and Cl. Perrault's own *Voyage à Bordeaux.*

18. On Mariotte, see Picolet, "Sur la biographie de Mariotte," 245–76, in *Mariotte, savant et philosophe,* who corrects previous biographies; Dorveaux, "L'autopsie"; Clair, *Rohault,* 59; Costabel, "Paradoxe"; Pelseneer, "Petite contribution"; Rochot, "Roberval, Mariotte"; Solovine, "Tricentenaire"; Brunet, "Méthodologie"; Condorcet, *Oeuvres,* 2: 23–33; Papillon, *Bibliothèque,* 2: 24–25; Mariotte, *Oeuvres;* Leibniz, *Lettres,* 40, 42. Although Mariotte is traditionally listed among the earliest academicians (*Mémoires,* 10: 307; Gauja, "L'Académie," 299; *IB*), the minutes, the financial record, and the earliest history of the Academy make it clear that he joined the Academy well after January 1667. He did not hear Perrault read his "projet pour la botanique" that month, he first

received a pension in 1668, and Du Hamel himself observed that Mariotte joined the Company late. Watson, "Early Days," 568, identifies Mariotte as the academician wearing a calotte and arguing with Perrault and Pecquet in plate 3.

19. On Duclos, see Todériciu, "Sur la vraie biographie de Samuel Duclos"; *Nouvelles de la république des lettres* (1685); Maindron, *L'Académie,* 4.

20. On Dodart, see Clair, *Rohault,* 59–60; Bourdelot, *Conversations;* Racine, *Oeuvres,* 6: 586–87; Saint-Simon, *Mémoires,* 15: 319, cited in Racine, *Oeuvres,* 6: 562–63; AdS, dossier "Denis Dodart"; AdS, Reg., 8: 2r–7v, 18: 144r–v; Stroup, *Royal Funding,* 25–26; Leibniz, *Oeuvres,* 1: 217, 239, 244, 259. Some of Dodart's notes for the natural history of plants survive in BMHN MSS. 447–51.

21. On Bourdelin, see Dorveaux, "Grands pharmaciens. 1. Bourdelin"; Bedel, "Conceptions"; Éloy, *Dictionnaire de la médecine,* 1: 433; Nicéron, *Hommes illustres,* 7: 98–101, 101–12; Oldenburg, *Correspondence,* 6: 143. The chemist's manuscripts survive in AdS, Cartons 1666–1793, 1–3, and BN MSS. n. a. fr. 5133–49; for Tournefort's account of his technique, see BMHN MS. 259; AdS, Reg., records contributions made by Bourdelin at nearly every seventeenth-century meeting. Mariotte and Homberg used Bourdelin's analyses differently: Stroup, "Wilhelm Homberg."

22. On Nicolas and Jean Marchant, see *BU,* 26: 486; Roger, *Sciences de la vie,* 169 n. 35; Bréchot et al., "Note bibliographique"; Laissus and Monseigny, *"Les Plantes du Roi";* Paul, *Science and Immortality,* 129, 176 n. 14; AdS, Reg., 18: 144v (28 Feb. 1699). Manuscripts at BMHN include the inventory of their libraries (MS. 2253) and their notes for the natural history of plants (MSS. 447–51). *JdS,* 1: 606–7, reviewed Jean Marchant's *De Febre Purpurata.*

23. On Homberg, see Stroup, "Wilhelm Homberg," and *Royal Funding,* 15 n. 6; Cap, *Études biographiques,* 2: 214–32; Nicéron, *Hommes illustres,* 14: 151–67; AdS, Reg., 12: 60r–v (7 May 1687); Leibniz, *Lettres,* 98; Régis, *Cours,* 1: 642.

24. On Tournefort, see Nicéron, *Hommes illustres,* 4: 354–71, 10A: 154–55, 10B: 146–47; *Tournefort;* Clair, *Rohault,* 60; Callot, "Système"; Leroy, "Tournefort"; Laissus and Laissus, "Tournefort et ses portraits"; Laissus and Monseigny, *"Les Plantes du Roi";* Lister, *Journey,* 62; and BMHN MS. 76, on his demonstrations of plants at the Jardin royal. Many of his manuscripts survive at the BMHN.

25. On Borelly, see Chabbert, "Jacques Borelly," and "Pierre Borel"; Clair, *Rohault,* 59; and Furetière, *Recueil des factums,* 2: 174. Although Borelly became an academician in 1674, he worked in the laboratory in 1672: BN MS. n. a. fr. 5134: 272–73.

26. On Charas, see Salomon-Bayet, "Opiologia"; Cap, *Études biographiques,* 1: 117–29; Dorveaux, "Grands pharmaciens. 2. Charas"; Nicolas, *Histoire littéraire de Nîmes,* 1: 357–61; Bouvet, "Apothicaires royaux"; *JdS,* 2: 475–79; Denis, *Recueil,* bound with *JdS,* 3: 74, 81–90.

27. On Du Verney, see Nicéron, *Hommes illustres,* 25: 350–57; Schiller, "Laboratoires"; Lister, *Journey,* 65, 69 (italics removed); Clair, *Rohault,* 60.

28. On Tauvry, see Hahn, *Anatomy,* 34; Roger, *Sciences de la vie,* 171–73; AdS, Reg., 17: 188r, 220r–v, 341r–45r; 18: 1r–4v, 45v–52v, 141v, 198v–99r,

274r (30 Apr., 24 May, 6 Aug., 12 Nov., 10 Dec. 1698, 28 Feb., 1 Apr., 6 May 1699).

29. AdS, Reg., 14: 204r (7 Dec. 1695); Lister, *Journey,* 79–80; Stroup, *Royal Funding,* 49, 127.

30. On Huygens, see his *Oeuvres;* Bos et al., *Studies on Christiaan Huygens;* Lister, *Journey,* 112; Nicéron, *Hommes illustres,* 19: 214–31; Stroup, "Christiaan Huygens"; Busson, *Religion des classiques,* 85–87.

31. On La Hire *père,* see Nicéron, *Hommes illustres,* 5: 335–46, 10B: 160–81; Huygens, *Oeuvres,* vols. 8, 9, and 10, contain correspondence of La Hire and Huygens and references to La Hire in other letters; Wolf, *Observatoire;* La Hire, "Description d'un tronc de palmier petrifié."

32. On La Hire *fils,* there is little information; Fontenelle wrote no eulogy but referred in passing to him: *Éloges,* 276.

33. On Sédileau, see Wolf, *Observatoire;* Condorcet, *Oeuvres,* 2: 90–91; Huygens, *Oeuvres,* 9: 378 n. 17.

34. On Cassini, see his *Anecdotes,* 257–62; Ch. Perrault, *Mémoires;* Nicéron, *Hommes illustres,* 7: 287–322, 10B: 236–48; *BU,* 133–36; Wolf, *Observatoire,* 6–7, 23, 71–72, 78–80, 151, 161, 206, 279; Stroup, *Royal Funding,* documents VII and VIII.

35. On Gallois, see Nicéron, *Hommes illustres,* 8: 153–60; Clair, *Rohault,* 59; Brunot, *Histoire,* 4, 1: 23; Sgard et al., *Dictionnaire; NBU; DBF;* AdS dossier "Jean Gallois"; Ch. Perrault, *Mémoires,* 45–46; AdS, Reg., 10: 80r-v, 111v; *Mémoires,* 10: 130–38 (31 July 1692); Neveu, "Vie," 464–65; Leibniz, *Lettres,* 84, 93, 96.

36. On Du Hamel, see *NBU; DBF;* Nicéron, *Hommes illustres,* 1: 265–74, 10A: 46–47, says that the additions to the 1701 edition of *Historia* were translated from Fontenelle's history as published in the annual *Histoire et mémoires.*

37. On Fontenelle, see Marsak, *Fontenelle;* Paul, *Science and Immortality; NBU; DBF;* Leibniz, *Lettres,* 195–235; AdS, Reg., 15: 253r (9 Jan. 1697); Musée du Conservatoire, *Histoire et prestige,* 59, oversimplifies the relationship between Du Hamel's and Fontenelle's histories.

38. On La Chapelle, see Stroup, *Royal Funding,* 103n. e; Huygens, *Oeuvres,* 8: 479n. 1; vol. 9 contains correspondence between La Chapelle and Huygens; correspondence in other volumes contains references to La Chapelle.

39. Stroup, *Royal Funding,* chaps. 5–6; he kept abreast of research and attendance by reviewing the minutes, which bear his signature.

40. On Bignon, see Saisselin, *Literary Enterprise,* 86; Clarke, "Bignon," 217, 222; BN MS. fr. 22225; BN MS. Clairambault 566: 186–94; Leibniz, *Oeuvres,* 1: 289, 308.

41. Saunders, *Decline and Reform,* 116–18, 124–27.

42. On Colbert, see Cole, *Colbert and Mercantilism;* Collas, *Chapelain;* Ministère de la Culture, *Colbert, 1619–1683;* Saunders, *Decline and Reform,* 30; Ch. Perrault, *Hommes illustres,* 1: 37–39, and *Mémoires de ma vie;* Colbert, *Lettres;* Lister, *Journey,* 128–29, describes Colbert's library; Bertrand, *L'Académie et les académiciens,* 39–40; Wolf, *Observatoire,* 115. On Du Hamel's book, his *Philosophia vetus et nova,* see Busson, *Religion des classiques,* 72, 75–76. Du Hamel included revision of the book in his annual reports of the Academy's activities: AdS, Reg., 10: 45v–46r, 73v–74r; for the third edition, he sought a contribution

by Huygens on the magnet: Huygens, *Oeuvres*, 8: 479. For a retrospective appreciation in 1691 of Colbert's preference for natural philosophy and mathematics over theology or history, see BN Archives de l'ancien régime 53: 10r.

43. On Louvois, see *Histoire*, 1: 386–87, 2: 132–33; Saunders, *Decline and Reform*, 26–130; Bertrand, *L'Académie et les académiciens*, 41–44.

44. On Pontchartrain, see Berger, "French Administration," "Pontchartrain and the Grain Trade," and "Rural Charity"; Frostin, "L'organisation," and "La famille ministérielle"; Saunders, *Decline and Reform*, 132–52; Stroup, *Royal Funding*; Leibniz, *Oeuvres*, 1: 289; Saint-Simon, *Mémoires*, 1: 52n. 2, and 6: 268–91; and Bignon's papers at the BN.

45. On Louis XIV, see Lavisse, *Louis XIV*; Goubert, *Louis XIV and Twenty Million Frenchmen*; Cassini, *Anecdotes*, 3, 94, 117–19, 289, 291, 292; Wolf, *Observatoire*, 118–20; Frostin, "L'organisation." Lister, *Journey*, 214, reported that the king "pleases himself in Planting and Pruning the Trees with his own Hand."

3: MODELS FOR A COMPANY OF SCIENTISTS

1. Martin, *Livre*, 965; Brunot, *Histoire*, 6, 1: 406–17.
2. Martin, *Livre*, 652–53, 964; Barber, *Bourgeoisie*; Fontenelle, *Éloges*, 44.
3. *Histoire... 1699*, 3–11; *JdS*, 2: 387–92; AdS, Reg., 4: 99r.
4. Goubert, *The Ancien Régime*, 215; Auvray, "Affiches," 203; Cipolla, "The Professions." For definitions of *académie* and *compagnie*, see Berthelin, *Abrégé du Dictionnaire de Trévoux*; Académie française, *Dictionnaire*; Furetière, *Dictionnaire*; Bourdelot, *Conversations*, 12–13, 14–21; Dubois and Lagane, *Dictionnaire de la langue française classique*; Le Roy de la Corbinaye, *Traité de l'orthographie*; Quemada, *Matériaux*; Huguet, *Petit glossaire*; Godefroy, *Dictionnaire de l'ancienne langue française*; Brunot, *Histoire*; Moréri, *Dictionnaire*; Sommer, *Lexique de Madame de Sévigné*, 13: 173–74; Livet, *Lexique de Molière*, 1: 438; compare OED.
5. Huygens, *Oeuvres*, 5: 41; McKeon, "Lettre"; Sealy, *Palace Academy*; George, "Seventeenth Century Amateur," and "Genesis"; Fauré-Fremiet, "Les origines"; Evans, "Learned Societies"; Gauja, "Les origines"; Adams, "Social Responsibilities"; Yates, *Academies*, chap. 12; Hahn, *Anatomy*, 4–7; Bigourdan, "Les premières réunions savantes"; Brown, *Scientific Organizations*; Ornstein, *Rôle of Scientific Societies*, 139–45; Taton, *Origines*.
6. Ranum, *Artisans of Glory*, 19; Ch. Perrault, *Mémoires*, 44–46; *Histoire ...1720*, 123; Paul, *Science and Immortality*. On false modesty, see La Bruyère, *Caractères*, 8: 42, 44.
7. On Descartes and his influence, see Descartes, *Discours*, pt. 6; Salomon-Bayet, *L'institution de la science*, 101; Dubarle, "The Proper Place of Science," 405–7; Paul, *Science and Immortality*. Compare Yates, *Academies*, 307–8, who interprets Fontenelle's *Entretiens* within a sixteenth-century tradition. On Bacon and his influence, see AdS, Reg., 1: 22–38; Bacon, *Works*, 3: 156–57; Huygens, *Oeuvres*, 19: 268–70; see also his correspondence during the 1660s; Hahn, *Anatomy*, 11, 15; Purver, *The Royal Society*, pt. 1; Zilsel, "The Sociological Roots

of Science," 558; Olmsted, "Scientific Expedition," 127; Juillard, "Société Royale," 80–82. On medical utility, see Nicéron, *Hommes illustres*, 7: 102.

8. Ultee, *Abbey*, 21–37.

9. King, *Science and Rationalism;* Nef, *Industry and Government*, gives background; Huygens, *Oeuvres*, 6: 212; Ch. Perrault, *Mémoires*, 45–47; Martin, *Livre*, 860, 923; Delorme, "Correspondance de Chapelain."

10. Colbert, *Lettres*, 5: 513.

11. Cole, *Colbert and Mercantilism*, chap. 6; Couton, "Effort publicitaire"; Collas, *Chapelain*, 360; Colbert, *Lettres*, 5: 499, 514–15.

12. Cole, *Colbert and Mercantilism*, 1: 349; see also 20–24, 346; Coleman, *Revisions in Mercantilism*, 71, 196–97.

13. Wilson, "Trade, Society and the State," 530; Heckscher, *Mercantilism*, 1: 24; Sée, *Esquisse*, 276; Aston, ed., *Crisis;* Coleman, ed., *Revisions in Mercantilism*.

14. Colbert, *Lettres*, 5: 514; Roger, *Sciences de la vie*, 173.

15. Colbert, *Lettres*, 5: 559.

16. Collas, *Chapelain*, 386–87, 375. See also Colbert, *Lettres*, 5: 513, 535–36, 550–51; Cole, *Colbert and Mercantilism*, 1: 315–17.

17. Collas, *Chapelain*, 355, 390–93.

18. Colbert, *Lettres*, 5: 593–94.

19. Ibid., 600–601, 603, 609.

20. Ibid., 513, 514, 515; Collas, *Chapelain*, 369–75, 383–84, 387; Cole, *Colbert and Mercantilism*, 1: 24, 294–95; Huygens, *Oeuvres*, 7: 361, 8: 112, 196–99; *Histoire... 1699*, 10–11.

21. Sée, *Esquisse*, chaps. 6–8; Grassby, "Social Status"; Thirsk, *Economic Policy and Projects;* Braudel and Labrousse, *Histoire économique*, 2: 356–58; Cole, *Colbert and Mercantilism*, 1: 315, 454, 477, 2: 142–43, 147–48, 171, 180; Schaeper, *Economy of France*, 25–54.

22. Hahn, *Anatomy*, 18–19; Caullery, "La biologie au XVIIe siècle," 33; Roger, *Sciences de la vie*, 173.

23. Colbert, *Lettres*, 5: 403–4, 407–8, 514; *Histoire*, 1: 361–66; Huygens, *Oeuvres*, 4: 513–16; Stimson, *Scientists and Amateurs*, 16; Brown, *Scientific Organizations*, 76; Dubarle, "The Proper Place of Science"; Wolf, *Observatoire*, 3; Olmsted, "Scientific Expedition," 119.

24. Colbert, *Lettres*, 5: 293, 314, 315–16, 334, 336, 351, 404n., 421–22; Sée, *Esquisse*, 289; Cole, *Colbert and Mercantilism*, 2: 124–25, 134. On plans to launch a scientific expedition when overseas headquarters for the East India Company were established, see Olmsted, "Scientific Expedition," 119.

25. Oldenburg, *Correspondence*, 6: 502, 507; Saunders, *Decline and Reform*, 105–18.

26. Grassby, "Social Status," 19–22.

27. In its formative period, therefore, the institution combined functions performed in the modern era by research laboratories, universities, learned societies, scholarly publishers, and scientific consultants: Hagstrom, *Scientific Community*, 23, 26–27, 48; Pyenson, "'Who the Guys Were,'" 169. Academicians also refereed books and articles before publication, taking this function so

seriously that an academician might blame his colleagues for insufficient vigilance when he published inaccurate claims: Huygens, *Oeuvres,* 7: 253, 255.

4: THE MATERIAL BENEFITS OF MEMBERSHIP: PENSIONS AND QUARTERS

1. Chapelain, *Lettres,* 2: 348, 406; Collas, *Chapelain,* 383 n. 4, and 386 n. 2.

2. Stroup, *Royal Funding,* 61–63.

3. In addition to pensions, academicians received *jetons: CdB,* 1: 780, 1203, 1367. Information is so sketchy that they have been omitted from calculations of expenditure on the Academy. Memoranda from the end of the seventeenth century and beginning of the eighteenth discuss the history of *jetons* and propose ways of using them as incentives for the Académie française: BN MS. Clairambault 566: 191r, 193r–v, 198r–200v, 202r. In the eighteenth century, when up to forty *jetons* could be distributed at each meeting of the Academy of Sciences, their status was summarized on the *estats* for pensions: BN MS. fr. 22225: 35r.

4. Huygens, *Oeuvres,* 7: 87–88 n. 1; 8: 456–57; 18: 4 n. 5. Pensions were not secret, but many rumors about them were inaccurate.

5. Stroup, *Royal Funding,* chaps. 2–3.

6. Antoine Niquet did so: Blanchard, *Dictionnaire,* 561–62, and *Ingénieurs,* 64 n. 146, 68, 99–100, 126n. 43, 160, 213, 302, 340, 389, 453, 458; Colbert, *Lettres,* 5: 140, 155–56, 161–62, 167, 181, 190–91, 214–15, 230–31; AN G[7] 898 (Jan. 1697).

7. BN MS. Clairambault 814: 633–34.

8. Stroup, *Royal Funding,* chap. 2.

9. Bertrand, "Les Académies d'autrefois" (1867): 752, and Stroup, *Royal Funding,* chaps. 2–3.

10. Stroup, *Royal Funding,* 28–30, 71, and app. B and C.

11. BN Archives de l'ancien régime 53: 10v; *Histoire,* 1: 13; BMHN MS. 1278: 1v; Bertrand, "Les Académies d'autrefois" (1867): 757; Hahn, "Scientific Careers," and "Scientific Research."

12. Of sixty-two members, fifty-seven attended meetings; the others were honorary, associate, or corresponding members who lived in the provinces or abroad. For a list of the sixty-two members before the reorganization of 1699, see Stroup, *Royal Funding,* fig. 2.1. Some persons who became members in 1699 had already attended meetings or corresponded with the Academy. On de Beauchamp, see Bigourdan, "Observatoires de la région provençale," 257; on Renau, see BN MS. Clairambault 566: 251v, quoted in Saunders, *Decline and Reform,* 258. Those who did not attend were Leibniz, Tschirnhaus, Chazelles, Langlade, and Guglielmini. The Academy differed from the Accademia del Cimento, which drew on a small region for its members, and from the Royal Society, which had far more members, most of whom "never played more than a nominal part in the activities of the Society": Webster, *The Great Instauration,* 89; Middleton, *Experimenters;* and Hoppen, "The Nature of the Early Royal Society."

13. The English were especially struck by the small size of the Academy. Francis Vernon wrote to Oldenburg that Cassini had told him "that the Royall

Academie are not as ours in Engld a great assembly of Gentlemen, Butt only a few Persons wch are eminent, & not in number above 13, or 14": Oldenburg, *Correspondence,* 5: 507 (11 May 1669).

14. The Academy also dissected the painter Le Brun in 1690: *Histoire,* 2: 92.

15. On shared expenses, see table 12; Schiller, "Laboratoires," 105–6, 110, 113–14; AN O^1 2124.

16. Méjanès, "Le Cabinet du Roi et la Collection des planches gravées"; Porcher, "La création du Cabinet des Planches gravées"; payments to Nicolas Clément and Goitton in *CdB* and BN Archives de l'ancien régime 1 and 2; BN Archives de l'ancien régime 53: 32r–41r, 51r–58v, 72r–66r, 86r; Ranum, "Islands and the South in a Ludovician Fête."

17. Laissus and Monseigny, *"Les Plantes du Roi,"* 204, 206–10, 216–17; *Tournefort,* 211n. 2; Schiller, "Laboratoires," 110; Muséum National d'Histoire Naturelle, *Exposition;* AdS, Reg., 9: 166v; 10: 44r–v; 11: 114v, 118v, 129r–v, 150v; 18: 125r–v; BMHN MSS. 1556–62, and MS. 89, dossier 2; BN MS. n. a. fr. 5147: 64r, 102r. On the early history of the Jardin royal, see Howard, "Medical politics," "Guy de La Brosse," and *La bibliothèque et le laboratoire.*

18. BN MS. Clairambault 814: 632–33; AN O^1 1678^A: items 4 and 19; Brice, *Description,* 1: 73–84; Blegny, *Livre commode,* 1: 122n. 2; Huygens, *Oeuvres,* 17: 498n. 3; 21: 7–8; Wolf, *Observatoire,* 1–2; Hillairet, *Dictionnaire des rues,* 2: 654–57. When Lister visited the Library in 1698, he counted twenty-two rooms: *Journey,* 108. Only a garden separated Colbert's *hôtel* and the Library.

19. On holdings of the library pertinent to the Academy's work, see: table 12g; BN MS. Clairambault 566: 252r; BN Archives de l'ancien régime 53: 22r, 26r–28r, 47r, 48v, and passim; BMHN MS. 450: 127r; Brice, *Description,* 1: 76; Lister, *Journey,* 109–11. Simone Balayé points out that eighteenth-century *encyclopédistes* were impressed by the Library's botanical works. Some academicians collected books for the Library: Franklin, *Anciennes bibliothèques,* 2: 179–80, 181; Delisle, *Cabinet des manuscrits,* 1: 278.

20. BN MS. Clairambault 566: 252r; BN Archives de l'ancien régime 1: 14r.

21. Wolf, *Observatoire,* 2, 136; *Histoire,* 1: 8, 66–67, 109; Huygens, *Oeuvres,* 22: 626; BN MS. n. a. fr. 5147: 15v, 106v, 111v (including payments for weeding the courtyard facing "la sale de lassemblée"); BN MS. Archives de l'ancien régime 1 and 2.

22. BN MS. Clairambault 566: 252v; Ch. Perrault, *Mémoires,* 49; Lister, *Journey,* 108; *Histoire,* 1: 319; *Histoire . . . 1699,* 14, 16; Oldenburg, *Correspondence,* 6: 401, 402; Maindron, *Académie,* 5. A *tapissier* repaired the door in 1685 for 1 lv. 1 s.: BN Archives de l'ancien régime 1: 21r.

23. Bacon took as his model for the House of Solomon a chemical laboratory, emphasizing procedure over theory: Salomon-Bayet, *L'institution de la science,* 262. On laboratories, see Howard, *La bibliothèque et le laboratoire;* Pelcher, "Boyle's Laboratory"; Eklund, *Incompleat Chymist;* Shapin, "House of Experiment."

24. In June 1670 Du Hamel wrote, "the laboratory is finished and they are hard at work": Oldenburg, *Correspondence,* 7: 33, 34. The best information about the size and contents of the laboratory is in Bourdelin's notebook of expenditure, BN MS. n. a. fr. 5147, and his inventory dated November 1688 of

Notes to pp. 39–43   293

the laboratory, BN MS. n. a. fr. 5149: 21r–34r. The latter describes an upstairs laboratory with three furnaces, a downstairs laboratory with seven furnaces, and a kind of pantry equipped with armoires and tables for storing supplies and equipment; see also BN MS. n. a. fr. 5134: 147, 181, 257 (1672). Bourdelin sometimes had to interrupt a distillation to allow Borelly to work: ibid., 272–73 (Sept. 1672). See also Ch. Perrault, *Mémoires,* 49, 54; Stubbs, "Chemistry at l'Académie"; Schiller, "Laboratoires."

25. Some of the equipment was expensive: a covered alembic made of red copper cost nearly 80 lv. and a large iron press 100 lv.; the instrument-maker Hubin sold the Academy dozens of aerometers at 2 lv. apiece and repaired several for 15 s.; Masselin, a royal master-coppermaker, earned 8 lv. by repairing and adding a red copper base to a round *bain de vapeur,* or steam bath; charcoal cost 500 to 650 lv. a year from 1672 to 1682: BN MS. n. a. fr. 5147: 1r, 60r–109r, 110r, 112r. Bourdelin recorded who broke what and kept the laboratory locked: ibid., 15v–16r, 66r, 102v, 108r. In November 1671, he outfitted Jean Richer with the medicines Duclos recommended for the forthcoming trip to Cayenne: ibid., 52v. In the eighteenth century, Sébastien Truchet bought from academician-chemist Geoffroy the drugs he needed for his trip to the Auvergne: AN M 851.

26. The taxidermist Colson sometimes assisted during dissections; he and the surgeon La Beurthe mounted the remains for display: *CdB,* 1: 270; 2: 536; AdS, Reg., 10: 108r (June 1681–July 1682). Additional payments, said to be for Versailles and the Jardin royal, may well have been for the Academy: *CdB,* 1: 631, 889, 947, 975, 1110, 1184, 1321 (1672–80).

27. See references to BN Archives de l'ancien régime 1 and 2 in table 6. Many of the expenses of the Academy's dissections were charged to the Jardin royal or the Bibliothèque du roi: Schiller, "Laboratoires," 103–5.

28. Lister, *Journey,* 65–66; cf. Huygens, *Oeuvres,* 6: 104; Schiller, "Laboratoires."

29. BN MS. n. a. fr. 5147: 12r; cf. 21v: "le 27e may [1669] lon a travaillé a un ours ou lon a mis dans ses entrailles a cause de sa tres grande puanteur 3 pintes deau de vie... plus lon a mis sur des mouchoirs de plusieurs de lassemblée bien trois onces de tres pur esprit de vin...." It cost ten sous a pint, and the anatomists submitted formal requests to Bourdelin for it: BN MS. n. a. fr. 5149: 35r–49v: "Je supplie tres humblement Monsieur Bourdelin de donner au present porteur six pintes deau de vie pour l'utilité des dissections anatomiques de l'Academie Royalle des Sciances, c'est de la part de son tres obeissant serviteur Mery." "Je vous prie, Monsieur, de m'envoyer six pintes d'Eau de vie, c'est pour renouveller les parties que ie conserve qui sont a sec. Vous obligerez sensiblement votre tres humble et tres obeist. servit. Du Verney."

30. Éloy, *Dictionnaire de la médecine,* 3: 508.

31. Huygens, *Oeuvres,* 6: 104; 7: 211; 10: 727; 22: 628; 19: 88; Lister, *Journey,* 108–9, 112; *CdB,* 1: 552.

2. Huygens, *Oeuvres,* 6: 91. This was before Louis XIV cracked down on gambling: Riley, "Police and the Search for *Bon Ordre,*" and "Louis XIV: Watchdog."

33. Huygens, *Oeuvres,* 7: 80, 84, 86, 100–101, 107–8, 113, 170, 172, 359 (July 1671–Oct. 1673). The two academicians were reconciled by the late

1670s, when Huygens experimented with Carcavi's magnet: AdS, Reg., 10: 41r–v. Carcavi helped get Huygens appointed to the Academy: Schiller, "Laboratoires," 102–3.

34. Todériciu, "Sur la vraie biographie de Samuel Duclos," 66.

35. See plate 5b; Wolf, *Observatoire;* Hirschfield, *Académie,* chap. 4; Ch. Perrault, *Mémoires,* 50–51; Colbert, *Lettres,* 5: 515; Oldenburg, *Correspondence,* 6: 147–49. In 1669, Cassini rented a house in nearby Ville-l'Évêque and Couplet moved next door to help him with observations; Cassini also observed from the gardens of Saint Martin des Champs (now the Conservatoire des arts et métiers) in 1671: Cassini, *Anecdotes,* 304, and Wolf, *Observatoire,* 65. Because Cassini disliked the astronomers working together and using the same instruments, they staked out separate territories, with Roemer working from one of the towers: *CdB,* 1: 1243. From 1677 until his death in 1682, Picard probably observed from his house in the rue des Postes; until he moved into Picard's apartment at the Observatory, La Hire observed primarily near the porte Montmartre: Wolf, *Observatoire,* 100.

36. Wolf, *Observatoire,* 28–39, 74; Hillairet, *Dictionnaire des rues,* 2: 442–43; AN O^1 883: 206–7; AN O^1 1691; AN O^1 1678^A.

37. AdS, Reg., 3: 77r–78v (3 July 1668); *CdB,* 1: 503, 505, 543, 600, 642, 647, 659, 686, 712, 723, 874, 928, 990, 994, 1089, 1208, 1211; Wolf, *Observatoire,* 62, 66.

38. Several workers were injured or killed during its construction: *CdB,* 1: 387, 567–68, 651, 715–16, 889.

39. *Histoire,* 2: 23; Wolf, *Observatoire,* 103–5, 109–10.

40. Designed by Sédileau, this map of the world was ready for the visit of Louis XIV in 1682; La Faye retouched it before the visit in 1690 of James II of England; in 1698 Lister admired its "accurateness and neatness": Wolf, *Observatoire,* 62–65, 83–84; Brown, *Story,* 218–19; Lister, *Journey,* 54–55; *Histoire,* 2: 96. The map has been preserved in an engraving by J. B. Nolin: Pelletier, "Les globes de Louis XIV."

41. *CdB,* 1: 1126, 1243; Locke, *Travels,* 151 and n. 6; Cassini, *Anecdotes,* 297; Lister, *Journey,* 53–55; Wolf, *Observatoire,* 13–14, 26, 53–58, 94–97, 115–16, 129. Brice, *Description,* 2: 99–103, and Blegny, *Livre commode,* 1: 122, are mistaken about which academicians had apartments at the Observatory.

42. On the Academy's instruments, see Wolf, *Observatoire,* chaps. 10–12; Cassini, *Anecdotes,* 304–5. Gosselin and Lagny lived and worked in a house owned by the king: AN O^1 1678^A.

43. On Roemer's designs, see Huygens, *Oeuvres,* 8: 343; 9: 262, 263–64; 22: 700. Academicians showed Butterfield's planisphere to James II in 1690: *Histoire,* 2: 100. Mahoney, "Christiaan Huygens: The Measurement of Time and of Longitude at Sea."

44. Paul, *Science and Immortality,* 71–72; *Histoire . . . 1725,* 137. Gossip had it that Carcavi hoped to marry his daughter to Cassini: Ch. Perrault, *Mémoires,* 45.

5: RESEARCH SUBVENTIONS AND
MINISTERIAL CONTROL

1. For tests of clocks, see Olmsted, "Voyage of Jean Richer"; *CdB,* 2: 236.
2. *Histoire,* 1: 199; Martin, *Livre,* 669; Wolf, *Observatoire,* 96–97; Saun-

ders, *Decline and Reform,* 91–99. The assistant called Pasquin, Pasquine, Paquin, or Pasquier helped in this work (table 1a, ii).

3. Perrault, *Abrégé;* BN MS. fr. 15189: 171v (7 May 1673); Auzout later attempted a more faithful translation: Leibniz, *Briefwechsel,* 1: 594. On Blondel and Perrault, see Hirschfield, *Académie,* 87–89.

4. Roland, "Alexis-Hubert Jaillot"; Brown, *Story,* 217–24, 245–49; *Neptune françois;* Bourgeois and André, *Sources,* 1: 51, 56–57, nos. 152, 177, 181; Saunders, *Decline and Reform,* 77–84, 164–65; Stroup, *Royal Funding,* 54–55, and documents VII and VIII; *Histoire*... *1700,* 120–24; *Mémoires*... *1718,* 2: 3 (from *De la grandeur et de la figure de la terre*). Richer, Varin, Des Hayes, and Deglos supplied data from both sides of the Atlantic: AdS, Reg., 9: 229r–v (July 1682 to June 1683).

5. *CdB,* 1: 1343. See also BN MS. n. a. fr. 5147: 107r–8r; BN MS. n. a. fr. 5136: 256–57, 271–83, 360.

6. In 1676, for example, the astronomer Richer constructed canal locks in Fère: Blanchard, *Ingénieurs,* 112–13.

7. BN MS. n. a. fr. 5147; see 108r, 112r, for notary's fees. Stroup, *Royal Funding,* table 5 (1693, 1695, 1697, 1698). Some reimbursements to Bourdelin are probably included in the small expenses of the Library (table 12c–d).

8. Saunders, *Decline and Reform,* 67.

9. BN MS. Clairambault 566: 252v; *CdB,* 2: 759, 1009; Saunders, *Decline and Reform,* 159.

10. Saunders, *Decline and Reform,* 89.

11. Compare *Histoire,* 1: 386–87, 2: 132–33, and *Histoire*... *1699,* 2–3, 14.

12. *Histoire,* 1: 408–19, 432–41; 2: 32, 56–58, 71–73.

13. Three missions cooperated with the Academy. The China missions included Fontaney, Tachard, Vissedelou, Bonnet, Le Comte, and Gerbillon, who left in 1685, and Avril, Beauvollier, and Barabé, who were to enter China via Grand Tartary as part of a diplomatic mission in 1688. Fourteen Jesuits went to Siam and India, twelve of them mathematicians; these included Antoine Thomas, Richaud, de Beze, and François Noel: "Observations... de Siam," and "Observations... des Indes et de la Chine," in *Mémoires,* 7, 2: 605–875. On their work, see AdS, Reg., 11: 114v, 115r–16v (16, 20 Dec. 1684, 17, 20 Jan. 1685); *Mémoires,* 10: 130–31; Saunders, *Decline and Reform,* 187; Mungello, *Curious Land,* 255–56. Other Jesuit missionaries also reported to the Academy, which in November 1685 examined information from Goa in January: AdS, Reg., 11: 144r, 148r–50r. The Jesuits also took with them a fashionable remedy for the poor, on which they had been authorized to spend 500 lv.: AN G^7 884 (June 1685). They were pensioned at 400 lv. a year while they were abroad, and their funds were usually channeled through père Verjus: *CdB,* 2: 660, 855, 870, 913. Payments continued during the 1690s: Stroup, *Royal Funding,* 45n. 20.

14. *Mémoires,* 7, 2: 743.

15. Ibid., 605–875; AdS, Reg., 12: 104v–5r (13, 20 Nov. 1688); *Mémoires,* 4: 325–33. The Jesuits sent other descriptions of eastern animals, printed in *Mémoires,* 3, 2: 251–88, with Du Verney's reflections.

16. AdS, Reg., 9: 124r–27r (29 Nov. 1681).

17. Saunders, *Decline and Reform*, 66–77, stresses Louvois's exasperation with academicians.

18. *Histoire*, 1: 395–404, 429–31; 2: 6–8, 23–25, 44–48, 63–66, 89–93.

19. AdS, Reg., 10: 95v (15 Apr. 1682).

20. BN MS. n. a. fr. 5147: 120v. Bourdelin recorded his out-of-pocket expenses: 365 lv. for the assistant's wages, at least 40 lv. for new bedding and laundry, 150 lv. for rent of a shop and cellar in which to work, and unnamed sums for tables, cabinets, etc.

21. AdS, Reg., 11: 121v–22r, 125r–v, 128v, 130r, 136v–38r, 139r–40r (7, 10–19, 21 Mar., 19 May, 6 June, 28 July, 4, 8 Aug. 1685); *Histoire*, 1: 442–45, 448; 2: 4–5, 14–15, 42, 59–60, 63, 87, 368–69; *CdB*, 2: 780; 3: 114; BN Archives de l'ancien régime 1: 26v–27r (16 Mar. 1685). See also AdS, Reg., 4: 123v–25r (4 Aug. 1668). The translation of Frontinus was never printed: BN MS. Clairambault 566: 252v. Picard published a treatise on surveying in 1684. In his eulogy of La Hire, Fontenelle made the reimbursement for 600 lv. of expenses incurred in this work (table 4c) the subject of an anecdote about Louvois's disdain for detail: *Histoire . . . 1718*, 81. Academicians discussed hydraulic machines and treatises with James II of England when he visited the Observatory in 1690: *Histoire*, 2: 103; McKie, "James, Duke of York, F.R.S." Trying to divert the Eure to Versailles was enormously expensive and cost nearly 6 million lv. between 1685 and 1687: *CdB*, 2: 1312–13.

22. *Mémoires*, 10: 29–36, 251–52, 325–39; AdS, Carton 1667–1699, pochettes 1692, 1693, and 1694.

23. Saunders, *Decline and Reform*, 198.

24. Ibid., 91–99.

25. AN O[1] 1934[B] 14 explains that his pension was "en considération de plusieurs machines qu'il a inventées."

26. *Histoire*, 1: 422–23, 448; 2: 14–15, 22, 33–35, 39.

27. For a detailed account of the financial record of the Academy during the 1690s, see Stroup, *Royal Funding*.

28. On the effects of the war on publishing and manufacturing, see BN MS. fr. 7801: 62r, and Lister, *Journey*, 80, 138, 145, 162–63.

29. Saunders, *Decline and Reform*, chaps. 5, 6, and 7; Stroup, *Royal Funding*, chap. 5.

30. Stroup, *Royal Funding*, chap. 3, and app. A, B.

31. AN G[7] 902, paid Aug. 1698.

32. See treatises in *Mémoires*, 7, 2; *Mémoires*, 10; *Histoire*, 2: 155–63, 189–200, 218–29, 259–65, 285–92, 300–331, 340–44; Stroup, *Royal Funding*, 50–51.

33. From 1693 the Academy's share of the expenses of the *petit jardin* cannot be separated from the Jardin royal's share; academicians continued to use the Jardin royal even after Fagon removed the *petit jardin* from Marchant's control in 1694.

34. Lister, *Journey*, 82, reported the cost of Tournefort's plates. The number of plates engraved during the 1690s has been inferred from the total engraved by 1701 and the number said in memoranda from the early 1690s to be completed: BN MS. Clairambault 566: 252r, and BN MS. fr. 22225: 36r, state that 250

plates have been engraved; see also Laissus and Monseigny, *"Les Plantes du Roi";* Bréchot et al., "Note bibliographique."

35. *Histoire,* 2: 135–39, 142–47, 150–52, 175–83, 205–7, 209–18, 238–57, 278–79, 281–84, 298–99, 335–39; *Historia,* 301–8, 311, 325–31, 374–76, 380–84, 408–9, 414–17, 440–47, 453, 455, 483–85, 494–500; *Histoire... 1715,* 82; Lister, *Journey,* 79–80; Stroup, *Royal Funding,* 39–40; Saunders, *Decline and Reform,* 164.

36. AdS, Reg., 14: 73v; 15: 43r–46r; *Histoire,* 2: 133–35, 141, 164–67, 173–74, 201–2, 204, 228, 259–62, 334; Stroup, *Royal Funding,* chap. 6; Saunders, *Decline and Reform,* 197–99.

37. Salomon-Bayet, "Préambule"; Stroup, *Royal Funding,* 57–60; AdS, Reg., 18: 79r–81v (14 Jan. 1699) for Homberg's talk on "Essays pour corriger la matiere des lettres de l'imprimerie."

38. By comparison the Gobelins and Savonnerie cost the crown on average 135,000 lv. a year from 1664 through 1690: BN MS. fr. 7801: 60v–62r.

39. Oldenburg, *Correspondence,* 5: 498; Brown, *Scientific Organizations,* 159.

40. In the seventeenth century, as now, there was considerable conformity within the scientific community about the relative importance of various problems and techniques: Hagstrom, *Scientific Community,* 52.

6: THE NATURAL HISTORY OF PLANTS: RIVAL CONCEPTIONS

1. On the transformation of botany in early modern Europe, see the works by Arber, Callot, Clark-Kennedy, Crestois, Davy de Virville, Delaporte, Eriksson, Greene, Henrey, Oliver, Raven, Reeds, von Sachs, and Webster, cited in the bibliography. Dodart was asked to explain sensitivity in the Academy's description of the mimosa: BMHN MS. 451: 417v.

2. Cailleux, "Progression du nombre d'espèces de plantes," 44; Raven, *John Ray,* 254.

3. Raven, *John Ray,* 191–92.

4. Huygens, *Oeuvres,* 6: 95–96 (1666); 19: 270–71; Bertrand, *L'Académie et les académiciens,* 8–10; Bacon, *Works,* 4: 251–52, 265–70.

5. AdS, Reg., 1: 30–38.

6. The word *la botanique* appeared in Randle Cotgrave's *Dictionarie of the French and English Tongues,* published in 1611. Although John Ray may have used *botany* to refer to the general study of plants in the 1690s, the French kept the more restricted usage: OED; Tournefort, *Élémens de botanique,* 520; Furetière, *Dictionnaire;* Bloch and Wartburg, *Dictionnaire étymologique de la langue française,* 79; Dauzat, *Dictionnaire étymologique de la langue française,* 99; Littré, *Dictionnaire,* 4: 1719.

7. The duke's 18-volume "histoire naturelle des oyseaux et des plantes peints en miniature" was acquired by the Bibliothèque du roi: BN Archives de l'ancien régime 53: 2v–3r. It is now part of BMHN Rés., *Vélins;* see also Dorst and Laissus, *Nicolas Robert et les Vélins;* Robert et al., *Vélins du Muséum.*

8. AdS, Reg., 1: 30–31, 36, 37–38. See Whitmore, *The Order of Minims,* on Plumier's botanical work.

9. Justel knew of the Academy's plans to use the duke's paintings and was familiar with Perrault's recommendations: Oldenburg, *Correspondence,* 4: 256–57 (Mar. 1668); Brown, *Scientific Organizations,* 157. Comparison of the vellums (BMHN Rés., *Vélins*) with the drawings (BMHN Rés., *Recueil des plantes*) shows that Perrault's suggestions were put into practice. In several drawings of large plants the scale is indicated by a leaf portrayed in its actual size. The drawing is sometimes reversed in the engraving: figs. 2, 5, 11 (BMHN Rés., *Recueil des plantes,* 230, 257, 293).

10. AdS, Reg., 4: 48r–55r (9 June 1668).

11. Duclos used the word *l'herbe,* which rarely occurs in the writings of other academicians. This is consistent with the pattern identified by Prévost, "Sur la sémantique des mots herba et herbe," and "Sémantique de 'herbe' et 'plante.'"

12. The solutions Duclos mentioned were *vitriol de mars, sel de plomb,* and *noix de galle.* Bacon had included the "Chemical History of Vegetables" in natural history: *Works,* 4: 254, 255, 299.

13. AdS, Reg., 4: 54r: "Et parce que l'on a dessein d'escrire cette histoire en langue françoise, Il seroit bon d'estre informé des noms que Le vulgaire des Principales Provinces de France donne a chaque Plante, pour les Joindre a ceux des autres Langues." For the distinction between "nom François" and "nom vulgaire," see BMHN MS. 448: 65r; cf. 109r. Conrad Gesner recorded vernacular names: Greene, *Landmarks,* 794.

14. Bréchot et al., "Note bibliographique," 374. There was discussion, however, about publishing the natural history with a bilingual—French and Latin—text: BMHN MS. 450: 6r.

15. AdS, Reg., 4: 48v, 54r, 54v–55r. Fontenelle's remark that the natural history was to be a catalogue of plants in France misinterprets Du Hamel's otherwise similar account for 1686: *Histoire,* 2: 1–2; *Historia,* 250. Tradition and Louis XIV's foreign policy encouraged the Academy to obtain numerous Mediterranean and middle eastern specimens; this is partial background to Tournefort's trip to the Levant. Dippy, interpreter and specialist in near eastern languages, helped academicians read the relevant literature: BMHN MS. 450: 418–49.

16. For information about the editions of *Mémoires des plantes,* see Nissen, *Die botanische Buchillustration,* 2: 48–49. Robert Hooke reviewed the book in *Philosophical Collections,* 1: 39–42.

17. Dodart, *Mémoires des plantes,* 125, 130, 132–33; Chevreul, "Recherches expérimentales," 114. The queries and drafts in BMHN MSS. 448–51 may shed light on the development of French botanical vocabulary; see for example MS. 450: 117r, item ii, or 300r.

18. Dodart, *Mémoires des plantes,* 123–25, 135, 136; AdS, Reg., 4: 49r; BMHN MS. 450: 47r, 56r, 103r, item ii. Dodart used the phrases "la plante parfaite" (the mature plant) and "la plante nait" (the plant first appears); he and Perrault used "la plante naissante" (the young plant). Botanical vocabulary, therefore, drew on zoological language to speak of the "naissance" or "birth" of plants. The drawing on which fig. 5 is based lacks the young shoot and shows a longer stem on the main plant (BMHN Rés., *Recueil des plantes,* 293). Bourdelin

sometimes supplied materials required by the engravers: BN MS. n. a. fr. 5147: 16r.

19. Dodart, *Mémoires des plantes,* 241; Davy de Virville, "De l'influence des idées préconçues," 112.

20. Dodart, *Mémoires des plantes,* 138; AdS, Reg., 1: 30, 31, 33–34; BMHN MS. 449: 154r–55v.

21. Dodart, *Mémoires des plantes,* 138; cf. Bacon, *Works,* 4: 261.

22. Dodart, *Mémoires des plantes,* 129, 137–38.

23. Ray had similar ideas during the 1650s and 1660s: Raven, *John Ray,* chaps. 4 and 5.

24. AdS, Reg., 1: 32: "Jardin Academique."

25. Laissus and Monseigny, *"Les Plantes du Roi,"* 204.

26. AdS, Reg., 7: 124v (4 Sept. 1677); 10: 17r; 11: 116v–17r, 124r, 125v (20 Jan., 11, 25 Apr. 1685); Marchant, *Descriptions de quelques plantes nouvelles;* Olmsted, "Voyage of Jean Richer," n. 84. Compare Raven, *John Ray,* 215.

27. AdS, Reg., 8: 215r–v (Apr. 1678–June 1679); 10: 44v, 72r–v, 83r, 109r, 152v (June 1679–June 1683); *Histoire,* 1: 307, 374, and passim. See also Bourdelin's notebooks in AdS, Cartons 1666–1793, 1–2, and BN MSS. n. a. fr. 5133–49.

28. AdS, Reg., 11: 118v (27 Jan. 1685).

29. "L'Academie a-t-elle confronté cette description, et toutes les autres avec la nature? Car cela est de l'ordre et d'une necessité absolue": BMHN MS. 451: 139r.

30. AdS, Reg., 1: 254, 256 (11 Feb., 9, 17 Mar. 1668); 4: 16v–17v, 21r–v, 30r–v, 48r–55v (5, 12, 19 May, 2 June 1668).

31. Ibid., 4: 54r; 8: 215v (Apr. 1678–June 1679); 10: 72r–v (Aug. 1680–June 1681); 11: 116v–17r (20 Jan. 1685); *Histoire,* 1: 307.

32. Bourdelin analyzed plants at his own initiative on 4 June 1668, five days before Duclos suggested the work: BN MS. n. a. fr. 5133: 14 (from back of volume).

33. BMHN MSS. 448–51 contain examples and advice about method. Academicians were supposed to read all existing descriptions of a plant first, then compare them with the plant itself, before drafting their own description. Plant descriptions often started with the root and worked up. They disputed earlier claims, listed the uses, cultivation, and source of the plant, named the persons who had brought rare specimens to the Academy, and often collated names; they rarely included chemical analyses but might describe the flavor of the plant. To discover this, the Marchants chewed parts of the plant slowly and noted the flavor at various stages. More plants were described and drawn than were engraved. See BMHN MS. 450: 52r, 55v, 62r, 322r. For some of Dodart's descriptions, see AdS, Reg., 8: 114r, 130r–v (26 May, 16 June 1677); 7: 134v (11 Dec. 1677); 8: 156r, 215r–v (June 1677–June 1679). See also *Historia,* passim; *Histoire,* 1: 282, 328, 374, 431; 2: 10–11, 29, 53, 68, 93, 122, 188, 257–58, 280, and passim.

34. AdS, Reg., 10: 85r–v (5 Dec. 1681).

35. Raven, *John Ray,* 213.

36. Chastillon got the right to attend all meetings of the Academy in 1707: BA MS. 4624: 68.

37. AdS, Reg., 10: 80v, 111v (3 Dec. 1681, 5 Aug. 1682, Marchant).

38. According to the minutes, *Histoire,* and BMHN MS. 89, dossier 2, Chastillon and Joubert drew the parts of plants and academicians compared the drawings with the plants in 1692, 1693, and 1695. But these activities cannot be correlated with the records of the royal treasury. Many drawings in BMHN Rés., *Recueil des plantes,* reflect revisions during the 1680s and 1690s; on them are glued smaller pieces of paper with a name, seed, flower, or leaf. For an explanation of the several strikes made from the Academy's engravings, see Nissen, *Die botanische Buchillustration,* 2: 48–49; Bréchot et al., "Note bibliographique"; and Laissus and Monseigny, *"Les Plantes du Roi,"* 214–36.

39. BMHN Rés., *Recueil des plantes,* 69–74, 81, 91, 147.

40. AdS, Reg., 8: 182v, 223r (20 July 1678, 26 July 1679); 11: 131v, 133r, 167r (20 June, 4 July 1685, 21 Mar. 1686); 12: 31v, 35r, 38r, 48r, 59v, 135r (19 Feb., 19 Mar., 28 May, 23 July 1687, 30 Apr. 1688, 1 June 1689); 13: 126r (21 Jan. 1693); 14: 69r (9 Mar. 1695). Tournefort followed the same procedure when he read his *Élémens de botanique* to the Academy in order to obtain permission for its publication: AdS, Reg., 13: 129v (11 Mar. 1693). Dodart occasionally brought drawings of parts of plants which had not yet been described or engraved: ibid., 8: 156r, 215v (June 1677–June 1679). For discrepancies between the illustration and the description, see BMHN MS. 450: 331v–32v.

41. BMHN MS. 450: 82r. Bosse's engraving is faithful to Robert's painting (BMHN Rés., *Vélins,* 22: 1), except that he elaborated the rock, earth, and grass in the drawing (BMHN Rés., *Recueil des plantes,* 130) and engraving.

42. "Je ne scay pourquoy ce rejetton a droite!... ny ce pot... il faudroit l'effacer & mettre la racine au lieu": BMHN MS. 450: 95r. A comparison of the vellum (BMHN Rés., *Vélins,* 27: 26), drawing (BMHN Rés., *Recueil des plantes,* 160), and engraving shows that the pot was elaborated and that the branch on the right was not in the vellum or the drawing.

43. "Pourquoy *Mas?* est ce a cause du chicot... mentionné dans la description? Dioscoride fait la distinction de masle et femelle et ne la tire pas de la.... La figure ne la represente pas masle par la racine mais plustost femelle et avec une affectation ridicule. Il la faut corriger. On ne doit pas donner dans ces visions": BMHN MS. 450: 129r, 179r, and 292r. The engraving was based on and elaborated Robert's vellum, which was one of a pair; the other, Mandragora foemina, inspired the Academy's Mandragora flore sub caeruleo purpurascente (fig. 8), from which the engraver discreetly omitted the root: BMHN Rés., *Vélins,* 23: 25, 26, and *Recueil des plantes,* 233, 234. Jean Marchant discussed mandrake plants and their "pretendues vertus" in 1721: BN MS. 89, dossier 3.

44. BMHN MS. 450: 109r, 266r. The drawing (BMHN Rés., *Recueil des plantes,* 183) and engraving both show this fault, but the drawing lacks the seed. Errors of proportion showed up in other cases as well: BMHN MS. 450: 154, is marked "il faut alonger la tige denviron de trois doits."

45. BMHN MS. 450: 45r–47r, 107r, 124r, and passim.

46. The butterflies are not in the drawing: BMHN Rés., *Recueil des plantes,* 84.

47. Cymbalaria (fig. 3), lacked the names of Caspar and Jean Bauhin; in desperation Dodart suggested, "... il suffira, peut-estre, de l'adjouster a l'im-

primé. Il semble qu'on ne se peut dispenser de mettre le nom de J B a toutes les Plantes quand le nom q[ui]l donne est different de celuy que son Frere a donné": BMHN MS. 450: 82r. The drawings reveal some confusion about the identities of plants: BMHN Rés., *Recueil des plantes,* 26–30.

48. AdS, Reg., 11: 55v–56r (15 Apr. 1684, Dodart).

49. BMHN MS. 89, dossier 2: "Memoires des dessins de plusieurs fleurs et autres parties de plantes données a Mr. de Chastillon lan 1692" and a similar list for 1693.

50. See n. 32, above. Duclos complained that he had no laboratory: AdS, Reg., 4: 58r. Chemical analysis of plants was discussed before June 1668: ibid., 1: 36–37, 203, 249 (15 Jan., 12 Feb. 1667, 21 Jan. 1668). Bourdelin's earliest full notebooks of regular experiments on plants date from 1672: BN MSS. n. a. fr. 5134–35.

51. References are too numerous to cite; see, for example, AdS, Reg., 11: 117r–v (20 Jan. 1685), for a list of the plants and animals Bourdelin distilled from Mar. to Sept. 1684. In 1680 he analyzed 90 plants: *Histoire,* 1: 307–8. For Bourdelin's notebooks, see AdS, Cartons 1666–1793, 1–2; BN MSS. n. a. fr. 5133–46, 5148.

52. Dodart, *Mémoires des plantes,* 241.

53. Jean Marchant catalogued 300 plants not yet analyzed that could be found within 20 *lieues* of Paris, indicating the times of flowering to assist Bourdelin in planning his analyses: AdS, Reg., 8: 215r–v (Apr. 1678–June 1679). Marchant later gave Bourdelin another catalogue of 200 plants: ibid., 10: 13r (20 Mar. 1680).

54. For the supply of plants to the laboratory, see *CdB,* 1: 781 (1674); AdS, Reg., 10: 44v, 72r–v (June 1679–June 1681); BN MS. n. a. fr. 5149: 11r, 12r, 13r, 16r, 17r, 18v (1678–86); BN MS. n. a. fr. 5147: 103r, 107r, 110r, 111r–v, 113v, 115v, 118v, 119r, and passim. Some payments were for gathering and transporting plants from the Jardin royal.

55. AdS, Cartons 1666–1793, 3. See also AdS, Reg., 8: 63r, 74r (6 Nov. 1675, 29 Jan. 1676); 10: 72r, 82v, 109r, 152v (Aug. 1680–June 1683). Dodart, *Mémoires des plantes,* 185–86.

56. Duclos completed the research for his *Observations* in 1671, and he was busy writing the book until 1674 or 1675.

57. Table 1a: Dodart received a pension in 1671 "because of his profound knowledge of natural philosophy, and since he has attended for nine months the meetings of the Académie des Sciences"; in 1672 it was for his work in "phisique" or natural philosophy. These entries belie Fontenelle's statement that Dodart did not become an academician until 1673: *Histoire,* 2: 364; *Histoire . . . 1707,* 186.

58. Huygens, *Oeuvres,* 7: 11–12 (25 Feb. 1670); AdS, Reg., 8: 4v–5r (23 Jan. 1675).

59. AdS, Reg., 7: 201r–9v (3 Sept. 1678); 8: 204r–v (24 May 1679); 10: 19v, 22r–25v, 47v (22 May, 12 June, 4 Sept. 1680). Puech-Milhau, "Interview on Canada." See also Dodart's nonbotanical publications.

60. AdS, Reg., 8: 5r–7v, 122r (1674, 2 June 1677); AdS, Cartons 1666–1793, 3; BMHN MS. 450; Jean Marchant's copies of Dodart's papers: BMHN

MS. 449: 154r–55v, 168r–69v, 188r–89r; Marchant asked Dodart to consult Morison's book (BMHN MS. 451: 200v).
61. BN MS. fr. 1333: 42v–44r.
62. BMHN MS. 1278: 2r, 4r, 7v, 8r, 10v–11r, 11v. Dodart, however, sought the formal approval of the Academy and Colbert for his "Avertissement" to the book: AdS, Reg., 7: 19v (7 Sept. 1675); La Hire referred to the Academy's mandate to Dodart to write the book: Huygens, *Oeuvres,* 9: 264 (3 Mar. 1688).
63. BMHN MS. 1278: 10v–11v; Dodart, *Mémoires des plantes,* 157–58, 160. For Dodart's efforts to incorporate the results of chemical analysis into the descriptions of plants, see AdS, Cartons 1666–1793, 3; BMHN MS. 450: 5r, 21r–31v, 42r; BMHN MS. 447, dossier 4, "Catalogue de quelques plantes analysées."
64. The critical notes in BMHN MS. 448–51 show that Dodart also clashed with the Marchants over the style and content of descriptions and illustrations.
65. For Reneaume's positive assessment of the work, see *Tournefort,* 212.
66. For failures to answer objections or to act on suggestions, see MS. 450: 127r.

7: JUSTIFYING THE CHEMICAL ANALYSIS OF PLANTS

1. Charas, *The Royal Pharmacopoea;* Lémery, *Course of Chemistry;* Neville, "Christophe Glaser," and "'Pratique de chymie'"; Handford, "Chemistry at the Jardin du Roi," 37, 56. Rohault, *Traité de physique,* pt. 1, chap. 20, was skeptical of such methods.
2. Distinguished scholars of sixteenth- and seventeenth-century chemistry have mapped much of the terrain of chemical activity and thought during this period: Debus, "Sir Thomas Browne," "Solution Analyses," "Fire Analysis," and *The English Paracelsians;* Multhauf, "Significance of Distillation" and *The Origins of Chemistry;* Boas, *Robert Boyle,* and "Quelques aspects"; M. B. Hall, "Humanism"; and Metzger, *Les doctrines chimiques en France.* But most modern discussions of the Academy's chemical research have relied almost exclusively on the printed sources, while some have condemned without elucidating it: Ornstein, *Rôle of Scientific Societies,* chap. 5; Partington, *History of Chemistry,* 3: 12; Bertrand, *L'Académie et les académiciens,* p. 340; Académie des Sciences, *Troisième centenaire,* 2: 1; Stubbs, "Chemistry at L'Académie." Chevreul's "Recherche expérimentale" represents the best of early efforts to assess Dodart's *Mémoires des plantes* within a seventeenth-century context, and Holmes has perceptively treated the change from distillation to solution analysis, introducing some neglected manuscript sources into the discussion: "Analysis by Fire and by Solvent Extractions" and "Tradition and Invention."
3. AdS, Reg., 4: 48r–55v, see 51r: "leurs eaux distillées, leurs esprits tant acres et sulphurez qu'acides et mercuriels, leurs huyles et leurs sels fixes ou volatiles."
4. Ibid., 62r–63r.
5. *Histoire,* 1: 121, 167, 252–53; *Historia,* 88–90: Du Hamel included chemical analysis under the heading "De animalium et plantarum Anatome." The

Oxford Clubbe during the 1650s had also "merged the methods and the object of study of the chemist and the anatomist": Davis, *Circulation Physiology,* 29.

6. Davis, *Circulation Physiology,* 9, 24; Debus, *The English Paracelsians,* 61, 157, 179–80; see also 95, 129–30, n. 33; 134n. 117.

7. Debus, "Fire Analysis," 147; Le Febvre, *Compleat Body,* 1: 151, 223, 255–56, 257, 262–63; 2: 4–5, 13–14.

8. *Histoire,* 1: 167.

9. Dodart, *Mémoires des plantes,* 156–58, no. 4.

10. Ibid., 158, nos. 5 and 7.

11. Antoine de Jussieu made this point in 1738 or 1739 in his assessment of the project: BMHN MS. 2651, draft, 9r–10r.

12. Dodart, *Mémoires des plantes,* 158–59, no. 8, and 160.

13. Mariotte, *Végétation,* 127.

14. Dodart, *Mémoires des plantes,* 159, no. 9.

15. Debus, *The English Paracelsians,* 38–39, and "Fire Analysis," 145; Gregory, "Chemistry and Alchemy," 110; Boas, *Robert Boyle,* 112–13.

16. *Histoire,* 1: 57–58 (1668).

17. Mariotte, *Végétation,* 145–46, 125.

18. AdS, Reg., 4: 59v.

19. BMHN MS. 1278: 4r.

20. AdS, Reg., 14: 123r (1 June 1695).

21. Dodart, *Mémoires des plantes,* 155, no. 2; 156, no. 3; 158, no. 6; 168–69, no. 1.

22. Ibid., 169–76.

23. *Histoire,* 1: 121–22 (1670).

24. Le Febvre, *Compleat Body,* 1: 244–45, and passim.

25. *Histoire,* 1: 122; Dodart, *Mémoires des plantes,* 168.

26. Dodart, *Mémoires des plantes,* 164–65; AdS, Cartons 1666–1793, 1–3; BN MSS. n. a. fr. 5133–49; BMHN MS. 259; Tournefort, *Élémens de botanique,* 516, and *Histoire des plantes,* e x^{r+v}.

27. Dodart, *Mémoires des plantes,* 164; the retort was called a "cornuë."

28. Ibid., 164–65. Hooke found the chemical analyses of the Academy particularly interesting and cited this list in his review of the book in *Philosophical Collections,* 1: 40.

29. Dodart, *Mémoires des plantes,* 186–92. Bourdelin stocked the laboratory with "teinture de tornesol, des solutions de sublimé, et de sel de Saturne et de l'eau de vitriol": BN MS. n. a. fr. 5147: 119r, and passim. For his method of preparing the solution of turnsole, see BN MS. n. a. fr. 5149: 7r. On color indicators, see Baker, "History"; Nierenstein, "Early History."

30. Glaser, *Compleat Chymist,* 212–14.

31. Ibid., 228–30, 240–45.

32. BMHN MS. 259: 8r–v. Tournefort described Bourdelin's equipment and sketched the cucurbit resting in a water bath. Bourdelin's recipients had mouths large enough to permit entry of an arm for cleaning to ensure that distillants were pure.

33. AdS, Reg., 8: 2r–7v (23 Jan. 1675); Eklund, *Incompleat Chymist,* 15, 36.

34. Dodart believed that maceration freed but changed the constituents, and

that fermentation reduced the oil in plants, perhaps because essential oils were changed into flammable spirits: Dodart, *Mémoires des plantes,* 176–80; AdS, Cartons 1666–1793, 1, 1: 9–11 (30 Aug. 1672); 3, 9: 11, 20 (June, July 1673); AdS, Reg., 8: 2r–v, 102r, 104r–v, 114r, 190r–v (16 Dec. 1676, 10 Feb., 26 May 1677, 23 Nov. 1678); Eklund, *Incompleat Chymist,* 25.

35. AdS, Reg., 8: 6v. Borelly suggested rectifying distillants four times: ibid., 1r (2 Jan. 1675).

36. AdS, Reg., 8: 61r, 77v, 190v–91r, 123v–24v, 222r (2, 23 Jan., 14 Aug. 1675, 4 Mar. 1676; 1677; 23 Nov. 1678; 12 July 1679); 10: 21r, 22r, 27v, 28v–35v, 69v, 74r, 91r, 92r (5, 12 June, 3 July 1680, 11, 25 June 1681, 18 Feb., 11 Mar. 1682); 11: 168r, 169r (6 Apr. 1686); AdS, Cartons 1666–1793, 2, 7: 105 (16 Nov. 1678); 3, 3: 253r–v (4 Mar. 1682); BN MS. 5147: 103v (24 Feb. 1682); Huygens, *Oeuvres,* 9: 264 (3 Mar. 1688). By 1681 the Academy's three big furnaces downstairs were "bruslés et ruinés entierement" and had to be replaced: BN MS. 5147: 102r–v, 105r, 108v, 118v.

37. AdS, Reg., 4: 49r–51r (Duclos); Dodart, *Mémoires des plantes,* 153, 221–24. Academicians also discussed analyzing the *marc* of a plant and placing flowers in *eau de vie* in the sun to be "distilled": AdS, Reg., 16: 208v–10r (31 July 1697, La Hire *fils*).

38. Dodart, *Mémoires des plantes,* 163.

39. AdS, Reg., 4: 61v–62v: "des matieres qui sont fixes & qui ne peuvent estre bruslées"; "les menstrues resolutifs pour en discontinuer la masse et rendre separables les parties constitutives"; "ne s'esleve point"; "embrasement"; "absoluement immobiles au feu."

40. Ibid., 4: 127v–33v, 134r–66r, 167r–75r (11, 18, 25 Aug. 1668).

41. Dodart asked whether the solvent might be useful "non pour l'analyse, a laquelle il ne peut servir, mais pour les effets merveilleux attribuez aux estres des plantes": ibid., 8: 4r. On the uselessness of these solvents, even if they could be made: Dodart, *Mémoires des plantes,* 146–48, 152; for other evidence of Dodart's skepticism, see BN MS. fr. 17054: 422r–23v, and BMHN MS. 449: 154r–55v. See also Multhauf, "Medical Chemistry and 'the Paracelsians.'"

42. BMHN MS. 1278: 4r: "Ce seroit pourtant un moyen beaucoup meilleur que celuy du feu puisque ce dissolvant n'altere point les choses, qu'il les laisse et qu'il les reduit en leurs principes constitutifs avec conservation de leurs vertus et proprietés specifiques, ce que le feu ne peut faire." Duclos recounted the changes which distillation and its prelude introduced in the plant: ibid., 9r.

43. Ibid., 3v–4r. Duclos cited the passage on p. 152 of Dodart's *Mémoires des plantes* as a careless dismissal of a technique that merited further consideration, and he mocked Dodart—"l'auteur du projet qui n'a ni l'usage ni la connoissance ni l'experience de cette sorte d'analyse," i.e., with solvents—as lacking expertise.

44. BN MS. fr. 1333: 42v–44r.

45. AdS, Reg., 8: 94r–v (26 Aug. 1676); 10: 21r (5 June 1680); 11: 168r. Lémery was not one of the first advocates of wet analysis, or the use of solvents, as has been claimed: Bedel, "Conceptions en chimie biologique," 397–98.

46. Classification of plants according to their chemical constituents does not seem to have been a motivation, although Dodart mentioned that one aim of analyzing animals was to discover differences between genres [AdS, Reg., 8: 126v

(2 June 1677)], and mineral waters were already classified by their chemical contents: Ulyatt, "Further Studies in the History of Mineral Waters," 1–32; Duclos, *Observations*, 62–107. That academicians never intended to classify plants chemically, however, is clear from the way Perrault, Duclos, and Dodart discussed the problem of classifying plants: AdS, Reg., 1: 30–38 (1667); 4: 48r–55r (1668); 8: 173r–78r (1678); Dodart, *Mémoires des plantes*, 241.

47. AdS, Reg., 1: 36–37 (15 Jan. 1667); Perrault was active in the analysis of plants only briefly, in 1678 and 1679: ibid., 8: 215v–16r, 222r (Apr. 1678 to June 1679, 12 July 1679). On Perrault's corpuscularianism, see "Le pesanteur des corps" in his *Oeuvres*, 1: 3. On Lémery and Homberg, see Boas, "Acid and Alkali," 26.

48. Mariotte, *Végétation*, pt. 1.

49. AdS, Reg., 4: 49r; *Histoire*, 1: 57 (1668).

50. AdS, Reg., 8: 2v: "ce que les plantes sont" and "ce qu'elles peuvent faire"; Dodart, *Mémoires des plantes*, 154.

51. Dodart, *Mémoires des plantes*, chap. iv, esp. pp. 149–51, 153, 154, 186.

52. AdS, Reg., 8: 190v (23 Nov. 1678), AdS, Cartons 1666–1793, 2, 7: 105r (16 Nov. 1678, Bourdelin); Mariotte, *Végétation*, 146; Dodart, *Mémoires des plantes*, 182. Duclos stressed that the purpose of chemical analysis was "to know what plants are in themselves, and not what they can become," and that only a sound theory assured medical knowledge (BMHN MS. 1278: 9r, 10v–11r), but he doubted the value of distillation for explaining natural compounds: Duclos, *Dissertation*, 2–3.

53. This was the thrust of Homberg's assessment in 1692, although he also thought some facts about the constituents of plants had been established: AdS, Reg., 13: 116r; *Histoire*, 2: 148. Louis Lémery later believed that only the distillants obtained from plants redeemed Bourdelin's work: Metzger, *Les doctrines chimiques en France*, 357–58.

54. See chap. 13, below, and AdS, Reg., 8: 155r–v (June 1677–Apr. 1678).

55. Dodart, *Mémoires des plantes*, 150–51, 231–36.

56. Le Febvre, *Compleat Body*, 1: 255–57, believed that only chemistry could explain "why these aliments do nourish and sustain."

57. *Histoire . . . 1707*, 187–89; Dodart brought his sweat to Bourdelin for analysis: AdS, Cartons 1666–1793, 2, 6: 84v–85r (14 Apr. 1677).

58. Locke, *Correspondence*, 2: 464; Roemer gave Dodart's letter to Locke, but no reply is known.

59. AdS, Reg., 8: 126v–27r (2 June 1677).

60. This seemed plausible because the distilled fruit produced a lot of carbon that yielded few cinders after distillation: AdS, Reg., 8: 7v (23 Feb. 1674), 175r (18 May 1678).

61. Ibid., 1: 203 (12 Mar. 1667); 8: 103r, 124v–26r, 173r–78r, 190r (27 Jan., 2 June 1677, 18 May, 23 Nov. 1678); 10: 71v (Aug. 1680–June 1681).

62. Dodart, *Mémoires des plantes*, 229.

63. Joravsky, *The Lysenko Affair*, 204–5, shows that the same process was necessary before genetics could develop, and that Mendel's work was ignored by contemporaries who were more interested in the general, philosophical implications of genetics than in the mechanism of change.

64. AdS, Reg., 8: 2r–7v (1674, 1675); "Comme il n'y a guere d'apparence que les analyses nous fassent bien voir dans les produicts, ce que les plantes sont, et ce qu'elles peuvent faire, il faut au moins, qu'elles nous fassent voir ce qu'on en peut faire, par quelque voye, que ce soit" (2v).

65. Dodart, *Mémoires des plantes,* 182.

8: MINISTERIAL INTERVENTION AND AN UNEXPECTED OUTCOME

1. Dodart, *Mémoires des plantes,* 239–42. His plans were adopted, for after 1676 the minutes mention the second part or continuation of the natural history: AdS, Reg., 8: 155v (June 1677–Apr. 1678); 10: 44r–v, 72r, 82v, 84v, 152v (June 1679–June 1683).

2. AdS, Reg., 8: 122r: "que l'on doit donner les premieres au public, comme sont la coriandre, la laictue, la Chicorée tant sauvage que Domestique, le Cresson, &c."

3. Ibid., 8: 215v: "dans le dessein d'en faire les Descriptions pour servir a l'histoire generale des plantes."

4. Ibid., 10: 72r: "...tous ces traittez qui devoient composer un juste volume luy ayant esté volez en entrant a Paris, ou il les apportoit pour les faire mettre au net, et les donner a l'Imprimeur; et toutes les diligences qu'il a faites pour les recouvrer luy ayant esté inutiles, il a esté obligé de refaire les deux plus importants de ces traittez, et de recueillir dans ses memoires tout ce qu'il a pû retrouver pour retablir les autres ouvrages."

5. Ibid., 10: 82v.

6. Ibid., 10: 84v; Dodart's other treatises were: "Examen de quinz a seize cents experiences de Medecine des Remedes Royaux distribuez par Monr. Pellisson," "L'Usage de la Raison en Physique, et en Medecine," "L'Histoire de la Medecine premiere partie du Regime, et des exercices," "Experiences sur le feu," "Traitté de la Transpiration," and "De la maniere de nourrir les malades." His history of diet and his practical manual on how to live a healthy life by eating and exercising properly were mentioned by Du Hamel in the annual report for 1678–79: AdS, Reg., 8: 214v–15r.

7. Ibid., 10: 109r, 112r–v, 152v (12, 19 Aug. 1682, July 1682–June 1683); cf. *Histoire,* 1: 374 (1683). Dodart's notes in BMHN MS. 450 seem to date from the 1680s.

8. AdS, Reg., 11: 114v (13 Dec. 1684). Reneaume claimed that Tournefort's Iberian herborizations were done for the Academy: *Tournefort,* 229.

9. AdS, Reg., 11: 113r (22 Nov. 1684); *Histoire,* 1: 405 (1684).

10. AdS, Reg., 11: 115v, 133r, 152v (10 Jan., 4 July, 19 Dec. 1685); 12: 21v (4 Dec. 1686).

11. Ibid., 11: 116v–17v, 124r (20 Jan., 11 Apr. 1685); *Histoire,* 1: 431; Huygens, *Oeuvres,* 9: 10 (23 May 1685).

12. Koenigsberger, "Republics and Courts," stresses the ill effects of self-interested patronage in music, art, and science.

13. Saunders, *Decline and Reform,* 99–102.

14. AdS, Reg., 11: 157r: La Chapelle's words were "recherche curieuse," "un

ieu," and "un amusement des Chymistes"; he requested instead "recherche utile ce qui peut avoir rapport au Service du Roy et de L'Estat." The connotation of "curieux" is unclear. According to Livet, *Lexique de Molière,* and the *Grand Larousse de la langue française,* 2: 1097, Descartes used the word to refer to the pseudo-sciences. Coming after Duclos's deathbed recantation and given La Chapelle's condemnation of the philosopher's stone, therefore, this could be an attack on the Academy for dabbling in alchemy. But Furetière, *Dictionnaire,* cites "curieux," "sciences curieuses," and "chimiste curieux," using "curieux" to convey a taste for experiment and discovery, or to suggest the contemporary English notion of the interests exhibited by virtuosi.

15. AdS, Reg., 11: 157v: "L'autre Recherche plus convenable à cette Compagnie et qui seroit plus du goust de Monseigneur de Louvois regarde tout ce qui peut illustrer la Physique et servir a la Medecine, ces deux choses estant presque inseparables parceque la medecine tire des Consequences et profite des nouvelles decouvertes de la Physique." Saunders, *Decline and Reform,* 254–55, has published the entire text of this speech.

16. He had previously done so in the case of the abbé de Lannion, who was rebuked and expelled from the Academy: AdS, Reg., 11: 162v (27 Feb. 1686).

17. Huygens, *Oeuvres,* 9: 164 (1 June 1687).

18. AdS, Reg., 8: 155v–56r, 195r–v, 216r–17v (June 1677–Apr. 1678, 11 Jan. 1679, Apr. 1678–June 1679); *Histoire,* 1: 50–51, 162–67, 198, 282, 320–21, 387–89 (1668, 1673, 1675, 1679, 1681, 1684).

19. AdS, Reg., 11: 158r.

20. Ibid., 11: 166v, 167r, 168r–69r (13, 27 Mar., 3, 6, 17 Apr. 1686).

21. Ibid., 12: 66v, 68v–69r, 134v (7, 21 Jan. 1688, 21 May 1689), and passim.

22. From 28 January 1688 until the end of 1689, only four descriptions of plants were recorded in the minutes, and only one engraving was verified: ibid., 12: 89v, 107r, 130v, 135r (9 June, 4 Dec. 1688, 30 Mar., 1 June 1689).

23. Ibid., 12: 89v–90r, 131v (12 June 1688, 20 Apr. 1689); *Histoire,* 2: 53, 68 (1688, 1689); cf. Hunt, *Catalogue,* 1: 369, on Jan Commelin.

24. Huygens, *Oeuvres,* 9: 481.

25. AdS, Reg., 12: 88r, 129v, 131v (26 May 1688, 16 Mar., 20 Apr. 1689); 13: 39r (15 Nov. 1690); *Histoire,* 2: 68 (1689).

26. AdS, Reg., 13: 14r–15v, 17r, 40r, 41r, 42v (17, 28 June, 1, 26 July, 29 Nov., 20 Dec. 1690, 13, 17 Jan. 1691), and passim; BMHN MS. 451: 131r–32r (29 Nov. 1690).

27. AdS, Reg., 13: 19v (12 Aug. 1690).

28. *Histoire,* 2: 10–11, 29, 53, 62, 63, 66, 68, 92–93, 116, 122 (1686–91).

29. External scholarly competition, an obvious source of botanical influence, had little effect one way or the other. Several foreign botanists were already working on similar projects—from the idiosyncratic books of Paolo Boccone to the specialized treatises of Nehemiah Grew and Marcello Malpighi—but the work of John Ray presented the most likely challenge. Ray's monumental *Historia plantarum* was to be a complete compendium of contemporary botanical knowledge and Tournefort praised it as a "botanical library": *Élémens de botanique,* 19. Like the academicians, Ray described plants and their chemical analysis; al-

though he could not illustrate his work, he surpassed the Academy's project in scope by discussing plant nutrition and the flow of the sap and proposing an impressive solution to the problem of classification. Ray supplemented his *Historia plantarum* with other Latin studies of British and European flora, but his prodigious output from 1686 through 1694 did not discourage academicians from resuscitating their natural history of plants as soon as their protector permitted. Jean Marchant owned Ray's books: BMHN MS. 447. See chap. 15, below; Stevenson, "John Ray and Classification," 254; Raven, *John Ray,* chaps. 4, 8–11; Arber, "A Seventeenth-Century Naturalist."

30. AdS, Reg., 13: 43r, 59v, 71r–v, 72r, 73r (14 Feb., 18, 25 Apr., 12, 19, 22 Dec. 1691); *Histoire,* 2: 116. Two of the plants were as yet unnamed.

31. AdS, Reg., 13: 71v, 81v, 105v–6r (15 Dec. 1691, 27 Feb., 12 July 1692); 14: 2v (21 Nov. 1693); 17: 38r–39v (27 Nov. 1697); on ginseng, cf. *Histoire... 1718,* 41–45.

32. AdS, Reg., 13: 133r–v (6, 13 May 1693); 14: 118v (25 May 1695); *Histoire,* 2: 153–54, 188 (1692, 1693); *Mémoires,* 10: 101–3, 119–26.

33. AdS, Reg., 14: 204r, 218v (3, 7 Dec. 1695).

34. Ibid., 14: 18v, 65r–v (19 June 1694, 2 Mar. 1695); 16: 122r (8 May 1697); *Histoire,* 2: 116; *Mémoires,* 10: 10–14 (31 Jan. 1692). The Marchants' annotated copy of *Pinax* is now BMHN MS. 1061. They corrected and added names of plants, working from personal observation and from books by Jean Bauhin, Parkinson, and others.

35. AdS, Reg., 16: 79v (10 Apr. 1697); *Histoire... 1699,* 60–63.

36. Only Homberg's paper on a vegetable dye focused exclusively on the use of a plant: AdS, Reg., 15: 97r–100r (20 June 1696): "L'Usage des Fleurs de Cartame dans la Teinture."

37. There are too many descriptions between 1693 to 1699 to list; see for example, AdS, Reg., 13: 131v (15 Apr. 1693, Marchant); 15: 87v (6 June 1696, Dodart).

38. *Histoire,* 2: 154, 188, 257–58, 280 (1692, 1693, 1695, 1696). By April 1693 academicians were again reviewing Bourdelin's notebooks: AdS, Reg., 13: 132v (29 Apr. 1693).

39. Stroup, "Wilhelm Homberg."

40. But not Jean Marchant, to whom he displayed animosity, seeing to it that Marchant would lose the *petit jardin: Historia,* 419, 448; Laissus and Monseigny, "Les Plantes du Roi."

41. Tournefort, *Histoire des plantes,* aiiij[r].

42. Tournefort, *Élémens de botanique,* 5.

43. See *Tournefort,* 217–18, on the several volumes of plant descriptions Tournefort wrote during the 1690s.

44. Tournefort, *Histoire des plantes,* avij[v]-eiij[r]: "premieres qualitez des corps," "la configuration des parties." See also *Tournefort,* 100–101.

45. Laissus and Monseigny argue that publication of Tournefort's *Élémens de botanique* ruled out release of the Academy's natural history of plants; see *"Les Plantes du Roi,"* 209–10.

46. Tournefort, *Élémens de botanique,* eiiij[r]: "ni les figures entieres de chaque espece de plante ni leurs vertus."

Notes to pp. 113-122

47. Ibid., 12-13.
48. Tournefort, *Histoire des plantes*, aiiij^{r-v}.
49. Tournefort, *Élémens de botanique*, 3, 516, 558; *Histoire des plantes*, aiij, avj^{r-v}, ex^{r-v}, 6, 11, 15, 40, and passim; Fontenelle, *Éloges*, 110; *Tournefort*, 216, 229. This may be why Fontenelle incorrectly claimed that the natural history had always been intended as a catalogue of plants in France; see chap. 6, above.
50. Marchant defined his work for 1699 as completing the second volume of the natural history of plants "sur le dessein que la Compagnie s'est anciennement proposé": AdS, Reg., 18: 144v.
51. *Tournefort*, 209, 212, 227-36.
52. Cf. Bertrand, *L'Académie et les académiciens*, 44-46, 50-51, for a different assessment of corporate and individual projects.

9: ANALOGICAL REASONING: THE MODEL

1. Tournefort alone among academicians kept up-to-date with botanical literature; many academicians came to their studies of plants from other disciplines altogether. The Marchants looked on new botanical treatises as sources of information about particular plants rather than as models for a new style of studying plants. See chap. 8, n. 29, above, on how little Ray influenced the Academy.
2. Canguilhem, preface to Delaporte, *Second règne*, 8.
3. The texts that academicians produced on the subject exist in manuscript and in publications. The minutes of the Academy record the papers of Perrault, Mariotte and Duclos plus the experiments that Marchant demonstrated: AdS, Reg., 1: 35 (15 Jan. 1667); 4: 67v-68r, 71v-77v, 79r-90r, 92v-99r (23, 30 June, 7, 14, 21 July 1668); AdS, Carton 1667-1699, pochette 1668. In 1679 Mariotte published his *Végétation,* a treatise on the nutrition and growth of plants that incorporated his earlier research on the circulation of sap, and a year later Perrault published his *Circulation,* on the circulation of sap, which borrowed heavily from Mariotte's experimental data and also contained Duclos's rebuttal. All three academicians revised their opinions between 1668 and 1680. La Hire addressed the problem in the late 1670s and the early 1690s: AdS, Reg., 8: 218r-v; *Mémoires,* 10: 317-19; *Histoire,* 2: 184-86. The histories of the Academy by its first two permanent secretaries report the summer debate: *Historia,* 62-66, and *Histoire,* 1: 58-63.
4. Salomon-Bayet, *L'institution de la science,* 87. Davy de Virville, "De l'influence des idées préconçues," 119, distinguishes between preconceived and directive ideas.
5. Canguilhem, "Role of Analogies and Models," 507, 516.
6. Ibid., 513.
7. Hesse, *Models and Analogies,* 81-85.
8. Ibid., 86-91.
9. That is, when "it has not been possible to observe or to produce experimentally a large number of instances in which sets of characters are differently associated": ibid., 76.
10. Ibid., 76-77.

11. Canguilhem, "Role of Analogies and Models," 513.
12. Hesse, *Models and Analogies,* 79–80.
13. Canguilhem, "Role of Analogies and Models," 517.
14. Hesse, *Models and Analogies,* 162.
15. Canguilhem, "Role of Analogies and Models," 515.
16. Hesse, *Models and Analogies,* 163.
17. Canguilhem, "Role of Analogies and Models," 508, 513, 516.
18. Canguilhem, *Connaissance de la vie,* 22–23.
19. Delaporte, *Second règne,* 18.
20. I am grateful to Shirley A. Roe for this quotation.
21. Canguilhem, "Role of Analogies and Models," 514.
22. Harvey, *The Circulation of the Blood,* 187.
23. Ibid., 132.
24. Harvey, *Movement of the Heart and Blood,* 58–59.
25. Pagel, *Harvey's Biological Ideas,* 25–26, 43, 51–58.
26. Harvey, *Movement of the Heart and Blood,* 61.
27. See Riolan's letters in Harvey, *The Circulation of the Blood;* Riolan, *Manuel,* 706–49; articles in *DSB* on Harvey and Riolan; Pagel, *Harvey's Biological Ideas;* Berthier, "Le mécanisme cartésien," 3: 33–44.
28. Berthier, "Le mécanisme cartésien," 3: 51. Descartes, *Discours,* pt. 5.
29. Tardy, *Cours de medecine;* Chaillou, *Mouvement du sang;* Martet, *Abbregé . . . ensemble de la circulation du sang;* Betbeder (who plundered Chaillou and Martet), *Questions nouvelles;* Guiffart, *Lettre,* preface; Pagel, *Harvey's Biological Ideas;* Berthier, "Le mécanisme cartésien," 2: 53–55; *DSB,* 10: 477.
30. Harvey, *Movement of the Heart and Blood,* 58–59.
31. Ibid., 51.
32. Ibid., 39.
33. Salomon-Bayet, *L'institution de la science,* 86.
34. Harvey, *Movement of the Heart and Blood,* 13, 31, and *The Circulation of the Blood,* 145, 172.
35. Harvey, *Movement of the Heart and Blood,* 94, 101, and *The Circulation of the Blood,* 118–19, 150–51, 152, 169.
36. Pagel, *Harvey's Biological Ideas,* 56.

10: ANALOGICAL REASONING: THE THEORY

1. Gasking, *Investigations,* 44; Delaporte, *Second règne,* 27–31. See for example [Aristotle], *De Plantis,* 1: 3, 818a17–20; 1: 2, 817a31–35; 1: 1, 816b11–23; Aristotle, *Historia Animalium,* 5: 1, 539a16–20; *De Anima,* 1: 5, 411b19–30; 2: 1, 412b1. Tournefort, *Élémens de botanique,* 21, 23, 515–26, 561; Grew, *Anatomy Begun,* 44–45, preface, and epistle dedicatory; Arber, "Nehemiah Grew," 47; Webster, "Recognition of Plant Sensitivity"; Davy de Virville, "De l'influence des idées préconçues," 114.
2. Delaporte, *Second règne,* 31–32.
3. Savants often continued to assume what they needed to prove: Davy de Virville, "De l'influence des idées préconçues," 115.

4. For what academicians wrote about the circulation of sap, see chap. 9 n. 3, above.

5. Hooke, *Micrographia,* 114, 116, 120; Gunther, *Early Science in Oxford,* 6: 337; Le Febvre, *Compleat Body,* pt. 2, 3–4; *Histoire,* 1: 3; Grew, *Anatomy of Plants;* Malpighi, *Plantarum Anatome;* Arber, "Grew and Malpighi." Régis, *Système,* 467–86, relied on Grew and Malpighi for the anatomy of plants and on Perrault for the circulation of sap.

6. He exemplifies the "medical style" of explanation characterized by King, ed., Hoffmann, *Fundamenta Medicinae,* xx–xxi. He also represents a process discussed by Ginzburg in "Morelli, Freud," by which savants raid popular culture for their scientific conjectures.

7. Bugler, "Précurseur," presents Mariotte as a precocious experimental genius.

8. Harvey, *Movement of the Heart and Blood,* 58–59, chap. 15; Perrault, *Circulation,* 72–73, 82, 105–6, 110–11; AdS, Reg., 4: 80r.

9. Harvey explained to Robert Morison why he could not accept Pecquet's theory that the lacteal veins were filled with chyle: *The Circulation of the Blood,* 193–200; cf. 129. Academicians were modified Harveians, and Mariotte disagreed with Harvey on the lacteal veins: "Il semble que comme les veines lactées qui sont dans le mesentere reçoivent le chyle et le portent dans les veines, d'ou il passe dans le Coeur, et du coeur dans les poumons, d'ou il est porté derechef dans le coeur et ensuitte dans les arteres pour servir a la nourriture de toutes les parties du corps, et le surplus repasse dans les veines qui le reportent au coeur." He compared the ends of roots with lacteal veins: "Vraysemblablement les extremitez des racines s'imbibent de l'humidité qui est dans la terre, et la portent dans le Corps de la racine, d'ou elle passe dans des petits Canaux qui sont dans la tige d'ou elle se distribue.... (AdS, Reg., 4: 79v).

10. AdS, Reg., 4: 79v. The beginning of the quotation appears in n. 9, above; it continues: "... d'ou elle se distribue dans les branches et jusques auz extremitez des feüilles, & le surplus est reporté par d'autres petits canaux vers la Racine pour s'y perfectionner par une Espece de cohobation et devenir un suc bien digeré et propre a la nourriture des fleurs et des fruicts...."

11. AdS, Reg., 4: 79v–80r; Perrault, *Circulation,* 77–80; *Mémoires,* 10: 194–95; Grew, *Anatomy of Plants,* 48–49; Dedu, *De l'âme des plantes,* 297–98.

12. AdS, Reg., 4: 79r, 80r–v, 81v–83r, 85r–v; Mariotte, *Végétation,* 133–34; Perrault, *Circulation,* 81–82, 84–85, 87–96; *Historia,* 62–63.

13. On the controversy with Duclos over the formation of dew, see AdS, Reg., 4: 83r–v, 89r–v; and Duclos's portion of Perrault's *Circulation.* Mariotte's and Perrault's views on the role of leaves: AdS, Reg., 4: 82r–83r, 85v–86r, 94r–v; Perrault, *Circulation,* 83–86, 91–93.

14. Harvey, *Movement of the Heart and Blood,* 93; Perrault, *Circulation,* 85–86; Mariotte, *Végétation,* 133.

15. Harvey, *Movement of the Heart and Blood,* 58–59.

16. AdS, Reg., 4: 72v–74v, 79r, 80r–v, 85r–v; compare Duclos's report of his experiments, ibid., 87r–88r. The Academy later identified the two saps in an aloe as "suc crud" and "suc nourissier": BMHN MS. 451: 88r (6 May 1671).

17. Perrault, *Circulation*, 77–78, 94–95, 105, 116, 119–20. See also AdS, Reg., 4: 89v–90r; 10: 95v (15 Apr. 1682); 12: 130r (23 Mar. 1689).
18. Harvey, *The Circulation of the Blood*, 125; AdS, Reg., 4: 72r–75r, 85r–86r; *Histoire*, 1: 61 (Mariotte); Perrault, *Circulation*, 73–77.
19. Mariotte, *Végétation*, 130–31.
20. Perrault, *Circulation*, 74–77, 88–89, 93–94, 96–97.
21. Mariotte, *Végétation*, 129–30. Delaporte, *Second règne*, 44, incorrectly interprets Mariotte as affirming their existence.
22. For Hooke, see Gunther, *Early Science in Oxford*, 6: 337 (1668); for La Hire, see *Histoire*, 2: 183–86; *Mémoires*, 10: 317–19.
23. Grew, *Anatomy of Plants*, 21.
24. Harvey, *Movement of the Heart and Blood*, 98, 103; see also, 37–38, 99. Régis, *Système*, 465, argued thus.
25. Hooke, *Attempt*, 1; Brown, *Scientific Organizations*, 85–86; cf. Boyle, *Works*, 1: 80.
26. Partington, *History of Chemistry*, 2: 3, 505, 511; Middleton, *History of the Barometer*, 185; Huygens, *Oeuvres*, 22: 586–87, 3: 328–29. Locke heard capillary tubes discussed during his French visit: *Travels*, 101.
27. Hooke, *Attempt*, 26.
28. Hooke, *Micrographia*, 11, 20–21, 28.
29. For the confusion, see Middleton, *History of the Barometer*, 186, citing Honoré Fabre and J. C. Sturm. See also Millington, "Theories of Cohesion" and "Studies in Capillarity," 258, 267–68, who cites Rohault, Bernoulli, and Huygens. On Borelli see Partington, *History of Chemistry*, 2: 444, and Millington, "Studies in Capillarity," 264. A brief review of Borelli's 1670 *De motionibus naturalibus* in *Phil. Trans.* 73 (1671): 2210, pointed out that Borelli argued against the fear of a vacuum and attraction; it also mentioned his discussion of capillary tubes. His views on these subjects thus reached a wide audience.
30. Mariotte, *Végétation*, 130. It is not clear how La Hire explained capillary action, but by 1693 he stressed the mediocre heights to which capillary action could raise a liquid: AdS, Reg., 8: 218r–v (Apr. 1678–June 1679); *Mémoires*, 10: 317–19. While Perrault never cited the evidence of water rising in thin glass tubes, he did compare the absorption of liquids in a sponge with the rise of sap, as had Cesalpino a century earlier. Perrault, *Circulation*, 74–75. Cf. Malpighi and Ray; see Clark-Kennedy, *Stephen Hales*, 59–60; Guyénot, *Évolution*, 118.
31. AdS, Reg., 4: 74v–76v.
32. Pierre Perrault, *Origine des fontaines;* the preface was reprinted in Huygens, *Oeuvres*, 7: 287–97; quotation from p. 296.
33. Perrault, *Circulation*, 76–77, 113, 124. Régis, *Cours*, 1: 485–86, 491–92, had similar views. Compare Savery's pulsometer of 1698: Burstall, *Mechanical Engineering*, 193. For fermentation in Descartes's circulatory theory, see his *Discours*, pt. 5. Grew had a similar theory of the rise of sap: *Anatomy Begun* and *Anatomy of Roots* in *Anatomy of Plants*, 17–18, 22–26, 82–83; Arber, "Nehemiah Grew," 58.
34. Mariotte, *Traité des couleurs* in *Oeuvres*, 311, first published in his *Essais de physique*. Duclos misunderstood air pressure. In 1680 he adopted the theory that sap circulated but ascribed an "expulsive faculty" to the branches and trunks

of trees in order to explain how sap rose. He rejected air pressure as a cause, arguing that the weight of the air was too weak to force sap upwards in trees since it could not prevent a delicate plant from growing straight and tall. Duclos based his view on Huygens's experiment in which plants grew in a tightly stopped bottle, but he revived a discredited assumption (once used by critics of Torricelli and Pascal), even though by the 1670s it had been shown that the air pressure in a sealed container was the same as that of the atmosphere in which it had been sealed, unless the container had been evacuated by a pump. Boyle showed that "air in a bell jar on a plate can be at atmospheric pressure, even though the glass keeps the air above from pressing on that within" (Middleton, *History of the Barometer*, 66). Perrault refuted Duclos simply: "the enclosed air acts with the same force to create pressure as it does when it communicates with the other air." This was because it "acts according to the strength of its spring, which is proportional to the weight of the air that it had when it was enclosed." For Perrault this was proven by experiments with carps' bladders in an evacuated bell jar. He argued further that the air enclosed in soil was no different from the air in a sealed container and could therefore exert pressure on juices in the earth and on the roots of plants. See Perrault, *Circulation*, 109, 113–14, 119; Dedu, *De l'âme des plantes*, 284–85, repeated Duclos's error.

35. Clark-Kennedy, *Stephen Hales*, 60–66.

36. *Mémoires*, 10: 191.

37. *Histoire*, 2: 185; Justel wrote to Huet in 1671 about Borelli's book and about the difficulties of explaining capillarity: BN MS. fr. 15189: 160r.

38. AdS, Reg., 13: 39r (15 Nov. 1690); *Mémoires*, 10: 317–19; *Histoire*, 2: 184–85.

39. These were said to be connected by their lower parts to tubes which carried rising sap, and to be attached by their upper parts to tubes that transported descending sap. La Hire claimed to have observed very large valves of this sort in canes and reeds: *Histoire*, 2: 185–86.

40. Canguilhem, *Connaissance de la vie*, 34.

41. Canguilhem, "Role of Analogies and Models," 519, 517.

42. Ibid., 517.

43. As Delaporte points out in *Second règne*, 11.

11: CHEMICAL AND MECHANICAL EXPLANATION OF PHYSIOLOGICAL PROCESSES

1. Neither the strict dualism that pitted mechanistic against biological explanation (Kiernan, *Science and the Enlightenment*) nor the dispute between chemical and triturationist mechanism (Brockliss, *French Higher Education*, 400–408) emerged in the seventeenth-century Academy. For the scope of chemical research in England, see George, "Chemical Papers."

2. Mendelsohn, "Philosophical vs. Experimental Biology," and discussions by Roger and Plantefol; Gasking, *Investigations*, 37–69; Guyénot, *Évolution*, 214–15; Roger, *Sciences de la vie*, 325–54, 363, 391, 442.

3. AdS, Reg., 1: 33. Perrault cited Theophrastus on the *Causes of Plants*,

chap. 4. Compare Dodart, *Mémoires des plantes,* 138, who revived the question in 1678: AdS, Reg., 8: 189v. On growing plants from their ashes, see Le Febvre, *Compleat Body,* 156, Marx, "Alchimie et palingénésie," and Debus, "Further Note on Palingenesis." The view that plants grew from their salts was defended by Pinault in his *Traité du jardinage,* dedicated to the Academy and reviewed by Duclos in AdS, Reg., 6: 48r–57v; see 54v.

    4. AdS, Reg., 8: 135v, 151v (4 Aug. 1677, 23 Mar. 1678, Dodart); 7: 234r (14 Jan. 1679, Perrault); 11: 116v, 124r, 125v (20 Jan., 11 Apr., 25 Apr. 1685, Marchant), 133r–v (14 July 1685, La Chapelle). For descriptions see Marchant, *Descriptions de quelques plantes nouvelles;* BMHN MSS. 449–51. For Bourdelin's chemical analyses, see: AdS, Reg., 8: 198r–v, 200r–v, 201v–2r (15, 22 Mar., 26 Apr., 10 May 1679). See also AdS, Cartons 1666–1793, 2, 7: 119v–26r, 129r–32v, 135v–37r (1679); 2, 8: 116r–19r, 131v–32r, 146v–47r (1681, 1682); 1, 9: 104–6, 109–10, 127–32, 139–40, 381 (1684, 1686). See also Homberg's analyses in AdS, Reg., 14: 122r–25r, 200r (1 June, 23 Nov. 1695).

    5. AdS, Reg., 1: 34 (Perrault); *Mémoires,* 10: 120–22, 123; *Historia,* 309–10 (Tournefort).

    6. *Mémoires,* 10: 122, 124, 125. Grew had found seeds in the capsules of a hart's-tongue. Tournefort planted a hart's-tongue in a well, hoping for a natural and isolated environment to test Grew's hypothesis; a year later he observed several young plants growing on the opposite side of the well. Doubting at first that this was the same plant, because the seedlings had only a single leaf (the gametophyte) that was rounder than the leaves of the original plant, Tournefort continued to observe and eventually saw the characteristic leaves. Morison had made a similar experiment; he identified the prothallium as cotyledons of young ferns and believed he had proved that ferns were reproduced by seeds: Gunther, *Early Science in Oxford,* 3: 209; Arber, "Nehemiah Grew," 63.

    7. *Mémoires,* 10: 102, 125. In the 1670s Marchant studied mushrooms, bringing to a meeting "les premiers commencements de la formation des champignons qui sont dans les crottes de cheval, mises en une couche depuis un an, dans lesquelles il a fait remarquer de la moisissure, puis des fillets. . . .": AdS, Reg., 8: 155r. No mention was made of seeds. See also Plantefol's remarks following Mendelsohn, "Philosophical vs. Experimental Biology," 228–29.

    8. *Mémoires,* 10: 124, 411–12, 414, and plate 17; Mariotte, *Végétation,* 136, 137. Tournefort cited observations in England by Ray and in Provence, Poitou, and elsewhere by other naturalists.

    9. Gasking, *Investigations,* 63; Berthier, "Le mécanisme cartésien," 2: 88n. 2. In his synthesis of contemporary research on plants, Régis summarized the arguments for preformation and against spontaneous generation: *Système,* 465–66 (livre 6).

    10. Mendelsohn, "Philosophical vs. Experimental Biology," 216–17.

    11. Mariotte, *Végétation,* 138.

    12. Gasking, *Investigations,* chaps. 2–5, esp. 41–42, 57–59; Guyénot, *Évolution,* 273–76, 331.

    13. Roger, *Sciences de la vie,* 352–53.

    14. AdS, Reg., 8: 141r, 156r–v, 189v, 192r, 215v (17 Nov. 1677, June 1677–

Apr. 1678, 23 Nov., 7 Dec. 1678, Apr. 1678–June 1679); 7: 158r (14 May 1678); *Historia,* 157–58, 170.

15. Mariotte, *Végétation,* 137, 139. Guyénot, *Évolution,* 289–93, and Gasking, *Investigations,* 66–67, discuss how preformationists explained hereditary variations. See Huygens, *Oeuvres,* 9: 361 (7 Feb. 1690), for the problem of explaining grafts along preformationist lines.

16. Mariotte, *Végétation,* 137; Gasking, *Investigations,* 42; Bugler, "Précurseur," 247–48. Fontenelle accepted epigenesis: Marsak, *Fontenelle,* 26. Citing variation, Daniel Tauvry later objected to *emboîtement:* Guyénot, *Évolution,* 291.

17. Mariotte, *Végétation,* 138.

18. Mendelsohn, "Philosophical vs. Experimental Biology," 225.

19. AdS, Reg., 1: 33. Dodart, *Mémoires des plantes,* 137, referred to the germination of plants in a vacuum and the extraction and analysis of lixivial salts from soil, work that bore no resemblance to Harvey's or Malpighi's studies.

20. Hoppen, *The Common Scientist,* 141–42, 260n. 200.

21. Mariotte, *Végétation,* 128–29; AdS, Reg., 7: 158r–v (14 May 1678, Mariotte); 8: 151v (23 Mar. 1678, Dodart). Cf. *Historia,* 170.

22. Mariotte, *Végétation,* 139.

23. Ibid., 137; Perrault, *Circulation,* 106–8, 118–19, 123–24.

24. Mariotte, *Végétation,* 135, and *De la nature des couleurs,* in *Oeuvres,* 310–11.

25. Mariotte, *Végétation,* 138; *Mémoires,* 10: 120, 124, 125, 126. For Tournefort, the exceptional number of mushrooms after the London fire showed that an alteration in the earth's juices could induce a dormant seed to grow. He conjectured that "the juice that dissolved the debris of calcinated houses" must be an especially good medium for causing seeds "that had been in the earth perhaps for a long time" to germinate.

26. AdS, Reg., 17: 44r–49r (4 Dec. 1697): ". . . il y a dans la terre un sel que l'on peut appeler comme naturel, lequel est un mélange de sel marin, de nitre, de sel fixe, de sel ammoniac. On y peut ajouter l'alun et le vitriol. En examinant tous ces sels sans employer le feu, l'on trouve qu'ils donnent des indices d'acide et d'alcali" (49r). Ibid., 18: 4v–5v (12 Nov. 1698): "la terre qui se trouve dans les champs et dans les jardins contient considerablement du soufre" (4v). Cf. *Historia,* 445–56. See also the preface to Tournefort's *Histoire des plantes,* and the review of that book in *Phil. Trans.,* no. 245 (1698): 385.

27. Dodart, *Mémoires des plantes,* 137–38; AdS, Reg., 8: 3v, 92v–95r, especially 94v (23 Jan. 1675, 26 Aug. 1676). Other academicians proposed tests to learn whether earth increased in weight due to cooking and whether it was saltier in the spring: AdS, Reg., 8: 190r (23 Nov. 1678); 10: 12r (13 Mar. 1680). Earths associated with mineral waters were tested with a solution of turnsole in 1680: ibid., 10: 46v (28 Aug. 1680). Perrault and Duclos discussed chalk in 1683: ibid., 11: 6v (1 Sept. 1683). Borelly wanted to compare the acids of plants with those in minerals: ibid., 10: 96r (15 Apr. 1682). La Chapelle revived Dodart's earlier plan of comparing the salts in plants and soils: ibid., 11: 157v (30 Jan. 1686).

28. AdS, Reg., 8: 63r–v (6, 13 Nov. 1675), cf. Borelly, 73v–74r (22 Jan. 1676); 75r, 78r–v, 81v, 85r, 88v, 89r (12 Feb., 11, 18 Mar., 6 May, 10 June, 1, 8 July 1676). Cf. AdS, Cartons 1666–1793, 1, 4: as numbered from front of vol.,

101r–7v (1675); as numbered from back of vol., 12r–13v, 15r–v, 19v–20r (1675); 1, 5: 28r–30v, 40v–42r, 55r–60v, 75r–78v, 85r–v, 87v–88v, 108r–20v, 124v–26v, 137r–39r, 157r–58v, 168r–69r (1676–77); 2, 6: 73. In 1675 Bourdelin had identified only one earth that yielded an acid liquid and another that produced a very sour (*acre*) spirit similar to spirit of salt; marl effervesced with spirit of salt: *Histoire,* 1: 198.

29. AdS, Reg., 8: 94v–95r (26 Aug. 1676). Borelly suggested using common salt, salt of tartar, "ou autre alcali sel armoniac, ou autre de Nature pareille sublimé, et derechef depuré &c."

30. Ibid., 8: 93r: "pour en tirer tout le sel, et touttes les diverses substances ensemble dans leur Cahos."

31. Ibid., 8: 93v–94v.

32. AdS, Reg., 12: 21r–v (4 Dec. 1686); 18: 13v (19 Nov. 1698).

33. Since a large volume of water contained only a small quantity of mineral salts, plants had to imbibe great quantities. Mariotte calculated how much water evaporated daily from plants: *Végétation,* 135–36, 140–41.

34. AdS, Reg., 4: 79v–80r; *Mémoires,* 10: 195–97, 406–15. Cf. Perrault, *Circulation,* 77–80; for other mechanistic explanations of growth, see Grew, *Anatomy of Plants,* 48–49; Dedu, *De l'âme des plantes,* 297–98; Bugler, "Précurseur," 247.

35. AdS, Reg., 14: 122v (1 June 1695): ". . . les organes des jeunes graines ne contiennent qu'une Séve aqueuse et fort fluide, qui n'est pas encore bien digerée, dont les parties salines, terrestres et aqueuses se mêlant avec le temps plus parfaitement s'epaississent, et forment en partie cette huile, qui se forme peu à peu, . . ."

36. Ibid., 14: 123v: ". . . dans les jeunes graines le phlegme avec son sel et une partie de sa matiere terrestre composent avec le temps la quantité d'huile qui se trouve dans les graines meures. . . ."

37. Ibid., 14: 123r. Duclos held similar views on how oil is formed in plants: ibid., 4: 48r–54v.

38. Howe, "Root"; Webster, "Water as the Ultimate Principle of Nature," 97, 100–107; *JdS* (1671): 612–13, review of Du Hamel's *De corporum affectionibus.*

39. *Histoire,* 2: 133 (1692); *Historia,* 1: 321–22.

40. Dodart, *Mémoires des plantes,* 205–6. Boyle pointed out that Helmont could not prove that minerals were produced by the water: *Sceptical Chymist,* in *Works,* 1: 496–98. His own experiments gave ambiguous evidence: *Considerations and Experiments Touching the Qualities and Forms,* in *Works,* 3: 102–9. Earlier English scientists, stimulated by Palissy and influenced by a Paracelsian tradition that emphasized "nutritive water and a life giving salt," had already studied the issue: Debus, "Palissy, Plat, and English Agricultural Chemistry," quotation from p. 88. Their interests resembled those of seventeenth-century academicians: *Phil. Trans.* 1 (1665): 91–92, and 10 (1675): 293–96.

41. Mariotte, *Végétation,* 124, 125, 127; Perrault, *Circulation,* 72–73.

42. Perrault, *Circulation,* 105, 110, 122.

43. AdS, Reg., 17: 40r–v (27 Nov. 1697); cf. Hoffmann, *Fundamenta Medicinae,* 2: 66–67.

Notes to pp. 153–157 317

44. *Mémoires*, 10: 120, 124, 126. This position could be Aristotelian or Paracelsian: Pagel, *Paracelsus*, 97, on "growing water."
45. Guyénot, *Évolution*, 115–18; AdS, Reg., 6: 122r (13 July 1669); Perrault, *Circulation*, 89; for Homberg, see n. 43, above.
46. On eclecticism in the sciences, see Roger, *Sciences de la vie*, 164; Debus, *The English Paracelsians*, 149; Brockliss, *French Higher Education*, chaps. 7 and 8.

12: THE NEW INSTRUMENTS AND BOTANY

1. See Brown, *Scientific Organizations*, chaps. 4–6; Stroup, "Christiaan Huygens." Huygens, *Oeuvres*, vols. 3–8, 19, 22, and passim, reveal his work with Thuret, Papin, and others on air pumps, microscopes, and clocks. For an inventory of Picard's instruments, see AdS, Reg., 9: 198v–99r.
2. Gunther, *Early Science in Oxford*, 6: 5, 22, 76, and passim; Ornstein, *Rôle of Scientific Societies*, 107–10; Middleton, *Experimenters;* Mariotte, *Traitté de nivellement;* La Hire, *L'école des arpenteurs;* Auzout, *Traité du micromètre;* Roberval, "Nouvelle manière de Balance"; Huygens, "Extrait d'une lettre . . . touchant une nouvelle manière de barométre."
3. The Academy tested an aerometer adapted from a design of the Florentine academy: Dodart, *Mémoires des plantes*, 188–91; see Hooke's comment in *Philosophical Collections*, 1: 39–40. Duclos used an aerometer, along with a compound balance, to analyze mineral waters: *Observations*, 198–201; discussed by Stubbs, "Chemistry at l'Académie Royale des Sciences," 90. Homberg experimented with an aerometer in his air pump: AdS, Reg., 16: 209r (31 July 1697). The enameler Hubin supplied and repaired aerometers for the Academy and probably also made its thermometers: BN MS. n. a. fr. 5147 records purchases and repairs. For a definition of "aerometer," see Furetière, *Dictionnaire;* for an illustration, see *Historia*, plate facing p. 389, fig. 2a, p. 439; for Homberg on the aerometer, ibid., 438–40. Academicians again adapted a design of the Accademia del Cimento when they used a thermometer to regulate the heat of distillatory fires; Dodart discussed apparatus in 1674 and 1675, but the amanuensis was unfamiliar with the words so that the instruments he named are unknown: Dodart, *Mémoires des plantes*, 180–81; AdS, Reg., 8: 7r, 3r, 4v.
4. Borel, *Observationum microscopicarum centauria*, Obs. 6–8, 17, 19, 46, 63, 96, tried to identify atoms as proof of atomism and also examined the exteriors of plants; he was not interested in plant anatomy. Hooke described the cellular structure of plants, while Henshaw found vessels in the wood of walnut trees: Hooke, *Micrographia*, 106; von Sachs, *History of Botany*, 229. In 1681 Schrader, *De microscopiorum usu*, summarized botanical and anatomical microscopy, citing Borelli, Swammerdam, Grew, Malpighi, Hooke, and others; he discussed the circulation of sap, 18–19, without mentioning the Academy's work. On the history of the microscope and its uses, see Turner, *Essays*, and Clay and Court, *History*.
5. Roger, *Sciences de la vie*, 183–84. Dodart recommended using microscope and loupe to correct descriptions or illustrations of plants: BMHN MS. 450: 107r, 124.
6. AdS, Reg., 1: 33:

> Les Experiences sur la naissance des Plantes se feront en considerant les racines, et semences, et les examinant diligemment avec le Microscope soit avant que de les mettre en terre soit en les en tirant en divers temps pour considerer les differents Changements qui leur arrivent en la grandeur ou en la figure de leurs pores, en leurs sucs, pesanteur, couleur, odeur, saveur &c. Ensuitte on considera ce qui arrive a leurs germes quand ils commencent a pousser principalement a ceux qui sont enfermez au dedans des grandes semences comme on voit aux glands du chesne, ou on remarque la racine, le tronc et les Branches de tout l'arbre qui paroist desja formé et distingué avant que de sortir d'entre les deux parties esquelles le gland a acoustumé de se fendre.

Perrault used the terms "microscope" and "engyscope": ibid. and 36; see Furetière, *Dictionnaire,* for the latter.

7. AdS, Reg., 8: 218r–v (Apr. 1678 to June 1679). See chap. 10, above, for La Hire's later views about the rise of sap.

8. Ibid., 8: 156r (23 Mar. 1678): "du bled en herbe, dans lequel on voyoit tous les noeuds, et l'epic formé avec les grains commencez sur un tuyau de deux lignes de long."

9. Ibid., 11: 1r (23 June 1683), 168v (6 Apr. 1686).

10. Ibid., 4: 81r–v (7 July 1668); Mariotte, *Végétation,* 129, 130–32, 137–38, 143. He distinguished between fibers and filaments and argued that spongy matter adhered to the membrane; his plants and descriptions are different from Hooke's in *Micrographia,* 101–15.

11. *Mémoires,* 10: 101–3, 120–22, 191–97, 406–15; *Histoire,* 2: 153–54 (1692).

12. AdS, Reg., 7: 176r, 185r–v (16, 30 July 1678). On the history of spherical lenses, see Rooseboom, "History of the Microscope," 272; Daumas, *Scientific Instruments,* 45. Several savants described how to make and use the glass globules: see, for example, Huygens, "Extrait d'une lettre . . . touchant une nouvelle maniere de microscope," and *Oeuvres,* 8: 90–93, 96–97, 113–14, 122–25, 128–29, 131, 187–88; 13, 2: 520–27, 680–85; Hartsoeker, "Extrait d'une lettre . . . touchant la maniere de faire les nouveaux microscopes"; Hooke, *Lectures and Collections . . . Microscopium,* 92, 97–98; Locke, *Travels,* 250.

13. AdS, Reg., 7: 185r–v, 244v; Huygens, *Oeuvres,* 22: 269.

14. AdS, Reg., 7: 244r: in sunflowers it resembled balls with rays; in wood sorrel (*trifolium acetosum*) it was round and pierced in the center; and in jonquil it looked like physic nut or croton seed (*pignons d'inde*). Du Hamel noted observations with Huygens's microscope in his annual report to Colbert: ibid., 8: 219r (Apr. 1678–June 1679).

15. Huygens, *Oeuvres,* 8: 65, 106, 112 (quotation), 205, 213.

16. Ibid., 13, 2: 699–700.

17. A spherical lens was most appropriate for examining transparent objects, since any object had to be held very near the glass globule, making illumination difficult. Observers needed either a dark background or an oblique source of light, and Huygens developed interchangeable diaphragms to regulate the amount of light. A liquid was best seen when a drop of the liquid adhered to the glass lens. For holding other objects, Huygens preferred mobile glass slides to narrow tubes. See Hooke, *Lectures and Collections . . . Microscopium,* 98–99; Huygens, *Oeuvres,* 13, 1: cxlii; 13, 2: 520–26; 8: 64–65, 212; Rooseboom,

"Huygens et la microscopie," 61, 72nn. 22, 24–26. On the mounting of the glass globule, see Daumas, *Scientific Instruments*, 46–47.

18. AdS, Reg., 7: 200r (20 Aug. 1678); Huygens, *Oeuvres*, 8: 92–93, 106, 112, 114, 123–24, 128–29, 130–31, 187–88.

19. On the relative advantages of the various simple microscopes available at the end of the seventeenth century, see: Hooke, *Lectures and Collections... Microscopium*, 96–97; Rooseboom, "History of Microscope," 270–71; van Cittert, "The 'van Leeuwenhoek Microscope,'" "The Optical Properties of the 'van Leeuwenhoek Microscope,'" and "On the Use of Glass Globes as Microscope-Lenses"; Rooseboom, "Concerning the Optical Qualities of Some Microscopes made by Leeuwenhoek." Huygens's spherical lenses probably had a magnifying power of 40x: Rooseboom, "History of the Microscope," 272. I am indebted to Gerard L'E. Turner for discussing with me the problems of late seventeenth-century lenses and for pointing out that any improvements made before the development of achromatic lenses in the eighteenth century would have been very modest. He is skeptical of the view that spherical lenses provided better images than ground lenses.

20. The articles he wrote in his own and Hartsoeker's names for the *Journal des sçavans* stirred up interest, and he demonstrated the instrument outside the Academy to Colbert, his brothers, and "some learned men who live with them": Huygens, *Oeuvres*, 8: 91–92, 96–99, 100–103; *Historia*, 171.

21. On Huygens's interest in microscopes and their uses, see Rooseboom, "Huygens et la microscopie," 59; Huygens, *Oeuvres*, 4: 334; 7: 315–16, 400, 417; 8: 21n. 2, 58–63; 22: 553, 564, 595, 599, 686, 698, 702.

22. *Phil. Trans.* (11 Mar. 1666); printed in Boyle, *Works*, 3: 154–55; Gunther, *Early Science in Oxford*, 6: 245–46 (7 June 1665); 7: 493 (11 July 1678). See also 'Espinasse, *Robert Hooke*, 51, 171n. 21. Borelly thought the air pump had potential for chemical research: AdS, Reg., 11: 168v (6 Apr. 1686).

23. AdS, Reg., 1: 256, 259–60 (17 Mar., 7 Apr. 1668); 4: 10r–v (21 Apr. 1668); Huygens, *Oeuvres*, 19: 200, 207; 17: 332. When Dodart referred in his 1676 *Mémoires des plantes*, 137, to academic studies of germination in a vacuum, he had in mind these tests made by Huygens between 7 April and 12 May 1668. On Huygens's development of the air pump see Stroup, "Christiaan Huygens."

24. AdS, Reg., 4: 10v–11r (21 Apr. 1668); Huygens, *Oeuvres*, 19: 209, 211–12. Huygens reported his observations three weeks later, but the experiment lasted only eight days. Huygens put into the bell jar a device intended to show whether all the air had been evacuated; this was a tube five to six *pouces* long that was filled with water and placed with its open end in the same container of water that held the branch.

25. AdS, Reg., 4: 19v–21r (12 May 1668); *Historia*, 58; Huygens, *Oeuvres*, 17: 312–14; 19: 211–12. Fontenelle elaborated Huygens's explanation. He asserted that all bodies contained air that could escape when external air pressure diminished. In an evacuated receiver, therefore, enclosed bodies would exhale an "artificial" air, whose characteristics varied according to its origin. Fontenelle supported his argument with the observation that a fallen column of mercury could rise in an evacuated receiver; to explain this, he cited the weight of newly

exhaled air: *Histoire,* 1: 46–47. See also Marsak, *Fontenelle,* 19–22; Dijksterhuis, *Mechanization of the World Picture,* 4: 261–82.

26. Huygens, *Oeuvres,* 22: 254; AdS, Reg., 8: 59v–60r (24 July 1675). Dr. N. B. Ward's similar observation of grasses in a sealed glass bottle from 1829 to 1833 led to the use of closed glass cases for oceanic transportation of rare plants; see Lemmon, *Golden Age,* 183.

27. Boyle inspired Huygens to work on air pumps, and Huygens's first machine resembled the one Hooke had built for Boyle; Homberg's earliest inspiration was von Guericke, and in 1683 he used a pump made by Dalancé which was an improvement of von Guericke's: *Mémoires,* 10: 648, reprinted from *JdS* (1683); *Histoire,* 1: 361 (1683); see also Middleton, *History of the Barometer,* 355. By 1692, however, Homberg had made his own air pump, which resembled the machines developed by Hooke, Boyle, and Huygens: *Mémoires,* 10: 215, 256, 281 (fig.); compare *Histoire,* 2: 138.

28. *Mémoires,* 10: 319–23.

29. Ibid., 348–54; see AdS, Reg., 13: 135r–v (13, 17 June 1693).

30. *Mémoires,* 10: 349–51.

31. Ibid., 353–54; see also *Histoire,* 2: 187–88 (1693); *Historia,* 324–25 (1693).

32. *Mémoires,* 10: 351–52.

33. Ibid., 352.

34. Ibid., 319–23, 353. For Homberg, "vapeur" was a mixture of ethereal matter with particles of water.

35. Ibid., 354, 283, 259. Compare Dodart, *Mémoires des plantes,* 209; *Histoire,* 1: 47; 2: 170–72; *Mémoires,* 10: 529–36, reprinted from *JdS* (1672).

36. Willis performed his experiment in 1669 and Plot published it in his *Natural History of Oxfordshire* (1677): Gunther, *Early Science in Oxford,* 3: 207–8.

37. Huygens, *Oeuvres,* 3: 383–84; Birch, *History of the Royal Society,* 2: 29, 56, 419–21.

38. Perrault, *Circulation,* 113.

39. *Histoire,* 2: 207 (1694).

40. *Mémoires,* 10: 348.

13: MEDICAL MOTIVATIONS AND SOCIAL RESPONSIBILITY

1. Clave, *Cours de chymie,* 8. For debate on how medicine affected botany, see: Arber, *Herbals,* 6–7, and "Robert Sharrock," 5; Webster, "Recognition of Plant Sensitivity," 9, 22; Roger, *Sciences de la vie,* pt. 1, and pt. 2, chap. 1; Debus, "Paracelsian Doctrine in English Medicine," 21–22.

2. *Histoire,* 2: 66.

3. Dorveaux, "Grands pharmaciens. 1. Bourdelin," 292; Brygoo, "Les médecins de Montpellier," 12; Éloy, *Dictionnaire de la médecine,* 1: 433, 588; 2: 104, 318; 3: 159, 507–8; *DBF,* 15: 907.

4. Éloy, *Dictionnaire de la médecine,* 1: 432, 594; 2: 64, 554; 3: 280; Brygoo, "Les médecins de Montpellier," 14; Dorveaux, "Apothicaires membres.

3. Boulduc"; *IB;* there is no entry for Langlade in Hazon, *Notice,* Éloy, *Dictionnaire de la médecine,* or the standard biographical encyclopedias.

5. Huygens, *Oeuvres,* 7: 11, 17; compare 8: 541 (22 Sept. 1684).

6. Hazon, *Notice,* 151; Éloy, *Dictionnaire de la médecine,* 2: 121–24.

7. Hazon, *Notice,* 176, 191; Éloy, *Dictionnaire de la médecine,* 2: 396; 4: 365–66, 415–19; Brygoo, "Les médecins de Montpellier," 16–17; *Tournefort,* 17, 20. Tournefort corresponded with Martin Lister about surgery he had performed: Bodleian MS. Lister 2: 155–56, no date.

8. See for example: *Histoire,* 1: 27–35, 36–39, 123–24, 198–99, 250–52, 370–73, 2: 51–52, 92.

9. See for example: Charas, *Pharmacopée royale,* "Nouvelle preparation du quinquina," and "Relation de l'accident arrivé en maniant les vipéres." Dodart, "Lettre... touchant quelques grains," and his posthumous *Medicina statica Gallica;* AdS, Reg., 10: 72r, 84v (1681); Fontenelle, *Éloges,* 101; Le Clerc, *History of Physick,* a. Tournefort, *Materia medica,* and *Histoire des plantes.* Duclos, *Observations;* Lémery, *Traité universel;* Jean Marchant, *Méthode nouvelle pour guerir la fievre maligne;* Tauvry, *Pratique des maladies croniques, Nouvelle pratique des maladies aigues,* and *Traité des medicamens et la maniere de s'en servir.*

10. Hoppen, "The Nature of the Early Royal Society," 255, 270n. 89.

11. Roger, *Sciences de la vie,* 169n. 35.

12. AdS, Reg., 11: 157v (30 Jan. 1686). Louvois also hoped that academicians would persuade physicians to abandon any "recherche inutile du remede universel qui est comme la pierre philosophale," a subject that had interested Duclos and Bourdelin: BN MS. n. a. fr. 5133: 45–58.

13. Brown, *Scientific Organizations,* 18–30, 195, 263; Roger, *Sciences de la vie,* 173, 175; Howard, "Medical Politics"; Brygoo, "Les médecins de Montpellier"; Whitmore, *The Order of Minims,* 228; Handford, "Chemistry at the Jardin du roi," 19–20, 47–50; Partington, *History of Chemistry,* 2: 172, 173, 269, 289; Multhauf, *The Origins of Chemistry,* 264–67.

14. AdS, Reg., 1: 30–31, 36–38 (Jan. 1667); 8: 117r–20r, 141r–v (2 June, 17 Nov. 1677); 10: 96v (22 Apr. 1682); 11: 24r, 64r–66r (17 Nov. 1683, 27 Apr. 1684); BN MS. n. a. fr. 5133: 29–31 (1667); *Histoire,* 1: 161–62.

15. Dodart, *Mémoires des plantes,* 140–42.

16. Ibid., 143. Other academicians also discussed poisons and their antidotes: AdS, Reg., 11: 163r–64v (2 Mar. 1686); 14: 24v (1 Sept. 1694); *Histoire,* 2: 182–83 (1693).

17. BN MS. n. a. fr. 5133: 29–30; Bertrand, "Les Académies d'autrefois" (1866): 345, and *L'Académie et les académiciens,* 14–15. See also Metzger, *Les doctrines chimiques en France,* 354–55; Salomon-Bayet, "Opiologia," 126–28. Bourdelin, however, obtained and tested the urine of the "petits garçons" and "petites filles de St Esprit": AdS, Cartons 1666–1793, 2, 6: 185v–81v [sic] (1676).

18. AdS, Reg., 14: 14v (12 May 1694); 15: 43r–46r, 194r–97r (5 Sept. 1696); BMHN MS. 450: 52r; Dodart, *Mémoires des plantes,* 142–43, 151, 232–36; *Mémoires,* 10: 244–47; Salomon-Bayet, "Opiologia," 142–50.

19. AdS, Reg., 10: 35r (3 July 1680); cf. 97r–v (6 May 1682); Oldenburg, *Correspondence,* 1: 225, 227, 229; also cited by A. R. Hall, "Henry Oldenburg et

les relations scientifiques au XVIIe siècle," 294. Oldenburg said Duclos practiced Paracelsian spagyric medicine; Whitmore, *The Order of Minims,* 227, points out that in English usage "spagyrical" meant "alchemical" but in France it referred to the use of antimony. Borelly wanted to test antimony and mercury: AdS, Reg., 11: 164r. See also Poynter, ed., *Chemistry,* 44.

20. In addition to previous references, see: AdS, Reg., 8: 134r, 135v, 174r–v, 224v (30 June, 28 July 1677, 18 May 1678, 23 Aug. 1679); 7: 244v (13 May 1679); 10: 149v (July 1682–June 1683); 11: 24r, 163r–64v (17 Nov. 1683, 2 Mar. 1686). *Histoire,* 1: 373; 2: 49–50, 68. Compare 'Espinasse, *Robert Hooke,* 151; Hoffmann, *Fundamenta medicinae,* 115–42.

For familiar remedies: AdS, Reg., 10: 4r, 26r, 69r, 86r, 113r (13 Dec. 1679, 19 June 1680, 4 June, 10 Dec. 1681, 26 Aug. 1682); 12: 1v–2r, 116v, 143v–44r (8 May 1686, 22 Dec. 1688, 3 Sept. 1689); 13: 140v (12 Aug. 1693); 14: 3v–4r, 16r (2 Dec.1693, 2 June 1694); 15: 107v (27 June 1696); 17: 38r–39v (27 Nov. 1697); *Histoire,* 1: 329 (1681), 2: 182, 183 (1693).

For less common remedies: AdS, Reg., 8: 174r (18 May 1678); 11: 126v, 129v–30r (2, 23, 26 May 1685); 12: 88v, 116v, 131r–v (2 June, 22 Dec. 1688, 13, 20 Apr. 1689); 13: 3r, 14r, 39v, 63v, 130v (15 Feb., 15 June, 15 Nov. 1690, 30 May 1691, 1 Apr. 1693); 14: 3r, 16r, 22v, 83v, 144r–v, 196r (25 Nov. 1693, 2 June, 4 Aug. 1694, 6 Apr., 6 July, 19 Nov. 1695); *Histoire,* 1: 427.

On quinine, see *Mémoires,* 10: 92–98 (31 May 1692); on opium, see n. 18, above, and AdS, Reg., 8: 192v (14 Dec. 1678); 14: 22v (4 Aug. 1694). On Dodart's interest in nutrition, see chap. 7.

21. Brockliss, *French Higher Education,* chap. 8, esp. sect. iv. Duclos, however, deplored the empirical approach: BMHN MS. 1278. The Academy's views were often controversial: Roger, *Sciences de la vie,* 179–81.

22. For the last, see the review in *JdS* (1671): 616, of Du Hamel's *De corporum affectionibus.*

23. Roger, *Sciences de la vie,* 444.

24. AdS, Reg., 4: 51v–52r (9 June 1668).

25. Boas, "Acid and Alkali," 14–18; Multhauf, "J. B. van Helmont's Reformation of the Galenic Doctrine of Digestion"; Mendelsohn, *Heat and Life,* 18–19.

26. Multhauf, *The Origins of Chemistry,* 218, 222–23; Webster, *The Great Instauration,* 274; Tournefort, *Histoire des plantes,* aiiij^v.

27. Dodart, "Lettre... touchant quelques grains," in *Mémoires,* 10: 561–66; parts of this section on ergotism have been published in Stroup, "Some Assumptions." I am grateful to Martinus Nijhoff Publishers for permission to reprint those passages.

28. Greulach and Adams, *Plants,* 50; Alexopoulos and Mims, *Introductory Mycology.*

29. Barger, *Ergot and Ergotism,* 10–13, 40–60, 65–70, 83; Bové, *Story of Ergot,* 137–44; Brothwell and Brothwell, *Food in Antiquity,* 145–55; von Hilden, *Gründlicher Bericht vom heissen und kalten Brand;* Thal, *Sylva Hercynia;* Bauhin, *Pinax,* 23, "Secale luxurans"; his *Theatri botanici,* 1, 4, xvii: 433–34, includes what is said by Barger (p. 10) to be the earliest illustration of ergot.

30. Éloy, *Dictionnaire de la médecine,* 1: 304–5; *NBU,* 14: 850; *DBF,* "Dubé." Dodart read the letters of Dubé and Chatton about spurred grain to the assembly

Notes to pp. 175–178   323

on Wednesday, 31 July 1675: AdS, Reg., 8: 60r–v. See also, *Mémoires,* 10: 562, 564, 565; Chatton, "Extrait"; Stroup, "Some Assumptions."

31. *Mémoires,* 10: 561–62, 564–65.

32. Above quotations from *Mémoires,* 10: 563. For an explanation of ardent and volatile spirits, see Eklund, *Incompleat Chymist,* 22, 40, 44.

33. *Mémoires,* 10: 565; in fact, the hallucinatory form of ergotism is prevalent in some regions, the gangrenous form in others.

34. Ibid., 562–63, 564.

35. Ibid., 563–65. During 1674 Bourdelin distilled rye, barley, and wheat, but his notebooks at the Academy for the period from 1674 through 1677 do not mention spurred rye: AdS, Cartons 1666–1793, 1, 2: 1–18, 33–52, 162–66, 204–8; 1, 3: 155–65, 171–76. In 1679, he distilled seeds and flour: ibid., 2, 7: 133r–35r, 163r–v.

36. Mariotte and Dodart studied barley, wheat, corn, and spurred rye: AdS, Reg., 8: 135v, 151v (4 Aug. 1677, 23 Mar. 1678); *Historia,* 170.

37. *Mémoires,* 10: 561–63; Bové, *Story of Ergot,* 23.

38. *Mémoires,* 10: 562, 563.

39. Ibid., 564–65. The experiment had not been performed when Dodart wrote.

40. Ibid., 564.

41. Hunault, *Discours physique sur les fievres,* 1, 54, 57; Dubé, *Medecin des pauvres,* 366-67, 374, and *Chirurgien des pauvres,* 69.

42. Pierre Goubert, "The French Peasantry," 68–69.

43. Bonnin, "À propos de la productivité agricole"; Hémardinquer, "Faut-il 'démythifier' le porc familial?"; Goubert, *French Peasantry;* Lebrun, *Les hommes et la mort.*

44. Goubert, *Louis XIV and Twenty Million Frenchmen,* 179; Barger, *Ergot and Ergotism,* chap. 2.

45. Tilly, "La révolte frumentaire," 742n. 32; Jean-Pierre Goubert, "Le phénomène épidémique," 1573–74; Barger, *Ergot and Ergotism,* 24–25.

46. Tilly, "La révolte frumentaire," 735, 749.

47. *Mémoires,* 10: 565–66. For the limited success of laws against the sale of ergot mixed with grain, see Barger, *Ergot and Ergotism,* 71, 75, 77; Jean-Pierre Goubert, "Le phénomène épidémique," 1574; Tilly, "La révolte frumentaire," 736.

48. The Academy continued this work in the eighteenth century: Fagon, "Sur le bled cornu"; Barger, *Ergot and Ergotism,* 31; Tillet, *Dissertation,* 42–45, 48; Diderot et al., *Encyclopédie,* 5: 906–7, "Ergot"; Wolff, *Vera causa,* and Lang, *Descriptio morborum ex usu clavorum secalinorum,* both discussed in *Acta eruditorum* (1718): 178–81, 309–16; Barger, *Ergot and Ergotism,* 62, 69–72.

49. Antoine, *Methode pour conserver la santé;* Belloste, *The Hospital Surgeon;* Bonet, *Bibliotheque de medecine et de chirurgie* and *A Guide to the Practical Physician;* Dubé, *Medecin des pauvres* and *Chirurgien des pauvres;* Fournier, *L'oeconomie chirurgicale* and *L'antiloimotechnie,* which includes his *Traicté de la gangrene;* Hecquet, *La medecine et la chirurgie des pauvres* and *Traité de la peste;* Hunault, *Discours physique sur les fievres;* Le Clerc, *The Compleat Surgeon* and *History of Physick;* Moreau, *De la veritable connoissance des fievres* and *Traité*

chymique de la veritable connoissance des fievres; Raynaud, *Traité des fievres malignes et pourprées;* Tardy, *Cours de medecine;* Wiseman, *Severall Chirurgicall Treatises.* See also Barger, *Ergot and Ergotism,* 70–77; Jean-Pierre Goubert, "Le phénomène épidémique," 1574; and Delamare, *Traité de la police.*

50. Dubé, *Poor Man's Physician,* 332–34; Antoine, *Methode pour conserver la santé,* 1, pt. 4, chaps. 3–6; Le Clerc, *The Compleat Surgeon,* 150–51. Peter, "Disease and the Sick at the End of the Eighteenth Century."

51. Dodart presented his study of purported remedies for the poor at several meetings: AdS, Reg., 10: 84v, 96v, 97r, 106r, 107r, 109r, 110v, 111v (5 Dec. 1681, 22, 29 Apr., 15, 22, 23, 29 July, 5 Aug. 1682). He was analyzing the medicaments discussed in such treatises as Sagot's controversial *Remedes des pauvres;* see also Denis, *Recueil... Quinzieme conference* (1674).

52. Fontenelle, *Éloges,* 102, or *Histoire... 1707,* 190–91.

53. Davis, *Society and Culture;* Darnton, *The Great Cat Massacre,* 9–72; Le Roy Ladurie, *Carnival in Romans;* Mousnier, *Peasant Uprisings;* Hanawalt, *Crime and Conflict.*

14: SCIENTIFIC PARIS AT THE END
OF THE CENTURY

1. Ross, "Scientist"; *OED,* 9: 221–23.
2. Stimson, *Scientists and Amateurs.*
3. Clair, *Rohault,* 59, citing Bourdelot, *Conversations,* 58–59.
4. Rudwick, "Charles Darwin in London." See also Allen, "Natural History and Social History."
5. AdS, dossier "l'abbé Jean Paul Bignon," assesses one of La Hire's memoirs in 1717 as not ready because "les recherches" were "trop fines et les experiences trop abstraites pour une assemblée publique," thereby indicating the Academy's view of its audience.
6. Stimson, *Scientists and Amateurs,* 55; A. R. Hall, "Introduction," to Birch, *History of the Royal Society,* 1: xix.
7. Dubarle, "The Proper Place of Science."
8. See, for example, Martin, *Livre,* on vernacular and popular scientific treatises in seventeenth-century Paris and the holdings of private libraries; Millburn, *Benjamin Martin,* on lecture demonstrations and popular journals in eighteenth-century England; Marion, *Recherches sur les bibliothèques privées à Paris;* Kaufman, *Borrowings, The Community Library,* and *Libraries and Their Users,* on patterns of borrowing from eighteenth-century English libraries.
9. Brunot, *Histoire,* 5: 21–24; 4, 1: 34n. 2; Nyrop, *Grammaire historique,* 1: 69, 72–74, 76, 77; Furetière, *Recueil des factums,* 1: 12, 15. For a complaint about words omitted from the dictionary of the Académie française, see BN MS. fr. 15189: 183r–v (1699).
10. Brunot, *Histoire,* 4, 1: 431, 432, 438–39, 45–46; Nyrop, *Grammaire historique,* 1: 76, 85.
11. AdS, Reg., 1: 30–38 (1667); Dodart, *Mémoire des plantes,* 132–33; Tournefort, *Élémens de botanique;* Brunot, *Histoire,* 4, 1: 430n. 1, 428n. 1; Furetière, *Recueil des factums,* 2: 174, 233.

12. Roberts, *Boisguilbert,* 104; Sedgwick, *Jansenism,* 141, 146, 154–55.
13. Cipolla, *Literacy,* 53, 60; Blegny, *Livre commode,* 1: 248–53; Stone, "Literacy and Education"; Huppert, *Public Schools,* and *Bourgeois Gentilshommes;* Brockliss, *French Higher Education.*
14. Burke, *Popular Culture,* 285; Ch. Perrault, *Histoires, ou contes du temps passé;* Daston and Park, "Unnatural Conceptions"; Thorndike, *History of Magic,* vol. 8; Kearns, *Ideas,* 23.
15. Martin, *Livre,* 926–57.
16. Blegny, *Livre commode,* 1: 204–16, 150–53, 156–57.
17. Martin, *Livre,* 926–57. The numbers of foreign language dictionaries suggest the greatest interest in Italian books, the least in English, making Mariotte's ability to translate Boyle's English treatises an uncommon asset: ibid., 938.
18. BMHN MSS. 447 and 2253 contain catalogues of the libraries of Nicolas and Jean Marchant.
19. Vauban estimated the number of houses in Paris in 1700: Avenel, *Histoire économique de la propriété,* 1: 476. On the air, see Colbert, *Lettres,* 5: 515; cf. Huygens, *Oeuvres,* 3: 398. For the description of Paris: Lister, *Journey,* 6–27, 232–34, 260n. 25; Scarron, Sonnet: "Un amas confus de maisons"; Blegny, *Livre commode,* 1: 241, 243–44; 2: 102–51, and passim for goods and services; Brice, *Description;* maps of Paris by de Fer and others point out the principal sites. On paving stones for Paris, see Locke, *Travels,* 269. Lister's book is a fairly reliable guide, for E. F. Geoffroy wrote to Lister that his book included "ce qui est de plus curieuse à Paris": Bodleian MS. Lister 2: 58. On Tournefort's death, see Nicéron, *Hommes illustres,* 4: 363–64.
20. Locke, *Travels,* 280; Locke's parentheses removed.
21. This simplified topography of scholarly Paris is based on: Bernard, *Emerging City;* Ranum, *Paris;* Hillairet, *Dictionnaire des rues;* Sainte-Beuve, *Port-Royal; Mémoires . . . 1722,* 139 (Fontenelle's eulogy of Varignon); Blegny, *Livre commode;* Brice, *Description;* Colletet, *La ville de Paris;* Michel, "Clergé et pastorale jansénistes"; Pedley, "The Map Trade in Paris"; Viguerie and Saive-Lever, "Essai pour une géographie socio-professionelle"; Ruestow, *Physics at Leiden,* 150n. 33. I am grateful to Armelle de Crépy for her assistance.
22. Lister, *Journey,* 23–24. Colletet's *Journal d'avis* was short-lived.
23. Martin, *Livre,* 670, 673–75, 720–27, 907–21.
24. Blegny, *Livre commode,* 1: 187–91. Shops that sold periodicals expected readers to browse and pitched their prices accordingly: a mail-order customer who wanted the journal posted on Wednesday paid more than a customer who waited until Saturday and thereby guaranteed the bookseller a larger stock for browsers: ibid., 1: 193n. 2.
25. Ibid., 1: 189, 190–91, and 2: 177; Neveu, "Vie," 501; Martin, *Livre,* 673–74. La Londe was "employé aux vérifications des toisés" at 2,000 lv. a year in 1685 and 1687 and died in 1688: *CdB,* 2: 955, 1271; Blanchard, *Ingénieurs,* 318.
26. Martin, *Livre,* 856–83. The printers included O. de Varennes, L. d'Houry, Sébastien Mabre Cramoisy, E. Michallet, J. Cusson, F. Léonard, Coignard, T. Moette, F. Le Cointe and D. Hortemels, Jacques Langlois, P. Rocolet, Jacques d'Allin, Barbin, and E. Martin. Publishers in Dijon, Geneva, Amster-

dam, Leiden, London, Oxford, Nuremberg, Leipzig, Ulm, and Milan also printed academicians' works in French, Latin, and English.

27. Pedley, "The Map Trade in Paris"; Blegny, *Livre commode*, 1: 149; plate 5 reproduces part of de Fer's eighth map in the *Traité de la police*.

28. Blegny, *Livre commode*, 1: 148. Butterfield, an emigrant from England, had become uncomfortable in his native language. He had not always been located so competitively on the quai de l'Horloge; Locke visited him in August 1677 in the rue neuve des Fossés of the faubourg Saint Germain, where his sign was "au Roy d'Angleterre": Butterfield's letters in Bodleian MS. Lister 2; Locke, *Travels*, 161–62.

29. AN O^1 1678^A, no. 6: 2v, and no. 14; AN O^1 1678, no. 9.

30. Blegny, *Livre commode*, 2: 363.

31. Brunot, *Histoire*, 4, 1: 407–8, lists several private observatories. Blegny, *Livre commode*, 1: 78, 282; 2: 73–74; Locke, *Travels*, 167; sundials, moondials, and a new pump were also for sale at rue Saint Pierre.

32. Hubin also sold glass eyes, as did Le Quin on rue Dauphine. Blegny, *Livre commode*, 1: 242; 2: 75; Mariotte, *De la nature des couleurs*, 316–17; Huygens, *Oeuvres*, 7: 261–62; *Histoire*, 1: 321–22; Stroup, "Christiaan Huygens." In 1674 and 1675 Hubin worked on a "machine des Fables d'Esope": *CdB*, 1: 804, 875; see also 934, 1010.

33. Blegny, *Livre commode*, 2: 76. La Hire wrote about Dalesme's "machine qui consume la fumée": *JdS* (1 Apr. 1686); Nicéron, *Hommes illustres*, 10: 180.

34. Blegny, *Livre commode*, 2: 75.

35. Locke, *Travels*, 161–62, 167.

36. Clair, *Rohault*, 45; McLaughlin and Picolet, "La bibliothèque et les instruments scientifiques du physicien Jacques Rohault"; Stroup, "Christiaan Huygens."

37. Ultee, *Abbey*, 17.

38. Blegny, *Livre commode*, 1: 241. Serious savants like Huygens enjoyed magic lanterns, and they also provided entertainment at *soirées*, for example at the *hôtel* de Liancourt.

39. Lister, *Journey*, 182–83. Louis Racine was frightened by an elephant at a fair: Racine, *Oeuvres*, 7: 294. Locke, *Travels*, 153.

40. Lister, *Journey*, 185–86, 190–98, 219–21, quotations on pp. 198, 220; Locke, *Travels*, 272–73; *État de la France* (1694), 1: 339; (1699), 330; *CdB*, 2: 1272, and passim (Beaulieu). Scudéry, *Entretiens*, 1: 265–336. In the eighteenth century Réaumur, casting about for sources of income for worthy savants and hoping to revitalize the *pépinerie*, suggested that its director be one of the Academy's botanists: Bertrand, *L'Académie et les académiciens*, 91–93.

41. Blegny, *Livre commode*, 1: 278–82.

42. Dodart, *Mémoires des plantes*, 139. La Quintinie's *Instructions pour les jardins fruitiers et potagers* appeared posthumously in 1690 and offered a calendar of monthly chores, advice about and pictures of gardening tools, methods of pruning trees, with illustrations, and information about cultivating orange trees. Ch. Perrault, *Hommes illustres*, 2: 83–84, gives a brief biographical notice. Davy de Virville, *Histoire*, puts La Quintinie into context.

43. Blegny, *Livre commode*, 2: 77, 97; Locke, *Travels*, 160n. 6.

Notes to pp. 195–197   327

44. Seneca, Letter 27; Blegny, *Livre commode,* 1: 216–36.
45. Blegny, *Livre commode,* 1: 227, 152n. 3; Huygens, *Oeuvres,* 4: 620.
46. Lister, *Journey,* 47–53, 94–96, 59–61; quotation on p. 60. In 1675 Locke saw Servière's museum in Lyon, which contained carved ivory, clocks, models of machines, a microscope, and other curiosities: Locke, *Travels,* 5–6.
47. Blegny, *Livre commode,* 1: 134n. 2 and 135n. 2; Fontenelle, "Éloge du P. Sébastien Truchet"; Lery, "Le P. Sébastien Truchet membre."
48. On the Academy's collection see: Wolf, *Observatoire,* 96–97, 129; Stroup, *Royal Funding,* 55–56. Huygens's apartment is described in his correspondence; for Tournefort, see chap. 2, n. 24, above. On Morin's museum, see Locke, *Travels,* 132. On Blondel, see Brice, *Description,* 2: 196–201.
49. Ultee, *Abbey,* 79–80; Saisselin, *Literary Enterprise,* 40.
50. Lister, *Journey,* 108.
51. Barthélemy d'Herbelot was renowned for the meetings of savants in his library, so that foreign visitors made a point of visiting him; sometimes he took them off to a coffee shop in the rue Mazarine, but finally stopped his meetings altogether after a series of thefts: Neveu, "Vie," 479.
52. Blegny, *Livre commode,* 1: 136n. 1, and refs. in n. 47, above; the library of the monastery of Saint Victor, where Louis Morin cloistered himself, was said by a French traveler to be small but excellent in 1733: Saisselin, *Literary Enterprise,* 40.
53. Blegny, *Livre commode,* 1: 137.
54. By 1733 the Bibliothèque du roi was "considered as holding first rank in Europe, especially for manuscripts": Saisselin, *Literary Enterprise,* 40. During the 1680s a recurring expense of the Bibliothèque du roi was the transport of books and manuscripts between it and other scholarly collections; see BN Archives de l'ancien régime 1.
55. Martin, *Livre,* 657–58; Clair, *Rohault,* 42–44, 59–60; Bigourdan, "Les premières réunions savantes," and "Les premières sociétés scientifiques." Bourdelot's Academy first met in the Hôtel de Condé and was occasionally attended during the 1640s by the two princes de Condé; later Bourdelot moved it to his own house, first on the rue de Rounon, later on rue Guénégaud; Bourdelot was widely mocked, Guy Patin and others derided him, the Condés beat him, and he was the buffoon of Queen Christina: Peumery, "Conversations médico-scientifiques," 130, 133. Denis's *conférences* began in 1664 and were held when the Academy met, on Saturday (later Wednesday) afternoons: Denis, *Recueil,* 156, 216, 240.
56. Blegny, *Livre commode,* 1: 123–24, 129, 227n. 1.
57. Roger, *Sciences de la vie,* 170–71. At least thirteen academicians— Auzout, Borelly, Carcavi, Cassini, Dodart, Du Verney, Gallois, Huygens, Homberg, Mariotte, Pecquet, Roberval, and Sauveur—attended private scientific academies: Bourdelot, *Conversations; Mémoires . . . 1731,* 93; Blegny, *Livre commode,* 1: 165–66n. 2; Stroup, *Royal Funding,* 57–60; Clair, *Rohault,* 42–60.
58. Sedgwick, *Jansenism,* 85–87, quotation from p. 86.
59. Dainville, *L'éducation des jésuites,* pt. 3.
60. Blegny, *Livre commode,* 1: 248–63.
61. Colletet, *Journal.*
62. Brockliss, *French Higher Education.*

63. Clair, *Rohault*, 25–26; Blegny, *Livre commode*, 1: 147–48; 2: 71, 342; most persons taught from their homes.

64. Blegny, *Livre commode*, 1: 146–47.

65. Blegny, *Livre commode*, 1: 142; the Collège royal was also outstanding for its teaching of oriental languages.

66. Ibid., 1: 124; see *CdB*, for the pensions paid to members of this Académie.

67. Contant, *L'enseignement;* Crestois, *L'enseignement;* Howard, "Medical Politics"; Lister, *Journey*. See catalogues of the Jardin royal prepared by Tournefort and Jean Marchant: BMHN MSS.1556–62. The frontispiece to the *Élémens de botanique* symbolized the Jardin royal: Académie des Sciences, *Troisième centenaire*, 2: 133–34.

15: ACADEMICIANS AND THE LARGER SCIENTIFIC COMMUNITY

1. Tournefort's correspondents, for example, wrote from Leiden, The Hague, Amsterdam, Barcelona, London, Venice, Rome, Bologna, Oxford, Martinique, Hamburg, Leipzig, Zurich, Basel, Palermo, Florence, and Lisbon, as well as France: BMHN MS. 253: 1–2r.

2. Mariotte, *Essai de logique*, in *Oeuvres*, 2: 612.

3. *Historia*, 6; Mariotte, *Essai de logique*, in *Oeuvres*, 2: 610.

4. Nicéron, *Hommes illustres*, 12: 96–102, quotation on p. 101.

5. Clair, *Rohault*, 46; Huygens never understood his own indebtedness to clockmakers: Leopold, "Christiaan Huygens and His Instrument Makers."

6. Raven, *John Ray*.

7. André and Bourgeois, *Recueil... Hollande*, 1: 410–11; *NBU*, 10: 169, 43: 379–80; Stroup, *Royal Funding*, 45n. 20, 56.

8. Mariotte, *Essai de logique*, in *Oeuvres*, 2: 610–12.

9. Mariotte, *Végétation*, 121. Lantin was *conseiller* to the Burgundian *parlement* and may have been the "Lentier" of the *parlement* of Dijon who participated in Bourdelot's scientific meetings, in which case he would also have known Dodart: Bourdelot, *Conversations*. See Picolet, "Sur la biographie de Mariotte," 246, 270n. 14, in *Mariotte, savant et philosophe;* Oldenburg, *Correspondence*, 5: 37–41, 74, 196–97; Leibniz, *Lettres*, 27, 44, 97, 111–18.

10. Le Febvre, *Compleat Body;* BN MS. n. a. fr. 1967: 191r–235v (1639–63), written from Montpellier and Paris.

11. BN MS. fr. 19658: 37r–38v (10 Apr. 1686).

12. Sainte-Beuve, *Port-Royal*, 5: 315–16n. 1, 328–29; 6: 156–66.

13. See n. 1, above, and BMHN MSS. 252; 998: f. 1, p. 389; 1105; 1391.

14. BN MSS. fr. 17051: 78r–82v, 84r–v, 153r–54v, 181r–82v, 192r–93v, 211r–12v; fr. 17052: 231r–v; fr. 17054: 206r–12v, 234r–35v, 259r, 260r, 263r, 288v, 292r–v, 302r, 311v–12r, 422r–23v, 453r, 487r, 491r–92v, 532r.

15. *Histoire*, 1: 12.

16. Lister became friendly with the Geoffroy family during his 1698 visit and corresponded with both father and son: Bodleian MS. Lister 2: 56, 58–59. Bourde, *Agronomie et agronomes*, 1: 81; Raven, *John Ray*, 209, 212, says that Sloane studied at Montpellier, where he met Tournefort; Sloane later conveyed

Tournefort's feelings of respect to the English botanist Ray. See also Lough, *France Observed,* and Cohen, "Isaac Newton, Hans Sloane."

17. *Mémoires,* 10: 84–90; Fontenelle's eulogy of Homberg; Stroup, "Wilhelm Homberg."

18. Vines and Druce, *Morisonian Herbarium,* xxv; Henrey, *British Botanical and Horticultural Literature,* 1: 119–27.

19. Huygens, *Oeuvres,* 3: 358.

20. Ibid., 3: 295.

21. Bodleian MS. Smith 52: 15–18. A. R. Hall, "Henry Oldenburg et les relations scientifiques au XVIIe siècle."

22. For Roemer, see AdS, Reg., 8: 220r–v, 221r (28 June, 5 July 1679); for Du Hamel, see his *De consensu veteris et novae philosophiae* (Oxford, 1669), and Bodleian MS. Rawlinson D. 398: 150, printed in Hart, *Notes on a Century of Typography,* 155. Thévenot's surviving correspondence in England treats mainly books and manuscripts, reflecting perhaps his responsibilities at the Bibliothèque du roi: Bodleian MS. Smith 130: 10, 11, and MS. Smith 11: 15r–v.

23. Fontenelle, *Éloges,* 107; Bodleian MSS. Lister 2: 153–54, 155–56 (1687), 137 (1698); MS. Ashmole 1816: 67 (1697). Lhwyd also corresponded with G. Roussel: Bodleian MS. Ashmole 1817 A: 364–68. BMHN MS. 1989 (J. Woodward, 15 Mar. 1696); Bodleian MS. Ashmole 1817 A: 450 (William Sherard, May 1701); see also Bodleian MS. Radcliffe Trust C. 2: 15r–v (Bobart to Richardson).

24. See Bodleian Sherard d. 84; Nehemiah Grew, *Anatomy of Plants,* preface; Harrison and Laslett, *Library of John Locke,* 125, no. 980 (now at the Bodleian). Locke owned books by other academicians; see ibid., nos. 789, 1380, 1381, 1907, 1908, 1908a, and 2259.

25. Locke, *Travels,* xxxix, xl, xlii, 251, 252, 254, 256, 261, 263, 275, 282. Auzout informed Locke about weights and measures, the Parisian bills of mortality, and medical remedies; Picard discussed pendulum clocks and a universal foot; Roemer demonstrated his model of Jupiter and its satellites, a level, and a tinder box; from Charas, Locke picked up medical ideas.

26. Ibid., 160–61, 272. This plant was included in N. Marchant's descriptions and engravings of rare flora printed with Dodart's *Mémoires des plantes.*

27. Juillard, "Société Royale," 85, 87. This may have followed Mariotte's request for help and perhaps explains how he came to have correspondence from Aberdeen on the subject of winds; see the treatise in his *Oeuvres.*

28. See, for example, the following Bodleian MSS.: Radcliffe Trust C. 3: 54 (Sherard to Richardson); Radcliffe Trust C. 4: 67, 68, 70, 84, 85; Radcliffe Trust C. 5: 23, 24, 106, 107, 112, 113 (Sherard to Richardson); English History C. 11: 14 (Jacob Bobart to E. Lhwyd, May 1698); Radcliffe Trust C. 1: 48, 69 (Sherard to Richardson); Ashmole 1816: 128 (M. Lister to Lhwyd). Like the Marchants, La Quintinie accumulated fruits from abroad: Bourde, *Agronomie et agronomes,* 1: 86–87.

29. Bourde, *Agronomie et agronomes,* 1: 82. La Quintinie traveled twice to England and was offered the patronage of Charles II.

30. AdS, Reg., 10: 110v (29 July 1682), communicated by Mariotte.

31. On Bishop Henry Compton, see Henrey, *British Botanical and Horticultural Literature,* 1: 144.

32. Bodleian MS. Rawlinson D. 371: 83. Dr. William Briggs wrote to Fagon that he hoped to discover microscopically "la texture la plus fine et la plus delicate des liqueurs et des parties solides, qui composent le Corps humain." He intended to publish an English account dedicated to the English king, and a French translation dedicated to Louis XIV.

33. For Huygens's correspondence with Boyle, see Maddison, "Studies in the Life of Robert Boyle." For Huygens's correspondence with English acquaintances on matters relevant to botany, see his *Oeuvres,* 3: 311, 384; 4: 201, 358; 5: 4, 58, 75; 7: 39, 473, 506, 528; 8: 311, 317. Huygens also wrote to other nonacademicians about botany: ibid., 2: 468; 3: 347–48; 4: 279; 9: 147–48; 10: 304.

34. Huygens, *Oeuvres,* 8: 38 (Oct. 1677); Dodart urged Huygens to express special gratitude to Leeuwenhoek, adding, "it seems to me that persons of this merit ought to receive a pension as external academicians." Leeuwenhoek became a corresponding member of the Academy in 1699.

35. AdS, Reg., 1: 248–49 (18 Jan. 1668).

36. See Huygens, *Oeuvres,* 7: 87–93 (1 Aug. 1671), for example. Auzout's correspondence with Oldenburg covers little more than the period during which he was also an academician, perhaps because he conveyed news in a quasi-official capacity: Oldenburg, *Correspondence,* 2–5 (from January 1665 until January 1669); *Histoire,* 1: 11; A. R. Hall, "Henry Oldenburg et les relations scientifiques au XVIIe siècle," 297.

37. Bertrand, "Les Académies d'autrefois," 339; Roger, *Sciences de la vie,* 179, identifies the emergence of "la vérité officielle, celle qu'établissent à Paris Messieurs de l'Académie des sciences, à Londres Messieurs de la Société royale."

38. AdS, Reg., 1: 200: "On a aussy arresté que toutes les choses qui seront proposées dans l'assemblée demeureront secrettes, que l'on ne communiquera rien au dehors que du consentement de la Compagnie" (19 Jan. 1667). Despite infractions, and recommendations during the 1680s that the Company publish extracts from its registers, the rule was not officially modified: ibid., 12: 19r (13 Nov. 1686).

39. Ibid., 12: 98v–99r (18 Aug. 1688). In 1691, however, when Pontchartrain asked the Academy to publish two articles a month, academicians found the burden too great, and were able to produce articles of sufficient merit for only two years before requesting a respite, on the grounds that they were too few to produce so much: ibid., 13: 71v–73r (19, 22 Dec. 1691); Saunders, *Decline and Reform,* 166–70; Stroup, *Royal Funding,* 50–51.

40. Brown, *Scientific Organizations,* 156.

41. Oldenburg, *Correspondence,* 4: 29, 31. Justel himself both criticized and praised the Academy in the 1660s and 1670s: BN MS. 15189: 141r and passim.

42. Stimson, *Scientists and Amateurs,* 70–96; Purver, *Royal Society,* passim.

43. Rohault has sometimes been taken as the model for one of M. Jourdain's teachers in Molière's *Le bourgeois gentilhomme;* Clair, *Rohault,* 33–36, argues against that view.

44. See, for example, Huygens, *Oeuvres,* 7: 253–54.

45. BN MS. fr. 1333: 1r–42v, contains Duclos's manuscript, 42v–44r, the

committee's decision. As Duclos recalled the incident on his deathbed, Du Hamel had opposed publication: *Nouvelles de la république des lettres* 4 (Oct. 1685), 1152–55; Stubbs, "Chemistry at L'Académie," 24. The Royal Society also examined the books of its members, but its policy, as Moray construed it, was to verify the factual claims of the author without judging whether the book merited publication. Corroboration of the evidence might delay, but not prevent the appearance of a book; thus Moray explained to Huygens why Digby's discourse on vegetation had not been printed seven months after its author had read it at Gresham College: Huygens, *Oeuvres,* 3: 285 (1 July 1661).

46. AdS, Reg., 10: 28v–29r (10 July 1680).

47. Ibid., 12: 98v–99r (18 Aug. 1688).

48. Saunders, *Decline and Reform,* 125–26.

49. *Historia,* 6; *Histoire,* 1: 15–16; cf. Hagstrom, *Scientific Community,* 12–16.

50. Middleton, *Experimenters,* 300.

51. *Histoire,* 1: 15.

52. Jean Marchant feared foreigners would copy the Academy's engravings of plants before the natural history could be finished: BN MS. fr. 22225: 62v.

53. BMHN MS. 89: dossier 2, draft of letter, probably late 1670s.

54. AdS, Reg., 12: 41v–42r (14 June 1687); cf. 22r, 23v, 45r (7, 14 Dec. 1686, 19 July 1687).

55. Purver, *Royal Society,* 13–14, 179.

56. Huygens, *Oeuvres,* 9: 165–66, 213. Compare G. A. Borelli's and M. Ricci's sentiment that the Accademia del Cimento should reveal "the conclusions found and demonstrated by us . . . withholding and keeping secret the arguments and demonstrations. In this manner . . . we can be certain that . . . priority . . . cannot be taken away." Quoted in Middleton, *Experimenters,* 301.

57. Middleton, *Experimenters,* 289, 291–92, 295; Neveu, "Vie," 461.

58. Sarton, *Six Wings,* 265n. 16.

59. Huygens, *Oeuvres,* 9: 91 (8 Sept. 1686); AdS, Reg., 10: 81v, 83v–85v (4, 5 Dec. 1681); Saunders, *Decline and Reform,* 89.

60. The death of the printer Cramoisy in June 1687 and La Hire's illness in the winter of 1687–1688 delayed publication: Huygens, *Oeuvres,* 9: 165–66, 262–63.

61. Articles by Huygens, Cassini, Mariotte, and Perrault were translated for the *Phil. Trans.,* and the books of Perrault, Mariotte, La Hire, Du Verney, Duclos, Huygens, Blondel, Tournefort, Cassini, Charas, and Du Hamel were reviewed there. Hooke reviewed favorably the 1679 edition of Dodart's *Mémoires des plantes* in his *Philosophical Collections,* 1: 39–42.

62. Urban Hiärne referred to the Academy's chemical laboratory in reporting on his own research in the royal chemical laboratory in Stockholm, after learning about the size and quality of the Academy's laboratory from Erik Odhelius, who wrote from Dijon in 1692. Hiärne was familiar with Dodart's chemical work and cited Dodart's results where they differed from his own. See Hiärne's *Actorum laboratorii Stockholmensis,* 3, 31–34; and Lindroth, "Urban Hiärne och Laboratorium Chymicum," 55n. 7.

63. The works of Mariotte, Perrault, and especially Tournefort were known

to the Swedish botanists. The *Propagatio plantarum* published in 1686 by Olof Rudbeck the younger is said to have been influenced directly by Mariotte and indirectly by Perrault, through the intermediary of Dedu's *L'âme des plantes*. Lars Robert, Magnus von Bromell, Jacob Ludenius, and others were influenced by Tournefort; Bromell attended Tournefort's lectures at the Jardin royal and joined his herborizations in 1702: Eriksson, *Botanikens historia i Sverige*, 80–81, 83, 113, 115–16, 124–29, 149, 160, 164, 166–69, 174–75. Tournefort's influence was also felt in Finland, where copies of his books have survived: Hjelt, *Naturalhistoriens studium vid Åbo Universitet*, 22, 66, 68, 90, 105, and *Naturhistoriens studium i Finland*.

64. Sachs, *History of Botany*, 403 (1890 ed.). On Tournefort's influence, see Bodleian MS. Radcliffe Trust C. 2: 15r–v (letter from James Bobart to R. Richardson); Bodleian MSS. Lat. Misc. C. 11, D. 25, and E. 28 and 31 (English manuscripts based on Tournefort's *Institutiones*); Laissus and Monseigny, "Les Plantes du Roi," 209; BMHN MSS. 10, 16, 17, 18, 797, 1032, 1093, 1146; Marion, *Recherches sur les bibliothèques privées à Paris;* BN MSS. fr. 21773: 299, and n. a. fr. 4732; Boccone, *Museo di piante*, 2, 61–62.

65. To avoid embarrassment, the Academy reviewed works by lesser figures, as for example when Duclos assessed Pinault's *Traité de jardinage:* AdS, Reg., 6: 48r–57v (9 Mar. 1669); *Histoire*, 1: 85; or when Dodart proposed that Gardrois be allowed to dedicate a book on natural philosophy to the Academy: AdS, Reg., 8: 15r, 37v (20 Feb., 13 Mar. 1675); see also nn. 96 and 116, below. When the Jesuit Gouye dedicated his "Theses de mathematique en forme de livre" to the Academy in 1686, academicians held a special session: ibid., 12: 6r, 8r–v (8 June, 3 July 1686). Mathurin Dissés dedicated his analyses of the mineral waters of Granssac and Fenayrols to the academician Jacques Borelly in 1686 and 1687: Chabbert, "Jacques Borelly," 226–27. Pardies dedicated his *Élémens de géométrie* to the Academy in 1671. Leibniz dedicated his *Hypothesis physica nova* (London, 1671) to both the Royal Society and the Academy, and became a member of each two and four years later, respectively; his book was reviewed in *Phil. Trans.*, 6: 22–23.

66. Evidence about the efforts of Leibniz, Hautefeuille, and Boccone is given in Stroup, "Louis XIV," n. 34. See also Michaud, *Biographie universelle*, 18: 556–57, and Locke, *Travels*, 250, on Hautefeuille; Huygens, *Oeuvres*, 8: 173, 218–19, and Leibniz's correspondence during 1690s, on Papin and Leibniz; AdS, Reg., 8: 160r, 218v, on Boccone; and Boccone's works.

67. AdS, Reg., 8: 183r (27 July 1678), printed in Huygens, *Oeuvres*, 22: 256; AdS, Reg., 11: 158v, 159r, 165v (6, 13 Feb., 6 Mar. 1686); 12: 3v, 5r, 7r, 14v–17v (22, 29 May, 15 June 1686); *Histoire*, 1: 448 (1685); 2: 14–15, 33–36, 110, 191 (1686, 1687, 1690, 1693); *Historia*, 256, 276–77, 337. The marine officer de Gennes brought "un modelle d'une machine pour faire de la toile par un simple mouvement des roües," probably the mechanical loom that he published in *JdS* the same year: AdS, Reg., 8: 172v–73r (18 May 1678); Daumas, ed., *History of Technology*, 2: 216–17; Ministère de la Culture, *Colbert, 1619–1683*, 172. Saint Hilaire and others proposed methods of desalinating water: AdS, Reg., 10: 75v, 76r–77r (30 July, 6 Aug. 1681); *Histoire*, 1: 320–21 (1681), *Historia*, 200–201; Colbert, *Lettres*, 3, 1: 238–39.

68. Videl de la Bavaniere, whose name was also given as de la Javaniere Videl, visited twice, the second time to discuss the tides of Saint Malo: *Histoire,* 1: 427 (1685), 2: 42 (1688). See also ibid., 1: 482, and *Historia,* 245 (1685), on de la Garouste or Carouze; *Histoire,* 1: 321–22 (1681), on Hubin.

69. Huygens, *Oeuvres,* 7: 253–54.

70. For remedies, see AdS, Reg., 13: 3r. For eclipses, see ibid., 14: 20r, and n. 85, below. For curious phenomena, see ibid., 13: 145v (26 Aug. 1693); *Historia,* 276, 310 (1692); *Histoire,* 2: 91, 140, 147 (1690, 1692). Academicians themselves were not exempt from the fascination for the curious: Dodart described the composition and appearance of floating islands at Saint Omer: AdS, Reg., 10: 47v (4 Sept. 1680). Perrault published a description of two unusual pears: ibid., 8: 40v (5 June 1675); *JdS* (1675): 166–67; *Mémoires,* 10: 552–54. This found its counterpart in an unsolicited letter to Sédileau describing a second such pear: AdS, Reg., 12: 89v–90r (12 June 1688). Thorndike, *History of Magic,* 8, chap. 30, emphasizes the enthusiasm for such reports of unrelated and bizarre phenomena during the seventeenth century; Daston and Park, "Unnatural Conceptions," show how that enthusiasm was transformed. For an experiment, see AdS, Reg., 10: 15r–v (3 Apr. 1678), which describes distillation of some matter from the bubo of a plague victim, and ibid., 8: 150r–v (1678), which gives a recipe for bread made with earth, sent by de Vinkeller.

71. As for example, the letter about a pear (n. 70, above) and a paper on ginseng (n. 93, below). A correspondent from Villefranche sent Borelly a paper "on the analysis of the nature of plants," recommending water of chalk (*eau de chaux*) as a solvent for extracting the sulphurous part of plants: AdS, Reg., 10: 21r (5 June 1680).

72. See tables 1 and 3–10 for payments made to these practitioners. On Deglos, Varin, and Des Hayes, see Colbert, *Lettres,* 5: 421, and n. 3; Wolf, *Observatoire,* 143–45; Olmsted, "Voyage of Jean Richer," n. 51.

73. Table 3; Stroup, *Royal Funding,* 43, 45n. 20, 56, 140–41.

74. Bodleian MS. Lister 3: 56–68; Wolf, *Observatoire,* 152, 154; Cassini, *Anecdotes.*

75. Ch. Perrault, *Mémoires,* 46. Charles Perrault unfairly includes Richer in this category, reflecting the Academy's disappointment with Richer after he returned from Cayenne. For a rehabilitation of this able astronomer, see Olmsted, "Scientific Expedition" and "Voyage of Jean Richer."

76. On Du Vivier, see AdS, Reg., 9: 110v (Aug. 1680–June 1681); *Histoire,* 1: 159, 199; Ministère de la Culture, *Colbert, 1619–1683,* 182.

77. Perrault, *Mémoires,* 47; for a summary of the duties of the usher (*huissier*) in 1714, see BA MS. 4624.

78. Chazelles helped to measure the earth in 1683 (see table 4), was professor of hydrography in Marseilles, sent measurements of latitude and longitude in the Mediterranean to Cassini, and became an academician in 1695: Stroup, *Royal Funding,* 54, 55, 77–78.

79. Bertrand, *L'Académie et les académiciens,* 5.

80. Oldenburg, *Correspondence,* 5: 507 (11 May 1669); cited also by Brown, *Scientific Organizations,* 158. This was only shortly after Cassini's arrival in Paris; Vernon also reports that Cassini told him the Academy met on Wednes-

days and Fridays, but in fact it met on Wednesdays and Saturdays. In June 1668, Lorenzo Magalotti wrote to Prince Leopold: "At the Royal Academy [of Sciences], which meets on Saturdays at the house of Monsieur Carcavi, His Majesty's librarian, I have found nobody who has offered to introduce me, and I have not recommended myself for admission at all, my ambition being extremely moderate in that direction." Quoted from Middleton, *Experimenters*, 32–33. Stubbs, "Chemistry at L'Académie," 27, says that visitors were not admitted until the 1670s.

81. Brown, *Scientific Organizations*, 159, states that Vernon was admitted to a meeting, but Vernon's letter of 12 June 1669 (Oldenburg, *Correspondence*, 6: 6) says only that he visited Huygens's apartment at the King's Library and while there had the chance to observe the dissection of a horse by Pecquet and Gaignan [sic for Gayant?], which Gallois recorded and Perrault drew. De Gennes described two experiments, one having to do with the vegetation of plants: AdS, Reg., 8: 172v–73r (1678); he later traveled to Africa and the Americas with Prozer, who showed the Academy drawings of plants the two had observed: *Historia*, 451 (1697); Goubert, *Louis XIV and Twenty Million Frenchmen*, 225. The Academy might hold special sessions for visiting princes and dignitaries — AdS, Reg., 12: 22v (ambassadors from Siam, 7 Dec. 1686); and *Histoire*, 2: 103 (James II, 1690) — thereby honoring both parties; Cosimo de' Medici seems merely to have visited the Library: Oldenburg, *Correspondence*, 6: 250 (1669).

82. *Historia* (1683, 1687); *Histoire*, 1: 361 (1683, with Mariotte), and 2: 20 (1687); AdS, Reg., 12: 60r–v (7 May 1687).

83. *Historia;* Papin had been a fellow of the Royal Society since November 1682.

84. *Histoire*, 2: 1–2 (1686).

85. Archives de l'Observatoire, Archives, B, 4, 9: 22, 28 June, 9 July 1694, 23 Nov. 1695, 18 May, 11 Nov. 1696, 7 Jan., 31 Oct. 1697, 16 Mar., 12 Apr., 27 Sept. 1699, 24 Feb. 1701. Letters from Gallet, Bonfa, and others are also found in full or summarized in AdS, Reg., from the 1680s, and in *Histoire et mémoires . . . 1699–1710*.

86. Dodart, *Mémoires des plantes*, 125, 124, 143. Informants were to be acknowledged in the description: BMHN MS. 450: 77r: "de qui la tenons nous? il ne faut jamais obmettre cela dans les plantes nouvelles." On ergotism and the *remède des pauvres*, see chap. 13, above. Bourdelin had a similar plan for encouraging physicians to send samples and information about mineral waters: BN MS. n. a. fr. 5133: 31.

87. AdS, Reg., 11: 114v–16v (16, 20 Dec. 1684, 17, 20 Jan. 1685).

88. *Mémoires*, 4: 325–33; 7, 2: 605–875.

89. Ibid., 10: 130. See also AdS, Reg., 13: 71v, 81v, 105v–6r; 14: 2v (15 Dec. 1691, 27 Feb., 12 July 1692, 21 Nov. 1693).

90. Colbert, *Lettres*, 5: 304, 314, 315–16, 320–21, 332, 336, 421, 425; AdS, Reg., 7: 255v (15 July 1679). Durasse, ambassador to Constantinople, wrote to Cassini in 1685 about "dactyles" he had seen in stones: ibid., 11: 128r (16 May 1685).

91. AdS, Reg., 10: 19v, 22r–25v (22 May 1680): letters from Antoine Galland to Dodart, written 24 Apr. 1680 in response to Dodart's request of 15

June 1679, and read by Perrault at a meeting. For an autobiographical sketch of Galland, see BN MS. fr. 15189: 78r–82r.

92. AdS, Reg., 16: 131v–32v (18 May 1697); *Historia,* 451; Prozer presented the drawings.

93. AdS, Reg., 17: 38r–39v (27 Nov. 1697); *Historia,* 451 (1697).

94. Nicolas Marchant, *Descriptions de quelques plantes nouvelles,* 247, 252, 256, 259, 276, 278, 284, 295, 309, 316, 321; AdS, Reg., 7: 234r; 8: 154v–55r; 10: 17r, 44r–v, 72r–v, 82v–83r, 109r; 11: 116v–17r, 124r, 125v; Huygens, *Oeuvres,* 8: 311, 317; Bodleian MS. Rawlinson C. 982: 27a and 28b. The plants Richer brought back are mentioned in AdS, Reg., 8: 40v, and 7: 124v.

95. Marchant, *Descriptions de quelques plantes nouvelles,* 245 ("Avertissement"): "These papers still lack several observations that the Company hopes to make this year [1676]. This delay may serve at least to provide able persons abroad with the time to send us their advice on all that we propose, before the Academy has produced anything."

96. *Histoire,* 1: 79 (1669); AdS, Reg., 8: 15r, 37r (13 Feb., 6 Mar. 1675, Needham); 10: 57v (8 Jan. 1681). G. A. Borelli's book on motion was selected for study: ibid., 14: 73v (12 Mar. 1695). Minor books reviewed during meetings included Pierre Le Givre's *Le secret des eaux minérales* (AdS, Reg., 1: 57–70; see Éloy, *Dictionnaire de la médecine,* 2: 104); see also n. 65, above. Surprisingly, the minutes do not mention Newton's *Principia,* and Newton's name appears there only rarely, once in connection with the visit of James II to the Observatory: Bertrand, "Les Académies d'autrefois," 427–28; Wolf, *Observatoire,* 129; see also Cohen, "Isaac Newton, Hans Sloane."

97. The translations cover the period from March 1668 to March 1670: Costabel, "Le registre académique 'Journaux d'Angleterre' et Mariotte," 321–25, in *Mariotte, savant et philosophe.*

98. Huygens, *Oeuvres,* 5: 283; 7: 11–12; 22: 700.

99. AdS, Reg., 4: 197r–v, 241r–46r, 252r–56v, 257r–59v, 295r–99r, 300r–309r, 318r–27v, 328r–32v; 6: 1r–6v, 7r–13r, 14r–20r, 21r–27r, 39r–47r (5, 12, 19, 26 Jan., 23 Feb. 1669); *Histoire,* 1: 79–81 (1669). BN MS. fr. 1333: 238r–62v, contains Duclos's "Remarques sur les Essais physiologiques de Boyle," with the date July 1668.

100. AdS, Reg., 8: 113v (19 May 1677). In August 1679 Mariotte presented excerpts he had translated from a "livre Anglois de Mr. Hook touchant le Ressort": ibid., 7: 256. Mariotte also knew German, for he read a book in that language by Caspar Horne on the habits and anatomy of elephants when the Academy dissected an elephant: ibid., 10: 72v (Aug. 1680–June 1681).

101. For *The Origine of Forms and Qualities,* see ibid., 1: 93–104, 107–16, 204–5 (26 Mar., 2, 16 Apr. 1667); *Histoire,* 1: 23–24 (1667); Multhauf, *The Origins of Chemistry,* 305. For *New Experiments Physico-Mechanical,* see AdS, Reg., 4: 27r (19 May 1668): Duclos reported on Boyle's experiments with the vacuum, Picard discussed those reported by the Accademia del Cimento, and then "la Compagnie a jugé que la matiere du vuide avoit esté suffisamment examinée et quil falloit passer a quelque autre matiere." For the *Aerial Noctiluca* (on phosphorus), see ibid., 10: 58r, 61r, 64r–v, 73r (22 Jan., 26 Feb., 16 Apr. 1681, Aug. 1680–June 1681); BN MS. n. a. fr. 5147: 100v, 103v, 104v (7 June 1681, 30

Mar., 5 May 1682); Homberg revived the study of phosphorus in the 1690s: Bertrand, *L'Académie et les académiciens,* 39; Leibniz, *Lettres,* 98, 107, 211–12. For the dissertation on desalinization, see *Histoire,* 1: 387-89 (1684). For an essay in *Phil. Trans.,* see AdS, Reg., 8: 77r (4 Mar. 1676), on the dissolution of copper by spirit of sal armoniac; 78v (18 Mar. 1676) on the increase of weight in certain substances. Du Hamel discussed Boyle in his *De corporum affectionibus: JdS* (1671): 614–16.

102. For individual experiments, see AdS, Reg., 11: 27r (1 Dec. 1683), on desalinating sea water; cf. AdS, Cartons 1666–1793, 1, 9: 89–91. In 1678 one of the proposals for future work was to investigate Boyle's experiments "dans les vaisseaux scellez hermetiquement": AdS, Reg., 8: 190v (23 Nov. 1678). Borelly analyzed Boyle's work on the hidden qualities of air: ibid., 10: 71r (Aug. 1680–June 1681).

103. *Histoire,* 1: 79 (1669); *Historia* (1698 ed.), 15. Partington, *History of Chemistry,* 2: 497, mistakenly assumes that Boyle's *Sceptical Chymist* was in question, but it was *Certain Physiological Essays (Tentamina Chymica);* see n. 99, above.

104. BMHN MS. 448: 40.

105. AdS, Reg., 8: 154v, 167v (30 Mar. 1678).

106. *Mémoires,* 10: 122, 124, 125; see chap. 11, above, for Tournefort's views. The Academy did not formally discuss Grew's *Anatomy of Plants* until 6 May 1699, when Geoffroy presented a copy of the book from Grew and read his own excerpt of it: AdS, Reg., 18: 274r. Correspondence at the Bodleian Library suggests Geoffroy's extensive English contacts; see n. 16, above.

107. Dodart, *Mémoires des plantes,* 241.

108. *Phil. Trans.,* 2 (1667–1668): 455, 797–99; 3–4 (1669): 853–62, 913–16, 963–65; 5 (1670–71): 1165–67, 1199, 2067–77; 6 (1671): 2119–28, 2144–49. Cf. AdS, Reg., 12: 130r (23 Mar. 1689); Mariotte, *Végétation;* Perrault, *Circulation.*

109. For academicians' studies of plant germination, see AdS, Reg., 8: 151v (23 Mar. 1678); 7: 158r–v (14 May 1678); *Historia,* 170; Mariotte, *Végétation,* 128–29, 137, 139; Perrault, *Circulation,* 89, 106–8, 118–19, 123–24; cf. Grew, *Anatomy of Plants;* discussed in chap. 11, above.

110. The difficulty of distinguishing between plants and animals was keenly felt in attempts to identify kermes and cochineal. For academicians' views, see: *Histoire,* 2: 206–7, 280; *Historia,* 339, 420; AdS, Reg., 14: 8r (20 Feb. 1694); 17: 176r (16 Apr. 1698, La Hire, with information from Guatemala). For the views of Martin Lister, John Ray, R. Reed, and Verchant (an apothecary in Montpellier), see *Phil. Trans.* 1 (1666): 362–63; 2 (1668): 796–97; 6 (1671): 2133, 2165–66, 2196–97, 2254–57, 2284–85; 7 (1672): 5059–60; AdS, Reg., 14: 201r–v (23 Nov. 1695, Verchant). On kermes, see also Locke, *Travels,* 43–44, 94, 95, 99, 100, 101; Locke knew Verchant during the 1670s, but another savant stimulated Locke's interest in kermes. Coral presented a similar problem: *Mémoires,* 10: 123 (Tournefort); cf. Boccone, "Account of some Natural Curiosities," and *Recherches,* 1: 1–46, 2: 43.

111. Perrault, *Circulation,* 90–91, 92, 122; La Hire's claims about valves are

reported in *Histoire,* 2: 184–86. Grew's *Anatomy... Begun* was reviewed in *Phil. Trans.,* 6 (1671): 3037–43.

112. Compare La Hire's paper of 1694 on the origin of springs, which examined Plot's work on the subject: *Histoire,* 2: 204; *Mémoires... 1703,* 56–69.

113. Lister noticed how isolated French savants were from both England and Italy: *Journey,* 74–75, 97, but cf. 132. See also Roger, *Sciences de la vie,* 174–77, on French ignorance of English and Dutch developments.

114. Stimson, *Scientists and Amateurs,* 75; A. R. Hall, *From Galileo to Newton,* 138.

115. A catalogue of Tournefort's library during the late 1680s survives: BMHN MS. 253. For the library of Nicolas and Jean Marchant, see BMHN MSS. 447 and 2253.

116. One scholar, a Philippe Billemet or Billemot who sent his unpublished "petit traité d'astronomie" to the Academy, asked for protection from "nostre illustre compagnie," which he compared to "une cour souveraine dans la republique des lettres": AN M 849, no. 18: 1. Pardies compared the Academy to a Chinese court of judges in his *Élémens de géométrie:* Ziggelaar, *Le physicien Ignace Gaston Pardies,* 51.

16: THE ACADEMY AS AN INSTRUMENT OF THE CROWN

1. Mariotte, *Essai de logique,* in *Oeuvres.*

2. Roche, "Milieux académiques," 98–108; reference to "personne morale," 107; confirmed by Duclos's comments in BMHN MS. 1278: 1v.

3. Stroup, "Louis XIV."

4. Bacon, *Works,* 4: 251; see also 265–70.

5. Greene, *Landmarks;* Sachs, *History of Botany.*

Bibliography

PRIMARY SOURCES: MANUSCRIPTS,
DRAWINGS, AND PAINTINGS

OXFORD. BODLEIAN LIBRARY.

MSS. Ashmole 1816 and 1817A: Correspondence.
MSS. Lat. Misc. C. 11, D. 25, E. 28, and E. 31: Botanical notebooks based on Tournefort's *Institutions*.
MSS. Lister 2 and 3: Correspondence.
MSS. Radcliffe Trust C. 1, 2, 3, 4, 5, and 11: Correspondence.
MSS. Rawlinson C. 982 and D. 398: Correspondence.
MSS. Smith 11, 52, 130: Correspondence.

PARIS. ARCHIVES DE L'ACADÉMIE DES SCIENCES.

Cartons 1666–1793, nos. 1–3: Papers and notebooks from 1666–1699, including Bourdelin's laboratory notebooks (cartons 1–2, 11 vols. organized chronologically) and Dodart's analyses, called *abrégés*, of Bourdelin's experiments (carton 3, 10 vols., organized topically or alphabetically and keyed by date to Bourdelin's notebooks; vol. 6 contains a key to chemical symbols and an assessment of the work: "Extrait des observations chymiques sur les plantes analysées selon la 2. maniere").
Carton 1667–1699: Pochettes de séances de 1667 à 1699, containing manuscripts of papers presented at meetings of the Academy, for example by Perrault and N. Marchant on botany, 1667, 1668, 1671; by La Hire, Sédileau, and Varignon on rainfall, 1692, 1693, 1694; by Morin on mineralogy, and Homberg on antimony and acid of sulphur, 1694; Dalesme on a bridge, 1695; Des Billettes on a canal lock, 1699.

Dossiers: "Bignon, l'abbé Jean Paul," "Couplet, Pierre," "Des Billettes, Gilles Filleau," "Dodart, Denis," "Gallois, Jean," "Jaugeon, Jacques," "Marchant, Nicolas," "Mariotte, Edme," "Morin de Toulon," "Perrault, Claude," "Tournefort, Joseph Pitton de," "Truchet, Sébastien."
Journal d'Angleterre. Translations into French of articles from *Philosophical Transactions*.
Registre des procès-verbaux des séances. Vols. 1–18 (1666–1699).

PARIS. ARCHIVES NATIONALES.

G^7 882–884: Trésor royal, feuilles mensuelles et états annuels des recettes et des dépenses, 1683–1685.
G^7 890: Calculs de recettes et dépenses. Consommation, 1689.
G^7 891–904: Trésor royal, feuilles mensuelles et états annuels des recettes et des dépenses, 1690–1699.
G^7 992: Pièces justificatives des états de distribution, 1690–1693.
G^7 1903: Mémoires de Desmaretz.
G^7 1905: Lettres, relations, plans de Paris, 1617–1743.
G^7 1906: Anciens inventaires du contrôleur général, n.d.
G^7 1907: Recueil par extraits des édits, arrêts... sur les affaires extraordinaires, 1689–1706 (lettres E, F, G).
M 802: Mémoire sur l'Imprimerie royale. États, règlements, arrêts et notes diverses concernant les libraires et imprimeurs. No. 1, "Mémoire historique sur l'Imprimerie royale (1789)."
M 803: Mémoires sur les principes de la navigation. Travaux du P. Truchet sur les voies navigables, les ponts et ouvrages à exécuter sur les différentes rivières, l'horlogerie, etc.
M 849: Papers of Truchet and Léonard. No. 3: Brouillons et desseins concernant un tableau de figures mouvantes que le P. Truchet fit pour le Roi, qui appelloit ce tableau, son "Petit Opera"; et dont il est parlé dans le *Dictionnaire* de Moréri, à l'article du P. Truchet. Another folder marked "Sciences" and called "Recueil de quelques memoires concernantes les nouvelles inventions dans les mechaniques, etc.," contains papers of père Léonard de Sainte-Catherine.
M 851: Papers of Gilles Filleau Des Billettes and Sébastien Truchet.
O^1 656: Pensions de la maison cassette du Roi; no. 1: "Gratifications aux gens de lettres, 1687."
O^1 883: Papiers du Grand écuyer. Personnels. Extraits mortuaires: pp. 206–7, Couplet de Tartreaux.
O^1 1678 and 1678^A: Descriptions of royal property in Paris.
O^1 1691: Observatoire de Paris.
O^1 1934^B 14: Information about pensions paid to members of the Academies of Painting and Sculpture, of Architecture, and of Rome.
O^1 2124, liasse 2: Papers relating to the Jardin royal, especially to its funding.

PARIS. ARCHIVES DE L'OBSERVATOIRE.

MS. B, 4, 9: Correspondence of Bonfa and Beauchamp to Cassini.

Bibliography

PARIS. BIBLIOTHÈQUE DE L'ARSENAL.

MS. 7464: Papers on the establishment and organization of the Academy.
MS. 4624: Regulations of the Academy of Sciences.

PARIS. BIBLIOTHÈQUE DU MUSÉUM
D'HISTOIRE NATURELLE.

MSS. 10, 16, 17, 18: Botanical notebooks of Plumier and de Jussieu, based on Tournefort.
MS. 252: Correspondence of Tournefort.
MS. 253: Tournefort's notes, especially about his library and correspondents.
MS. 259: Tournefort's notes about Bourdelin's method of distilling plants: "Observations physiques sur l'analyse des corps et principalement sur celle des plantes."
MS. 447: Notes of Nicolas and Jean Marchant about botany and catalogues of their library.
MSS. 448–451: Notes by Nicolas Marchant, Jean Marchant, and Denis Dodart for the Academy's natural history of plants. Catalogued as work of the Marchants, but many notes from the 1670s and 1680s are in Dodart's hand, known from correspondence in BN MSS. fr. 17051: 181r–82v, 211r–12v; 17054: 422r–23v, and from the abrégés in AdS, Cartons 1666–1793, 3.
MS. 797: Catalogue of plants according to Tournefort's system of classification.
MS. 998: Correspondence of Tournefort.
MS. 1032: "Catalogues des plantes du Jardin du Roy, avec la Methode de M de Tournefort, ou ses Elemens de botanique...1694."
MS. 1061: Caspar Bauhin's *Pinax* (1623), annotated by Nicolas and Jean Marchant.
MS. 1093: Vaillant's catalogue of plants, based on Tournefort.
MS. 1105: Notes of Tournefort.
MS. 1140: Antoine de Jussieu, "Reflexions sur quelques propositions."
MS. 1146: Botanical notebook of de Jussieu, based on Tournefort.
MS. 1278: "Remarques sur le projet de l'Histoire des plantes, dressé par M. Dodart, de l'Académie Royale des Sciences, et imprimé au Louvre, par M. Duclos, de la même Académie." This is Samuel Cottereau Duclos's private review of Denis Dodart's *Mémoires des plantes,* and it survives only in this copy made by Antoine de Jussieu.
MS. 1279: Guillaume Homberg, "Liste des plantes analysées au premier volume de l'Histoire des plantes."
MS. 1391: Correspondence of Tournefort.
MSS. 1556–1562: Catalogues of the Jardin royal by Jean Marchant and Tournefort, 1681–1699.
MS. 1989: Correspondence of Tournefort.
MS. 2253: Catalogue of the library of the Marchants.
MS. 2651: Antoine de Jussieu, "L'histoire et l'usage que l'on peut faire d'un histoire considerable de botanique, qui se met au nombre des volumes du Recueil d'Estampes du Cabinet du Roy," 1738 or 1739.

Rés.: *Vélins du Roy.*
Rés.: *Recueil des plantes dessinées par ordre du Roy,* by Nicolas Robert from his own miniatures, Louis Claude de Chastillon, and Abraham Bosse.

PARIS. BIBLIOTHÈQUE NATIONALE.

Archives de l'ancien régime 1, 2: Expenses of the Bibliothèque du roi, 1680s–1690s.
Archives de l'ancien régime 40: List made in 1727 of papers concerning the Bibliothèque du roi that had been in the possession of de Boze.
Archives de l'ancien régime 42: Papers about the seventeenth-century Bibliothèque du roi, with passing references to the Academy of Sciences.
Archives de l'ancien régime 50: Bibliothèque du roi, buildings, lodgings, finances, budgets.
Archives de l'ancien régime 53: Letters and documents related to acquisitions under the direction of Nicolas Clément (1682–1710). Includes "Mémoire touchant la bibliothèque envoyé à M. de C. le 30e juillet 1691," f. 8–11.
MS. Clairambault 566: Bignon's papers regarding the Academies. Fols. 251-52 have been published in Saunders, *Decline and Reform,* 256–59.
MS. Clairambault 814: États d'Officiers des Maisons des Rois Louis XIII, Louis XIV, Louis XV. Contains papers about the history and expenses of the Académie des Sciences.
MS. fr. 1333: Samuel Cottereau Duclos, "Dissertations physiques... faites en l'an 1677," and "Remarques sur les Essais physiologiques de Boyle," July 1688, from f. 238. The former is the original version of Duclos's *Dissertations physiques sur les principes des mixtes naturels,* as submitted to the Academy in an application to publish the book; the negative report of four academicians—Blondel, Du Hamel, Perrault, and Mariotte—appears on fols. 42v–44r.
MS. fr. 7801: Papers of the surintendants des bâtiments du roi, 1667–1739.
MS. fr. 13070: Papers seized from the père Léonard relating to the Academies of Medicine and Sciences.
MS. fr. 15189: Copies of the correspondence of Huet.
MS. fr. 17050: Correspondence of Mme de Sablé and docteur Vallant, including letters to and from academicians.
MSS. fr. 17051–17052, 17054: Medical, pharmacological, and natural historical notes, mostly in the hand of docteur Vallant, including letters to and from academicians.
MS. fr. 19658: Correspondence of Mabillon, including letters of Thévenot.
MS. fr. 21773: Contains extracts from the "Voyage de Tournefort."
MS. fr. 22225: Bignon's papers concerning the Academies.
MS. n. a. fr. 1967: Correspondence of Samuel Cottereau Duclos and Paul Ferry.
MS. n. a. fr. 4732: Contains notes on the "Voyage de Tournefort" (Lyon, 1727).
MS. n. a. fr. 5133–5149: Notebooks from the Academy's chemical laboratory, kept by Claude Bourdelin, 1666–1699.

PRIMARY SOURCES: PRINTED AND ENGRAVED

Académie française. *Le dictionnaire de l'Académie françoise.* 2 vols. Paris: Jean Baptiste Coignard, 1694.

Académie royale des sciences. *Estampes pour servir à l'histoire des plantes*. See Bosse et al., *Estampes*.

———. *Histoire de l'Académie royale des sciences, avec les mémoires de mathématique et de physique. Tirés des registres de cette Académie (1699–1790)*. Paris: Imprimerie royale, 1702–97.

———. *Histoire et mémoires de l'Académie royale des sciences depuis 1666 jusqu'à 1699*. 11 vols. Paris: Martin, Coignard, Guerin & La Compagnie des Libraires, 1729–33.

———. *Machines et inventions approuvées*. 7 vols. Paris: Martin, Coignard, Guerin, 1735, 1777. Vol. 1: 1666–1701.

———. See Godin, ed. *Table alphabétique*.

Acta eruditorum. Leipzig, 1682–1731.

Anonymous. "Autre machine pour faire travailler les invalides." *JdS* (13 July 1678): 282–84.

———. "Machine pour faire travailler les invalides." *JdS* (13 June 1678): 235–36.

Antoine, Dominique. *Methode pour conserver la santé . . . en faveur des pauvres*. Paris: B. Girin, 1699.

Aristotle. *Works*. Trans. and ed. J. A. Smith, W. D. Ross, et al. 12 vols. Oxford: Clarendon Press, 1908–52.

Auzout, Adrien. *Traité du micromètre*. Paris, 1667; repr. in *Mémoires*, 7.

Bacon, Francis. *Works*. Ed. James Spedding, Robert Leslie Ellis, and Douglas Dennon Heath. 7 vols. London: Longmans & Co., et al., 1879.

Bauhin, Caspar. *Pinax: Theatri botanici . . . sive index in Theophrasti, Dioscoridis, Plinii et Botanicorum qui a seculo scripserunt opera . . . .* Basel: L. Regis, 1623.

———. *Theatri botanici sive historiae plantarum ex veterum et recentiorum placitis propriaque observatione concinnate liber primus editus oper & cura Io. Casp. Bauhini*. Basel, 1658.

Bayle, Pierre. *Oeuvres diverses*. 5 vols. in 6. Hildesheim: George Olms Verlagsbuchhandlung, 1964–68. Repr. from edition published in 1727 at The Hague. Vol. 1: *Nouvelles de la république des lettres*.

Belloste, Augustin. *The Hospital Surgeon: Containing the Nature and Cure of the Following Diseases . . . .* London: J. Clark, c. 1745.

Berthelin, Pierre Charles. *Abrégé du dictionnaire universel françois et latin, vulgairement appellé Dictionnaire de Trévoux*. 3 vols. Paris: Chez les Libraires Associés, 1762.

Betbeder, Pierre de. *Questions nouvelles sur la sanguification et la circulation du sang. Ensemble, un traité des vaisseaux lymphées ou lymphatiques découvertes depuis peu*. Paris: Jean d'Houry, 1666.

Birch, Thomas. *The History of the Royal Society of London*. Ed. A. R. Hall and M. B. Hall. 4 vols. London and New York: Johnson Reprint, 1968.

Blegny, Nicolas de. *Le livre commode des adresses de Paris pour 1692 par Abraham du Pradel*. Ed. and annotated by Édouard Fournier. 2 vols. Paris: Paul Daffis, 1878.

Boccone, Paolo. "Account of Some Natural Curiosities presented to the Royal Society." *Phil. Trans.* 8 (1673): 6158–61.

———. *Icones & descriptiones rariorum plantarum Siciliae, Melitae, Galliae, & Haliae*. London: Robert Scott, 1674.

———. *Museo di fisica e di esperienza.* Venice: Io. Baptista Zuccato, 1697.
———. *Museo di piante rare della Sicilia, Malta, Corsica, Italia, Piemonte, e Germania.* Venice: Io. Baptista Zuccato, 1697.
———. *Recherches et observations curieuses sur la nature du corail . . . et autres choses . . . proposées et examinées . . . dans l'Académie de Mr. l'abbé Bourdelot.* 2 vols. Paris: Claude Barbin, 1671.
———. *Recherches et observations naturelles . . . touchant les plantes qu'on trouve dans la Sicile, avec quelques reflexions sur la vegetation des plantes.* Amsterdam: Jean Jansson à Waesberge, 1674.
Boislisle, A. M. de, ed. See Saint-Simon, *Mémoires.*
Bonet, Theophile. *Bibliotheque de médecine et de chirurgie contenant la maniere de guerir toutes les maladies. . . .* 4 vols. Geneva, 1708.
———. *A Guide to the Practical Physician . . . to which is added, an Appendix concerning the Office of a Physician.* London, 1684.
Borel, Pierre. *Observationum microcospicarum centauria.* The Hague: Adrian Vlacq, 1656.
Bosse, Abraham; Chastillon, Louis; Robert, Nicolas. *Estampes pour servir à l'histoire des plantes.* Paris: Imprimerie royale, c. 1701. Copies at New York Botanical Garden, Radcliffe Science Library of Bodleian Library, Oxford. Later series, c. 1750, at BN, Département des estampes.
Bourdelot, Pierre Michon, l'abbé. *Conversations de l'Academie de Monsieur l'abbé Bourdelot.* Ed. Pierre Le Gallois. Paris: Chez Thomas Moethe, 1672.
Boyle, Robert. *Works.* Ed. Thomas Birch. 6 vols. London: J. and F. Rivington, 1772.
Brice, Germain. *Description nouvelle de ce qu'il y a de plus remarquable dans la ville de Paris.* 2 vols. Paris: Nicolas Le Gras, 1684.
Cassini, Jean Dominique, I. *Anecdotes de la vie . . . rapportées par lui-même.* In J. D., comte de Cassini, *Mémoires pour servir à l'histoire des sciences et à celle de l'Observatoire royale de Paris,* 255–309. Paris: Bleuet, 1810.
———. *Observations astronomiques faites en Hollande, et en Angleterre, en 1697. & 1698.* In *Mémoires,* 7, pt. 2: 535–72.
Cassini, Jean Dominique, I and Cassini, Jean Dominique, II. *Observations astronomiques faites en France et en Italie en 1694. 1695. & 1696.* In *Mémoires,* 7, pt. 2: 461–533.
Chaillou, Jacques. *Questions de ce temps sur l'origine et le mouvement du sang.* Angers and Paris: G. Soly, 1664.
Charas, Moyse. "Nouvelle préparation du quinquina et la maniere de s'en servir pour la guérison des fièvres." *Mémoires,* 10: 92–98 (31 May 1692).
———. *Pharmacopée royale galénique et chymique. . . .* Paris: l'auteur, 1676.
———. "Relation de l'accident arrivé à M. Charas en maniant les vipéres, et de la maniere dont il s'est gueri." *Mémoires,* 10: 244–51 (31 Jan. 1693).
———. *The Royal Pharmacopoea, Galenical and Chymical.* London: John Starkey, Moses Pitt, 1678.
Chastillon, Louis Claude de et al. *Estampes pour servir à l'histoire des plantes.* See Bosse et al., *Estampes.*
Chatton. "Extrait d'une lettre écrite à Mr Denis par Mr Chatton Chirugien de Montargis, le 8. Novembre 1673." In Denis, *Recueil,* 292–93.

Clave, Estienne de. *Le cours de chimie*. Paris: O. de Varennes, 1646.
Colbert, Jean Baptiste. *Lettres, instructions et mémoires*. Ed. Pierre Clément. 8 vols. (including vol. of errata and table analytique). Paris: Imprimerie impériale, Imprimerie nationale, 1861–70, 1882.
Colletet, François. *Le journal de Colletet: premier petit journal parisien (1676)*. Ed. Arthur Heulhard. Paris: Le Moniteur du Bibliophile, 1878.
——. *La ville de Paris, contenant le nom de ses rues, de ses fauxbourgs, églises, monasteres, chapelles & colleges; le temps de leur fondation*. . . . Paris: Jean Musier, 1699 [Paris: Antoine de Rafflé, 1677].
Les comptes des bâtiments du roi sous le règne de Louis XIV. 5 vols. Paris: Imprimerie nationale, 1881–1901.
Condorcet, Jean Antoine Nicolas de Caritat, marquis de. *Oeuvres*. See under Secondary Sources.
Dedu, N. *De l'âme des plantes. De leur naissance, et de leur nourriture & de leurs progrez. Essay de physique*. Leiden, 1685.
Delamare, Nicolas. *Traité de la police*. 4 vols. Paris: Pierre Cot, 1703–38.
Denis, Jean Baptiste, ed. *Recueil des mémoires et conférences qui ont esté présentées à Monseigneur le Dauphin*. Paris: Frederic Leonard, 1672–74; Laurent d'Houry, 1682, 1693.
Descartes, René. *Discours de la methode pour bien conduire la raison, et chercher la verité dans les sciences. La dioptrique. et les meteores. Qui sont des essais de cette methode*. Paris: Henry Le Gras, 1658. Repr. in *Oeuvres*. Ed. Charles Adam and Paul Tannery. 12 vols. Paris: Léopold Cerf, 1897–1913.
Dodart, Denis. "Lettre. . . contenant des choses fort remarquables touchant quelques grains." *JdS* (1676): 69–71; *Mémoires*, 10: 561–66.
——. *Medicina statica Gallica*. In P. Noguez, ed., *Sanctorii Sanctorii de statica medicina*, vol. 2. Paris: Natale Pissot, 1725.
——, ed. *Mémoires pour servir à l'histoire des plantes*. Paris: Imprimerie royale, 1676; repr. in *Mémoires*, 4. Contains Nicolas Marchant, *Descriptions de quelques plantes nouvelles*. 2d ed., lacking Marchant's *Descriptions*, Paris: Imprimerie royale, 1679.
Dubé, Paul. *Le chirurgien des pauvres, qui enseigne le moyen de guerir les maladies externes par remedes faciles à trouver & preparer, en faveur de ceux qui sont éloignez des villes*. Dernière éd. Paris, 1686 [1669].
——. *Le medecin des pauvres*. . . . Dernière éd. Paris, 1686 [1669].
——. *The Poor Man's Physician and Surgeon*. . . *with an addition of the True Use of the Quinquina or Jesuites pouder*. Trans. from the 8th ed. printed at Paris. London, 1704.
Duclos, Samuel Cottereau. *Dissertation sur les principes des mixtes naturels, faite en l'an 1677*. Amsterdam: Elsevier, 1680. References are to the reprint of this work in *Mémoires*, 4: 1–40.
——. *Observations sur les eaux minérales de plusieurs provinces de France, faites en l'Académie royale des sciences en l'année 1670 et 1671*. Paris: Imprimerie royale, 1675; repr. *Mémoires*, 4: 41–120, to which references are keyed.
Du Hamel, Jean-Baptiste. *De consensu veteris et novae philosophiae libri duo*. Oxford: W. Hall, J. Grosley and Amos Curteyn, 1669.

———. *Philosophia vetus et nova ad usum scholae accommodata, in regia burgondia olim pertractata.* 2 vols. 3d rev. ed. Paris: Étienne Michallet, 1684.

———. *Regiae Scientiarum Academiae Historia.* Paris: Étienne Michallet, 1698. Rev. ed. Paris: J.-B. Delespine, 1701.

Estampes pour servir à l'histoire des plantes. See Bosse et al.

État de la France. Ed. Nicolas Besongne. Paris: various printers, 1661–1698. Ed. Louis Trabouillet. Paris: various printers, 1699–1718.

Fabry von Hilden. See Hilden.

Fagon, Guy-Crescent. "Sur le bled cornu appellé ergot." *Histoire...1710,* 61–64.

Fer, Nicolas de. *Atlas curieux ou le monde représenté.* Paris: chez l'auteur dans l'Isle du Palais sur le quay de l'orloge a la Sphére Royale, 1704.

Fontenelle, Bernard de. *Éloges des academiciens de l'Académie royale des sciences morts depuis l'an 1699. Oeuvres diverses de M. de Fontenelle,* vol. 3. New ed., with engravings by Bernard Picart le Romain. The Hague: Gosse & Neaulme, 1729. Reprinted from *Histoire...[date].*

———. "Éloge du P. Sébastien Truchet, Carme." *Histoire...1729,* 93–101.

———. *Histoire de l'Académie royale des sciences.* 2 vols. Paris: Martin, Coignard, Guerin, 1733. Vols. 1–2 of *Histoire et mémoires de l'Académie royale des sciences depuis 1666 jusqu'à 1699.*

Fournier, Denis. *L'antiloimotechnie.* Paris: Veuve J. Rebuffé, 1671.

———. *L'oeconomie chirurgicale pour le restablissement des parties molles du corps humain.* Paris: F. Clouzier, 1671.

———. *Traicté de la gangrene et particulierement de celle qui survient en la peste.* Paris: Veuve J. Rebuffé, 1670.

Furetière, Antoine. *Dictionnaire universel, contenant tous les mots françois tant vieux que modernes, et les termes de toutes les sciences et des arts; savoir la philosophie, logique, et physique; la medecine; ou anatomie; pathologie, terapeutique, chirurgie, pharmacopée, chymie, botanique, ou l'histoire naturelle des plantes, et celle des animaux, mineraux, metaux et pierreries, et les noms des drogues artificielles:...les arts, la rhetorique, la poesie, la grammaire, la peinture, sculpture, etc.* New ed. 2 vols. The Hague and Rotterdam: Chez Arnout et Reinier Leers, 1694.

———. *Recueil des factums d'Antoine Furetière de l'Académie françoise contre quelques-uns de cette Académie suivi des preuves et pièces historiques données dans l'édition de 1694.* Ed. Charles Asselineau. 2 vols. Paris: Poulet-Malassis et de Broise, 1859.

Gallois, abbé Jean. "Extrait du livre intitulé *Observations physiques & mathématiques envoyées des Indes & de la Chine.*" *Mémoires,* 10: 130–38.

Gennes, [Jean Baptiste?] de. "Nouvelle machine pour faire de la toile sans l'aide d'aucun ouvrier, presentée à l'Academie Royale...." *JdS* (8 Aug. 1678): 317–20.

Glaser, Christophe. *The Compleat Chymist, or, a New Treatise of Chymistry.* Trans. F. R. S. from the 4th French ed. London, 1677.

Godin des Odonais, Louis, ed. *Table alphabétique des matieres contenues dans l'Histoire et les Mémoires de l'Académie royale des sciences.* 2 vols. Paris: La Compagnie des Libraires, 1729–34.

Gouye, Thomas. *Observations physiques et mathématiques, pour servir à la perfec-*

tion de l'astronomie et de la geographie. Envoyées des Indes & de la Chine à l'Académie. In Mémoires, 7, pt. 2: 741–875.

———. Observations physiques et mathématiques, pour servir à la perfection de l'astronomie et de la geographie. Envoyées de Siam à l'Academie. In Mémoires, 7, pt. 2: 605–740.

Le grand vocabulaire français..., par une Société de Gens de Lettres. 30 vols. Paris: C. Panckoucke, 1767–74.

Grew, Nehemiah. *The Anatomy of Plants. With an Idea of a Philosophical History of Plants and Several Other Lectures read before the Royal Society.* [London]: W. Rawlins, 1682. Other contents are: *The Anatomy of Plants Begun* [1671], and *The Anatomy of Roots* [1672, 1673], *The Anatomy of Trunks* [1673, 1674], *The Anatomy of Leaves, Flowers, Fruits, and Seeds* [1676, 1677], and several chemical lectures related to plants.

———. *Anatomie des plantes.* Trans. Louis Le Vasseur. Paris: L. Roulland, 1675; 2d ed., Paris, 1679; Michallet published an edition of Grew with works by Boyle and Dedu in 1679, vander Aa did so in 1691. The plates of the French editions are not identical with each other or with the English originals.

Guiffart, Pierre. *Lettre à un docteur et professeur en medecine touchant la connoissance du chyle et de ses vaisseaux.* 2d ed. Rouen: François Vaultier, 1658.

Guiffrey, Jules, ed. See *Les comptes des bâtiments du roi.*

Haller, Albrecht von. *Bibliotheca botanica.* 2 vols. Vol. 1, *Tempora ante Tournefortium.* Zurich: Orell, Gessner, Fuessli, 1771–72.

Hartsoeker, Nicolas. "Extrait d'une lettre... touchant la maniere de faire les nouveaux microscopes, dont il a esté parlé dans le Iournal il y a quelques iours." *JdS* (29 Aug. 1678): 355–56.

Harvey, William. *Movement of the Heart and Blood in Animals: An Anatomical Essay (De motu cordis et sanguinis, 1628) and The Circulation of the Blood (De circulatione sanguinis, 1649).* Trans. Kenneth J. Franklin. In William Harvey, *The Circulation of the Blood and Other Writings.* New York: Dutton, Everyman Library, 1963.

Hecquet, Philippe. *La medecine et la chirurgie des pauvres.* Paris, 1758 [1740, 1742].

———. *Traité de la peste.* Paris: G. Cavelier fils, 1722.

Hiärne, Urban. *Actorum laboratorii Stockholmensis.* Stockholm: Michael Laurel, 1706.

Hilden, Wilhelm Fabry von. *Gründlicher Bericht vom heissen und kalten Brand, welcher Gangraena et Sphacelus oder S. Antonii- und Martialis Feuer genannt wird.* Prepared from 1603 edition and not the first edition, *De gangraena et sphacelo* (1593). Bern and Stuttgart: E. Hintzsche, 1965.

Hoffmann, Friedrich. *Fundamenta medicinae.* Trans. Lester S. King. London: Macdonald; New York: American Elsevier, 1971.

Hooke, Robert. *An Attempt for the Explication of the Phenomena, Observable in an Experiment published by the Honourable Robert Boyle, Esq: in the xxv. experiment of his Epistolical Discourse touching the Air.* London, 1661.

———. *Lectures and Collections... Microscopium.* London: J. Martyn, 1678.

———. *Micrographia: or Some Physiological Descriptions of Minute Bodies Made* by *Magnifying Glasses, with Observations and Inquiries thereupon.* London, 1665; repr. New York: Dover, 1961.

———, ed. *Philosophical Collections.* 7 nos. London, 1681–82.

Hunauld [Hunault], Pierre. *Discours physique sur les fievres qui ont regné les années dernieres.* Paris: Laurent d'Houry, 1696.

Huygens, Christiaan. "Extrait d'une lettre... touchant une nouvelle manière de baromètre qu'il a inventée." *JdS* (1672); repr. in *Mémoires,* 10: 540–44.

———. "Extrait d'une lettre... touchant une nouvelle maniere de microscope qu'il a apporté de Hollande." *JdS* (15 Aug. 1678): 331–32; *Mémoires,* 10: 608–9.

———. *Oeuvres complètes.* 22 vols. Amsterdam: Swets & Zeitlinger, N. V.; The Hague: Martinus Nijhoff, 1888–1950.

Journal des sçavans. Paris and Amsterdam, 1665–1699.

La Bruyère, Jean de. *Oeuvres.* Ed. G. Servois. 4 vols. Paris: Hachette, 1865–82.

La Hire, Philippe de. "Description d'un tronc de palmier petrifié et quelques refléxions sur cette pétrification." *Mémoires,* 10: 140–43.

———. *L'école des arpenteurs, ou l'on enseigne toutes les pratiques de géométrie, qui sont nécessaires à un arpenteur.* Paris: T. Moette, 1699.

La Hire, Philippe de, and Sédileau. "Description d'un insecte qui s'attache à quelques plantes étrangeres, et principalement aux orangers." *Mémoires,* 10: 10–14.

Lang, Christian N. *Descriptio morborum ex usu clavorum secalinorum cum pane.* Lucerne, 1717.

La Quintinie, Jean de. *Instructions pour les jardins fruitiers et potagers, avec un traité des orangers....* 2 vols. Paris: Claude Barbin, 1690.

Le Clerc, Daniel. *The Compleat Surgeon.* 3d ed. London, 1701.

———. *The History of Physick.* Trans. Dr. Drake and Dr. Baden. London, 1699.

Le Febvre, Nicaise. *A Compleat Body of Chemistry.* Trans. P. D. C. Esq. London, 1670.

Le Gallois, Pierre, ed. See Pierre Bourdelot, *Conversations de l'Académie de... Bourdelot.*

Leibniz, Gottfried Wilhelm. *Der Briefwechsel... mit Mathematikern.* Ed. C. J. Gerhardt. Berlin: Mayer & Müller, 1899.

———. *Lettres et opuscules inédits.* Ed. A. Foucher de Careil. Paris: Librairie Philosophique de Ladrange, 1854.

———. *Oeuvres.* Ed. A. Foucher de Careil. Lettres, 2 vols. Paris: Firmin Didot, 1859–60.

Lémery, Nicolas. *Course of Chemistry.* Trans. W. Harris. London, 1677.

———. *Traité universel des drogues simples.* Paris: Laurent d'Houry, 1698.

Le Roy de la Corbinaye, Charles. *Traité de l'orthographie françoise, en forme de dictionnaire....* New ed., rev. M. Restaut, Avocat au Parlement, et aux Conseils du Roi. Poitiers: Chez J. Felix Faulcon, 1787.

Lister, Martin. *A Journey to Paris in the Year 1698* (London: Jacob Tonson, 1699); references are to the 3d ed., edited by Raymond Phineas Stearns, unless otherwise noted. Facsimile reprint in The History of Science, 4. Urbana: University of Illinois Press, 1967.

Locke, John. *Correspondence*. Ed. E. S. de Beer. 7 vols. Oxford: Clarendon Press, 1976–82.

———. *Locke's Travels in France, 1675–1679, as Related in his Journals, Correspondence and Other Papers*. Ed. John Lough. Cambridge: Cambridge University Press, 1953.

Major, Johann Daniel. *Dissertatio botanica, de planta monstrosa Gottorpiensi . . . & circulatione succi nutritii . . . .* Schleswig: J. Carstens, 1665.

Malpighi, Marcello. *Anatome plantarum*. London: John Martyn, 1675, 1679.

Marchant, Jean. *Méthode nouvelle pour guerir la fievre maligne et pourprée*. Dijon: A. Michard, 1674.

Marchant, Nicolas. *Descriptions de quelques plantes nouvelles*. Paris: Imprimerie royale, 1676. Bound with Denis Dodart, *Mémoires des plantes* (1676) and repr. in *Mémoires*, 4.

Mariotte, Edme. *Essais de physique, ou mémoires pour servir à la science des choses naturelles. (1) De la végétation des plantes. (2) De la nature de l'air. (3) Du chaud et du froid. (4) De la nature des couleurs*. Paris: Estienne Michallet, 1679, 1681; repr. in Mariotte, *Oeuvres* (1717): 1: 117–320.

———. *Oeuvres*. 2 vols. Leiden: Pierre vander Aa, 1717; The Hague: Jean Neaulme, 1740.

———. *Traitté de nivellement, avec la description de quelques niveaux nouvellement inventez*. Paris: J. Cusson, 1672; repr. in Mariotte, *Oeuvres* (1740): 535–56.

———. *De la végétation des plantes*. See Mariotte, *Essais de physique;* references to text as repr. in Mariotte, *Oeuvres* (1717): 121–47.

Martet, Jean. *Abbregé des nouvelles experiences anatomiques, des veines lactées, reservoirs du Chyle . . . ; et comme le Chyle est porté au coeur pour faire le sang. Ensemble de la circulation du sang*. Paris: Charles de Sercy & Jean Guignard, 1664.

Ménage, Gilles. *Dictionnaire etymologique de la langue françoise, les origines françoises de M. de Caseneuve, les additions du R. P. Jacob, et de M. Simon de Valhébert, le discours du R. P. Besnier sur la science des etymologies, et la vocabulaire hagiologique de M. l'abbé Chastelain*. New ed., A. F. Jault, docteur en médecine et Professeur en langue syriaque au Collège royal. 2 vols. Paris: Chez Briasson, 1750.

Moreau, Jacques. *De la veritable connoissance des fievres continues, pourprées, et pestilentes*. A later edition of *Traité chymique* (1683). Paris, 1685.

———. *Traité chymique de la veritable connoissance des fievres continues, pourprées et pestilentielles*. Dijon: J. Ressayre, 1683.

Moréri, Louis. See Secondary Sources.

Le neptune françois. See Pène et al.

Nicéron, Jean Pierre. See Secondary Sources.

Nouvelles de la république des lettres. Ed. Pierre Bayle. Amsterdam, 1684–1689.

Oldenburg, Henry. *Correspondence*. Ed. A. Rupert Hall and Marie Boas Hall. 10 vols. Madison: University of Wisconsin Press, 1965–75.

Papillon, Philibert. *Bibliothèque des auteurs de Bourgogne*. 2 vols. Dijon: Philippe Marteret, 1742.

Pardies, Ignace Gaston, S. J. *Élémens de géométrie*. Paris: Sebastien Mabre-Cramoisy, 1671.

Pène, Charles de; Sauveur, Joseph; Cassini, Jean Dominique et al. *Le neptune françois, ou recueil des cartes marines, levées et gravées par ordre du roy. Premier volume. Contenant les costes de l'Europe sur l'Ocean, depuis Dronthem et Norvege jusques au Detroit de Gibraltar, avec la Mer Baltique.* Paris: Imprimerie royale, 1693; Paris: Hubert Jaillot, 1693. Two different printings.

Perrault, Charles. *Histoires, ou contes du temps passé, avec des moralitez.* Paris: Charles Barbin, 1697.

———. *Les hommes illustres qui ont paru en France pendant ce siècle.* 2 vols. Paris: Antoine Dezallier, 1696, 1700; repr., Geneva: Slatkine Reprints, 1970.

———. *Mémoires de ma vie* published with *Voyage à Bordeaux (1669) par Claude Perrault.* Ed. Paul Bonnefon. Paris: Librairie Renouard, H. Laurens, 1909.

Perrault, Claude. *Abrégé des dix livres de Vitruve.* Paris: Jean Baptiste Coignard, 1674. 2d ed., revised, corrected, augmented, Paris, 1684, with 70 plates and many woodcuts.

———. *La circulation de la sève*, in Claude and Pierre Perrault, *Oeuvres diverses*.

———. *Voyage à Bordeaux (1669).* See Charles Perrault, *Mémoires de ma vie.*

Perrault, Claude, and Perrault, Pierre. *Oeuvres diverses de physique et de méchanique.* 2 vols. Leiden: Pierre vander Aa, 1721.

Perrault, Pierre. *De l'origine des fontaines.* Paris: P. Le Petit, 1674.

Perrault, Pierre, and Perrault, Claude. See Claude Perrault and Pierre Perrault, *Oeuvres diverses*.

Philosophical Transactions. London, 1665–99; repr., New York, 1963.

Picard, Jean. *Traité de nivellement.* Paris: É. Michallet, 1684.

Racine, Jean. *Oeuvres.* Ed. Paul Mesnard. 8 vols. Paris: Librairie Hachette, 1887–1921.

Raynaud, François. *Traité des fievres malignes et pourprées.* Brussels, 1695; Carpentras: B. Ravasi, 1695.

Régis, Pierre Sylvain. *Système de philosophie.* 2 vols. Paris: De l'imprimerie de Denys Thierry, aux depens d'Anisson, Posuel et Rigaud Libraires à Lyon, 1690.

———. *Cours entier de philosophie, ou systeme general selon des principes de M. Descartes....* 3 vols. Amsterdam: Huguetan, 1691.

Riolan, Jean. *Manuel anatomique et pathologique ou abregé de toute l'anatomie.* Trans. Sauvin. Paris: Chez Gaspar Meturas, 1661.

Robert, Nicolas et al. *Vélins du Muséum. Peintures sur vélin de la collection du Muséum national d'Histoire naturelle de Paris.* Brussels: Bibliothèque royale Albert 1er, 1974.

———. *Estampes pour servir à l'histoire des plantes.* See Bosse et al., *Estampes*.

Roberval, Gilles Personne de. "Nouvelle maniere de balance." *JdS* (1670); repr. *Mémoires*, 10: 494–96.

Rohault, Jacques. *Traité de physique.* 2 vols. Paris: Veuve de Charles Savreux, 1671.

Royal Society of London. *Philosophical Transactions.* See *Philosophical Transactions.*

[Sagot]. *Remedes des pauvres. Cures extraordinaires. Relations des cures extraordinaires, et surprenantes, faites par les remedes des pauvres, dans les terres de*

Monseigneur le duc de Montausier, ou il en fait distribuer, en divers hospitaux de son gouvernement.... N.p., n.d., published anonymously.

Saint-Simon, Louis de Rouvroy, duc de. *Mémoires.* New ed., ed. A. de Boislisle. 41 vols. Paris: Hachette, 1879–1928.

Scarron, Paul. "Sonnet" [sur Paris]: "Un amas confus de maisons." In *Poésies diverses,* 2:136. Ed. Maurice Cauchie. 2 vols. Paris: Librairie Marcel Didier, 1948, 1948, 1960, 1961.

Schrader, Frideric. *Dissertatio epistolica de microscopiorum usu in naturali scientia et anatome.* Göttingen: Typis Johann Christoph Hampe: sumptibus B. Fuhrmanns, 1681.

Scudéry, Madeleine de. *Entretiens de morale.* 2 vols. Paris: Jean Anisson, 1692.

Sédileau. "Observations de la quantité de l'eau de pluye tombée à Paris durant près de trois années, et de la quantité de l'evaporation." *Mémoires,* 10: 29–36.

Sédileau, and La Hire, Philippe de. See La Hire.

Tardy, Claude. *Cours de medecine.* Paris, 1667.

Tauvry, Daniel. *Nouvelle anatomie raisonnée, ou les usages de la structure du corps de l'homme, et de quelques autres animaux, suivant les lois des mechaniques.* Paris: Estienne Michallet, 1690.

———. *Nouvelle pratique des maladies aigues.* Paris: Laurent d'Houry, 1698.

———. *Pratique des maladies croniques.* Paris: Laurent d'Houry, 1712.

———. *Traité des médicaments et la maniere de s'en servir.* Paris: Estienne Michallet, 1691.

Thal, Johann. *Sylva Hercynia, sive catalogus plantarum sponte nascentium in montibus, et locis vicinis Hercyniae.* Frankfurt on Main, 1588. Article on rye and ergot, 45–47.

Theophrastus. *Enquiry into Plants.* Trans. Sir Arthur Hort. 2 vols. Loeb Classical Library. London and New York: W. Heinemann, 1916.

Tillet, Mathieu. *Dissertation sur la cause qui corrompt et noircit les grains de bled dans les epis.* Bordeaux: Veuve de P. Brun, 1755.

Tournefort, Joseph Pitton de. "Conjectures sur les usages des vaisseaux dans certaines Plantes." *Mémoires,* 10: 191–97.

———. "Description d'un champignon extraordinaire." *Mémoires,* 10: 120–22.

———. *Élémens de botanique, ou methode pour connoître les plantes.* 3 vols. Paris: Imprimerie royale, 1694.

———. *Histoire des plantes qui naissent aux environs de Paris, avec leurs usages dans la médecine.* Paris: Imprimerie royale, 1698.

———. *Materia medica, or a description of simple medicines.* London: A. Bell, 1708.

———. "Observations physiques touchant les muscles de certaines Plantes." *Mémoires,* 10: 406–15.

———. "Réfléxions physiques sur la production du champignon dont il a été parlé dans les Mémoires du mois dernier." *Mémoires,* 10: 122–25.

Wiseman, Richard. *Severall Chirurgicall Treatises.* London, 1676. Treatise 1, on tumors, contains chap. 6, "Of an Erysipelas."

Wolff, Christian Friedrich von. *Vera causa multiplicationis frumenti admirandae, omnem plantarum vegetationem.* Halle, 1718; *Discovery of the True Cause of the Wonderful Multiplication of Corn....* London, 1734.

SECONDARY SOURCES

Académie des Sciences. See Institut de France or Musée du Conservatoire National des Arts et Métiers.

Académie Royale des Sciences, des Lettres et des Beaux-Arts de Belgique. See *Biographie nationale*.

Adams, Robert P. "The Social Responsibilities of Science in *Utopia, New Atlantis* and After." *Journal of the History of Ideas* 10 (1949): 374–98.

Alexopoulos, Constantine J., and Mims, Charles W. *Introductory Mycology*. 3d ed. New York: John Wiley & Son, 1979.

Allen, D. E. "Natural History and Social History." *Journal of the Society for the Bibliography of Natural History* 7 (1976): 509–16.

André, Louis, and Bourgeois, Émile, eds. *Recueil des instructions données aux ambassadeurs et ministères de France depuis les traités de Westphalie jusqu'à la révolution française*. 30 vols. Paris: various publishers, 1884–1983. Vols. 16, 17, and 18 contain *Recueil... Hollande*, 3 vols. Paris: E. de Boccard, 1922–24. Page references are to these three volumes, numbered 1, 2, and 3.

Arber, Agnes. *Herbals, Their Origin and Development: A Chapter in the History of Botany, 1470–1670*. New ed. Cambridge: Cambridge University Press, 1938.

———. "Nehemiah Grew (1641–1712)." In *Makers of British Botany*, ed. Oliver, 44–64.

———. "Nehemiah Grew (1641–1712) and Marcello Malpighi (1628–1694): An Essay in Comparison." *Isis* 34 (1942): 7–16.

———. "Robert Sharrock (1630–1684): A Precursor of Nehemiah Grew (1641–1712) and an Exponent of 'Natural Law' in the Plant World." *Isis* 51 (1960): 3–8.

———. "A Seventeenth-Century Naturalist: John Ray." *Isis* 34 (1943): 319–24.

Aston, Trevor, ed. *Crisis in Europe, 1560–1660*. Intro. by Christopher Hill. New York: Basic Books, 1965.

Aucoc, Léon, ed. *L'Institut de France. Lois, statuts et règlements concernant les anciennes Académies et l'Institut de 1635 à 1889*. Paris: Imprimerie nationale, 1889.

Auvray, L. "Affiches des cours du Collège Royal pour les années 1681, 1682 et 1683." *Mémoires de la Société de l'Histoire de Paris et de l'Île de France* 49 (1927): 194–238.

Avenel, Georges, le vicomte d'. *Histoire économique de la propriété, des salaires, des denrées et de tous les prix en général depuis l'an 1200 jusqu'en l'an 1800*. 6 vols. Paris: Ernest Leroux, 1884–1912.

Baker, A. Albert, Jr. "A History of Indicators." *Chymia* 9 (1964): 147–67.

Balteau, J. et al., eds. *Dictionnaire de biographie française*. 16 vols. Paris: Libraire Letouzey et Ané, 1933–81.

Barber, Elinor G. *The Bourgeoisie in 18th Century France*. Princeton: Princeton University Press, 1955.

Barger, G. *Ergot and Ergotism*. London: Gurney & Jackson, 1931.

Bedel, Charles. "Les conceptions en chimie biologique du XVIIe au XXe siècle." *Archives internationales de l'histoire des sciences* 12 (1959): 397–99.

Bénézit, Emmanuel. *Dictionnaire critique et documentaire des peintres, sculpteurs, dessinateurs et graveurs.* New ed. 10 vols. Paris: Librairie Gründ, 1976.

Berger, Patrice. "French Administration in the Famine of 1693." *European Studies Review* 8 (1978): 101–27.

———. "Pontchartrain and the Grain Trade during the Famine of 1693." *The Journal of Modern History* 48 (December 1976, on-demand supplement): 37–86.

———. "Rural Charity in Late Seventeenth Century France: The Pontchartrain Case." *French Historical Studies* 10 (1978): 393–415.

Bernard, Leon. *The Emerging City: Paris in the Age of Louis XIV.* Durham: Duke University Press, 1970.

Berthier, Aug. Georges. "Le mécanisme cartésien et la physiologie au XVIIe siècle." *Isis* 2 (1914): 37–89; 3 (1920): 21–58.

Bertrand, Joseph L.F. *L'Académie des Sciences et les académiciens de 1666 à 1793.* Paris: J. Hetzel, 1869; reprinted, Amsterdam, 1969.

———. "Les Académies d'autrefois. *L'ancienne Académie des sciences,* par Alfred Maury..., 1865. — Procès-verbaux inédits des séances de l'Académie des Sciences." *JdS* (1866): 337–53, 420–32, 576–93, 715–25, 758–69; (1867): 167–82, 752–66; (1868): 107–23.

Bigourdan, G. "Les premières réunions savantes de Paris au XVIIe siècle. L'Académie de Montmor." *CRA* 164 (1917): 159–62, 216–20.

———. "Les premières sociétés scientifiques de Paris au XVIIe siècle. Les réunions du P. Mersenne et l'Académie de Montmor." *CRA* 164 (1917): 129–33.

———. "Sur quelques observatoires de la région provençale au XVIIe siècle. L'observatoire d'Avignon." *CRA* 164 (1917): 253–59.

Biographie universelle, ancienne et moderne, redigée par une Société de gens de lettres. 52 vols. Paris, 1811–28.

Blanchard, Anne. *Dictionnaire des ingénieurs militaires 1691–1791.* Montpellier: the author, 1981.

———. *Les ingénieurs du "Roy" de Louis XIV à Louis XVI. Étude du corps des fortifications.* Montpellier: the author, 1979.

Bloch, Oscar, and Wartburg, W. von. *Dictionnaire étymologique de la langue française.* 4th ed. Paris: Presses Universitaires de France, 1964.

Boas, Marie. "Acid and Alkali in Seventeenth Century Chemistry." *Archives internationales de l'histoire des sciences* 9 (1956): 13–28.

———. "Quelques aspects sociaux de la chimie au XVIIe siècle." *RHS* 10 (1957): 132–47.

———. *Robert Boyle and Seventeenth-Century Chemistry.* Cambridge: Cambridge University Press, 1958.

———. See also Hall, Marie Boas.

Bonnin, B. "À propos de la productivité agricole: L'exemple du Dauphiné au XVIIe siècle." *Annales* 23 (1968): 368–75.

Bos, H. J. M.; Rudwick, M. J. S.; Snelders, H. A. M.; and Visser, R. P. W., eds. *Studies on Christiaan Huygens. Invited papers from the Symposium on the Life and Work of Christiaan Huygens, Amsterdam, 22–25 August 1979.* Lisse: Swets & Zeitlinger, 1980.

Bourde, André J. *Agronomie et agronomes en France au XVIIIe siècle.* 3 vols. Paris: S.E.V.P.E.N., 1967.
Bourgeois, Émile, and André, Louis. *Les sources de l'histoire de France: XVIIe siècle (1610–1715).* 8 vols. Paris: Auguste Picard, 1913–1935.
Bouvet, M. "Les apothicaires royaux (suite). 2: Les apothicaires de Louis XIV, XV, XVI." *Revue d'histoire de la pharmacie* (1930): 31–43, 76–83, 189–211, 242–62.
Bové, F. J. *The Story of Ergot.* Basel and New York: S. Karger, 1970.
Braudel, Fernand, and Labrousse, Ernest, eds. *Histoire économique et sociale de la France.* 4 vols. Paris: Presses Universitaires de France, 1970–82. Vol. 2. *Des derniers temps de l'âge seigneurial aux préludes de l'âge industriel (1660–1789),* by Ernest Labrousse, Pierre Léon, Pierre Goubert, Jean Bouvier, Charles Carrière, and Paul Harsin.
Bréchot, Jean-François; Laissus, Yves; and Guédès, Michel. "Note bibliographique sur les *Plantes du Roi.*" *Revue française d'histoire du livre* (1974).
Brockliss, L. W. B. *French Higher Education in the Seventeenth and Eighteenth Centuries: A Cultural History.* Oxford: Clarendon Press, 1987.
Brothwell, D., and Brothwell, P. *Food in Antiquity.* New York: Praeger, 1969.
Brown, Harcourt. *Scientific Organizations in Seventeenth Century France (1620–1680).* History of Science Society Publications, n. s. 5. Baltimore: Williams & Wilkins Co., 1934.
Brown, Lloyd A. *The Story of Maps.* Boston: Little, Brown and Company, 1949.
Brunet, P. "La méthodologie de Mariotte." *Archives internationales d'histoire des sciences* 1 (1947): 26–59.
Brunot, Ferdinand. *Histoire de la langue française des origines à 1900.* 13 vols. in 21. Paris: Librarie Armand Colin, 1905–1972.
Brygoo, Édouard. "Les médecins de Montpellier et le Jardin du Roi à Paris." *Histoire et Nature: Cahiers de l'Association pour l'Histoire des Sciences de la Nature* 14 (1979): 3–29.
Bugler, G. "Un précurseur de la biologie expérimentale: Edme Mariotte." *RHS* 3 (1950): 242–50.
Burke, Peter. *Popular Culture in Early Modern Europe.* New York: Harper, 1978.
Burstall, Aubrey F. *A History of Mechanical Engineering.* Cambridge: MIT Press, 1972 [1965].
Busson, Henri. *La religion des classiques.* Paris: Presses Universitaires de France, 1948.
Cailleux, André. "Progression du nombre d'espèces de plantes décrites de 1500 à nos jours." *RHS* 6 (1953): 42–49.
Callot, Émile. "Système et méthode dans l'histoire de la botanique." *RHS* 18 (1965): 45–53.
The Cambridge Economic History of Europe. See Postan and Habbakkuk, eds.
Canguilhem, Georges. *La connaissance de la vie.* 2d ed. Paris: Vrin, 1965.
———. "The Role of Analogies and Models in Biological Discovery." In *Scientific Change,* ed. Crombie, 507–20.
Cap, Paul-Antoine. *Études biographiques pour servir à l'histoire des sciences.* 2 vols. Paris, 1857–64.

Carus-Wilson, E. M., ed. *Essays in Economic History.* 3 vols. London: Edward Arnold (Publishers) Ltd, 1954–62.
Caullery, Maurice. "La biologie au XVIIe siècle." *XVIIe siècle* 30 (1956): 25–45.
Centre Méridional de Rencontres sur le XVIIe Siècle. *Le XVIIe siècle et la recherche.* Colloque de Marseille, 6. Marseille: C. M. R., 1977.
Chabbert, Pierre. "Jacques Borelly (16...–1689): Membre de l'Académie des Sciences." *RHS* 23 (1970): 203–27.
———. "Pierre Borel (1620?–1671)." *RHS* 21 (1968): 303–43.
Chevreul, Michel Eugène. "Recherches expérimentales sur la végétation par M. Georges Ville (Paris, Librairie de Victor Masson, 1853). Examen précédé de considérations sur différents ouvrages d'agriculture et sur différentes recherches relatives à l'agriculture et à la végétation des XVIIIe et XIXe siècles." *JdS* (1858): 108–28.
Cipolla, Carlo M. *Literacy and Development in the West.* Baltimore: Penguin Books, 1969.
———. "The Professions. The Long View." *The Journal of European Economic History* 2 (1973): 37–52.
Cittert, P. H. van. "On the Use of Glass Globes as Microscope-Lenses." *Proceedings of the Koninklijke Nederlandsche Akademie van Wetenschappen te Amsterdam,* ser. B, 57 (1954): 103–11.
———. "The 'van Leeuwenhoek Microscope' in possession of the University of Utrecht," and "The Optical Properties of the 'van Leeuwenhoek Microscope' in possession of the University of Utrecht." *Proceedings of Koninklijke Akademie van Wetenschappen te Amsterdam* 35 (1932): 1062–63; 36 (1933): 194–96; 37 (1934): 290–93.
Clair, Pierre. *Jacques Rohault (1618–1672). Bio-bibliographie avec l'édition critique des Entretiens sur la philosophie.* Recherches sur le XVIIe siècle, 3; Cahiers de l'Équipe de Recherche, 75. Paris: C. N. R. S., 1978.
Clarke, Jack A. "Abbé Jean-Paul Bignon 'Moderator of the Academies' and Royal Librarian." *French Historical Studies* 8 (1973): 213–35.
Clark-Kennedy, A. E. *Stephen Hales.* Cambridge: Cambridge University Press, 1929.
Clay, Reginald S., and Court, Thomas H. *The History of the Microscope.* London: Chs Griffin & Co., Ltd., 1932.
Cohen, I. Bernard. "Isaac Newton, Hans Sloane and the Académie Royale des Sciences." In *Mélanges Alexandre Koyré,* 1: 61–116. 2 vols. Paris: Hermann, 1964.
Cole, Charles Woolsey. *Colbert and a Century of French Mercantilism.* 2 vols. Hamden, Conn.: Archon Books, 1964 [1939].
Coleman, D. C., ed. *Revisions in Mercantilism.* London: Methuen, 1969.
Collas, Georges. *Jean Chapelain, 1595–1674: Étude historique et littéraire d'après des documents inédits.* Paris: Perrin & Cie, 1912.
Condorcet, Jean Antoine Nicolas de Caritat, marquis de. *Oeuvres.* 12 vols. Paris: Firmin Didot, 1847–49.
Conservatoire National des Arts et Métiers. See Musée du Conservatoire National des Arts et Métiers, Paris.

Contant, Jean-Paul. *L'enseignement de la chimie au Jardin royal des plantes de Paris.* Cahors: A. Coueslant, 1952.

Costabel, Pierre. "Le paradoxe de Mariotte." *Archives internationales d'histoire des sciences* 2 (1949): 864–86.

Couton, G. "Effort publicitaire et organisation de la recherche: les gratifications aux gens de lettres sous Louis XIV." In *Le XVIIe siècle et la recherche,* 41–55. See Centre Méridional de Rencontres sur le XVIIe Siècle.

Crestois, Paul. *L'enseignement de la botanique au Jardin royal des plantes de Paris.* Cahors: A. Coueslant, 1953.

Crombie, A. C., ed. *Scientific Change: Historical Studies in the Intellectual, Social, and Technical Conditions for Scientific Discovery and Technical Invention from Antiquity to the Present.* Symposium in the History of Science. University of Oxford, 9–15 July 1961. New York: Basic Books, Inc., Publishers, 1963.

Crosland, Maurice, ed. *The Emergence of Science in Western Europe.* New York: Science History Publications, 1976.

Dainville, François de. *L'éducation des jésuites (XVIe-XVIIIe siècles).* Ed. Marie-Madeleine Compère. Paris: Les Éditions de Minuit, for Institut National de Recherche Pédagogique, 1978.

———. *La géographie des humanistes.* Paris: Beauchesne, 1940.

Darnton, Robert. *The Great Cat Massacre and Other Episodes in French Cultural History.* New York: Basic Books, 1984.

Daston, Lorraine J., and Park, Katharine. "Unnatural Conceptions: The Study of Monsters in Sixteenth- and Seventeenth-Century France and England." *Past and Present* 92 (1981): 20–54.

Daumas, Maurice, ed. *A History of Technology and Invention: Progress Through the Ages.* Trans. Eileen B. Hennessy. Vols. 1–2. New York: Crown Pub., 1969, 1970.

———. *Scientific Instruments of the Seventeenth and Eighteenth Centuries and Their Makers.* Trans. M. Holbrook. New York: Praeger Publishers, 1972 [1953].

Dauzat, Albert. *Dictionnaire étymologique de la langue française.* Paris: Librairie Larousse, 1938.

Davis, Audrey B. *Circulation Physiology and Medical Chemistry in England, 1650–1680.* Lawrence, Kansas: Coronado Press, 1973.

Davis, Natalie Zemon. *Society and Culture in Early Modern France.* Stanford: Stanford University Press, 1985.

Davy de Virville, Adrien. "De l'influence des idées préconçues sur le progrès de la botanique du XVe au XVIIIe siècle." *RHS* 10 (1957): 116–19.

Davy de Virville, Adrien et al. *Histoire de la botanique en France.* Paris: Société d'édition d'enseignement supérieur, 1954.

Debus, Allen G. *The English Paracelsians.* London: Oldbourne, 1965.

———. "Fire Analysis and the Elements in the Sixteenth and the Seventeenth Centuries." *Annals of Science* 23 (1967): 127–47.

———. "A Further Note on Palingenesis: The Account of Ebeneezer Sibly in the *Illustration of Astrology* (1792)." *Isis* 64 (1973): 226–30.

———. "Palissy, Plat, and English Agricultural Chemistry in the Sixteenth and Seventeenth Centuries." *Archives internationales de l'histoire des sciences* 21 (1968): 67–88.

———. "Paracelsian Doctrine in English Medicine." In *Chemistry in the Service of Medicine,* ed. Poynter, 5–26.

———. "Sir Thomas Browne and the Study of Colour Indicators." *Ambix* 10 (1962): 29–36.

———. "Solution Analyses Prior to Robert Boyle." *Chymia* 8 (1962): 41–61.

Delaporte, François. *Le second règne de la nature: Essai sur les questions de la végétalité au XVIIIe siècle.* Preface by Georges Canguilhem. Paris: Flammarion, 1979. Trans. by Arthur Goldhammer as *Nature's Second Kingdom.* Cambridge: MIT Press, 1982.

Delisle, Léopold. *Le cabinet des manuscrits de la Bibliothèque impériale.* 3 vols. (Histoire générale de Paris.) Paris: Imprimerie impériale, 1868, 1874, 1881.

Delorme, Suzanne. "La correspondance de Chapelain et son intérêt pour l'historien des sciences du XVIIe siècle." *Proceedings of the 14th International Congress of the History of Science* (1975), 3: 385–88.

The Dictionary of Scientific Biography. See Gillispie, ed.

Dictionnaire de biographie française. See Balteau et al., eds.

Diderot, Denis et al., eds. *Encyclopédie ou Dictionnaire raisonné des sciences, des arts et des métiers.* Facsimile of 1751–80 edition. Stuttgart-Bad Canstatt: Fromann, 1966.

Dijksterhuis, E. J. *The Mechanization of the World Picture.* Trans. C. Dikshoorn. London: Oxford University Press, 1969 [1961].

Dorst, Jean, and Laissus, Yves. *Nicolas Robert et les Vélins du Muséum National d'Histoire Naturelle.* Paris: Henri Scrépel, 1980.

Dorveaux, Paul. "Apothicaires membres de l'Académie Royale des Sciences. 3. Simon Boulduc." *Revue d'histoire de la pharmacie* 1 (1930): 5–15.

———. "L'autopsie de Mariotte." *Mémoires de l'Académie des sciences, arts et belles-lettres de Dijon,* année 1937 (1938): xii–xiv.

———. "Les grands pharmaciens apothicaires membres de l'Académie Royale des Sciences. 1. Claude Bourdelin." *Bulletin de la Société de l'Histoire de la Pharmacie* 17 (1929): 289–98.

———. "Les grands pharmaciens apothicaires membres de l'Académie Royale des Sciences. 2. Moyse Charas." *Bulletin de la Société de l'Histoire de la Pharmacie* 17 (1929): 329–40, 377–90.

Dubarle, D. "The Proper Place of Science: Reflections on a Cartesian Theme concerning Humanity and the State as Audiences of the Scientific Community." *Minerva* 1 (1963): 405–27.

Dubois, J., and Lagane, R. *Dictionnaire de la langue française classique.* Paris: Librairie Classique Eugène Belin, 1960.

Eklund, Jon. *The Incompleat Chymist: Being an Essay on the Eighteenth-Century Chemist in His Laboratory, with a Dictionary of Obsolete Chemical Terms of the Period.* Smithsonian Studies in History and Technology, 33. Washington, D.C.: Smithsonian Institution Press, 1975.

Éloy, N. F. J. *Dictionnaire historique de la médecine ancienne et moderne.* 4 vols. Mons: H. Hoyois, 1778.

Eriksson, Gunnar. *Botanikens historia i Sverige intill år 1800.* Stockholm: Almqvist & Wiksell, 1969.

'Espinasse, Margaret. *Robert Hooke.* London: William Heinemann, 1956.

Evans, R. J. W. "Learned Societies in Germany in the Seventeenth Century." *European Studies Review* 7 (1977): 129–51.

Fauré-Fremiet, E. "Les origines de l'Académie des sciences de Paris." *Notes and Records of the Royal Society of London* 20 (1966): 20–31.

Fisher, F. J. "The Development of London as a Centre of Conspicuous Consumption in the Sixteenth and Seventeenth Centuries." In *Essays in Economic History,* ed. Carus-Wilson, 2: 197–207.

Forster, Robert, and Ranum, Orest, eds. *Biology of Man in History: Selections from the "Annales: Économies. Sociétés. Civilisations."* Trans. Elborg Forster and Patricia M. Ranum. Baltimore: The Johns Hopkins University Press, 1975.

Franklin, Alfred. *Les anciennes bibliothèques de Paris. Églises, monastères, collèges, etc.* 3 vols. (Histoire générale de Paris.) Paris: Imprimerie impériale, 1867, 1870, 1873.

Frostin, Charles. "La famille ministérielle des Phélypeaux: Esquisse d'un profil Pontchartrain (XVIe-XVIIIe siècle)." *Annales de Bretagne* 86 (1979): 117–40.

———. "L'organisation ministérielle sous Louis XIV: cumul d'attributions et situations conflictuelles (1690–1715)." *Revue historique de droit français et étranger* 58 (1980): 201–26.

Gasking, Elizabeth. *Investigations into Generation, 1651–1828.* Baltimore: The Johns Hopkins Press, 1967.

Gauja, Pierre. "L'Académie royale des sciences (1666–1793)." *RHS* 2 (1949): 293–310.

———. "Les origines de l'Académie des sciences de Paris." In *Troisième centenaire,* 1–51. See Institut de France; Académie des Sciences.

George, Albert J. "A Seventeenth Century Amateur of Science: Jean Chapelain." *Annals of Science* 3 (1938): 222–31.

———. "The Genesis of the Academy of Sciences." *Annals of Science* 3 (1938): 372–401.

George, Philip. "The Chemical Papers published in the Philosophical Transactions from 1664/5 until 1700." *Archives internationales d'histoire des sciences* 12 (1959): 47–56.

Gillispie, Charles Coulston, ed. *The Dictionary of Scientific Biography.* 16 vols. New York: Charles Scribner's Sons, 1970–1980.

Ginzburg, Carlo. "Morelli, Freud and Sherlock Holmes: Clues and Scientific Method." *History Workshop* 9 (1980): 5–36.

Godefroy, Frédéric. *Dictionnaire de l'ancienne langue française et de tous ses dialectes du IXe au XVe siècle.* 10 vols. New York: Kraus Reprint Corporation, 1961 [Paris, 1880–1902].

Goubert, Jean-Pierre. "Le phénomène épidémique en Bretagne à la fin du XVIIIe siècle (1770–1787)." *Annales* 24 (1969): 1562–88.

Goubert, Pierre. *The Ancien Régime: French Society, 1600–1750.* Trans. Steve Cox. New York: Harper and Row, 1973. Vol. 1 of *L'ancien régime.* 2 vols. Paris: Armand Colin, 1969–73.

———. *The French Peasantry in the Seventeenth Century.* Trans. Ian Patterson. Cambridge: Cambridge University Press, 1986.

Bibliography   359

———. "The French Peasantry in the Seventeenth Century: A Regional Example." *Past and Present* 10 (1956): 55–77.
———. *Louis XIV and Twenty Million Frenchmen*. Trans. Anne Carter. New York: Pantheon Books, 1970.
Grand Larousse de la langue française. See Guilbert et al., eds.
Grassby, R. B. "Social Status and Commercial Enterprise under Louis XIV." *The Economic History Review*, 2d ser. 13 (1960): 19–38.
Greene, Edward Lee. *Landmarks of Botanical History*. Ed. Frank N. Egerton. 2 vols. Stanford: Stanford University Press, for the Hunt Institute for Botanical Documentation, 1983.
Gregory, J. C. "Chemistry and Alchemy in the Natural Philosophy of Sir Francis Bacon, 1561–1626." *Ambix* 2 (1938): 93–111.
Greulach, V. A., and Adams, J. E. *Plants: An Introduction to Modern Botany*. New York: Wiley, 1967.
Guilbert, Louis; Lagane, René; and Niobey, Georges, eds. *Grand Larousse de la langue française*. 7 vols. Paris: Librairie Larousse, 1971–78.
Gunther, Robert W. T. *Early Science in Oxford*. 14 vols. Oxford: Oxford University Press, for the subscribers, 1920–1945. Vols. 6, 7, 8, 10, 13: *The Life and Work of Robert Hooke*. Oxford, 1930–38.
Guyénot, Émile. *L'évolution de la pensée scientifique: Les sciences de la vie aux XVIIe et XVIIIe siècles, l'idée d'évolution*. Paris: A. Michel, 1941.
Hagstrom, Warren O. *The Scientific Community*. Carbondale: Southern Illinois University Press, 1975.
Hahn, Roger. *The Anatomy of a Scientific Institution: The Paris Academy of Sciences, 1666–1803*. Berkeley, Los Angeles, London: University of California Press, 1971.
———. "Scientific Careers in Eighteenth-Century France." In *The Emergence of Science in Western Europe*, ed. Crosland, 127–38.
———. "Scientific Research as an Occupation in Eighteenth-Century Paris." *Minerva* 13 (1975): 501–13.
Hall, A. Rupert. *From Galileo to Newton, 1630–1720*. London: Wm. Collins Sons & Co. Ltd., 1963.
———. "Henry Oldenburg et les relations scientifiques au XVIIe siècle." *RHS* 23 (1970): 285–304.
Hall, Marie Boas. "Humanism in Chemistry." *Chymia* 8 (1962): 33–39.
———. See also Marie Boas.
Hallays, André. *Les Perrault*. Paris: Perrin, 1926.
Hanawalt, Barbara. *Crime and Conflict in English Communities, 1300–1348*. Cambridge: Harvard University Press, 1979.
Handford, J. R. "Chemistry at the Jardin du roi from D'Avisson to Macquer." M. Sc. thesis., University College, London, 1958.
Harrison, John R., and Laslett, Peter, eds. *The Library of John Locke*. Oxford Bibliographical Society Publications, n. s. 13. Oxford: Oxford University Press, 1965.
Hart, Horace. *Notes on a Century of Typography at the University Press, Oxford, 1693–1794*. Oxford: Oxford University Press, 1900.

Hazon, Jacques-Albert. *Notice des hommes les plus célèbres de la Faculté de Médecine en l'Université de Paris*. Paris: Benoît, 1778.
Heckscher, Eli Filip. *Mercantilism*. Authorized translation by Mendel Shapiro. 2 vols. Rev. 2d ed., ed. E. F. Söderlund. London: Allen & Unwin; New York: Macmillan, 1955 [1935].
Hémardinquer, J.-J. "Faut-il 'démythifier' le porc familial d'Ancien Régime?" *Annales* 25 (1970): 1745–66.
Henrey, Blanche. *British Botanical and Horticultural Literature before 1800*. 3 vols. London and New York: Oxford University Press, 1975.
Hesse, Mary B. *Models and Analogies in Science*. London: Sheed and Ward, 1966.
Hillairet, Jacques [pseud.]. *Dictionnaire historique des rues de Paris*. 3 vols. Paris: Les Éditions de minuit, n.d. [c. 1961].
Hirschfield, John Milton. *The Académie Royale des Sciences (1666–1683): Inauguration and Initial Problems of Method*. New York: Arno Press, 1981. Reprint of Ph.D. diss., University of Chicago, 1957.
Hjelt, Otto E. A. *Naturalhistoriens studium vid Åbo Universitet*. Åbo Universitets Lärdomshistoria, 8. Helsingfors: Tidnings. & Tryckeri-Aktiebolagets Tryckeri, 1896.
———. *Naturhistoriens studium i Finland under sjuttonde och adertonde seklet. 1. Tiden före Linné. 2. Tiden efter Linné*. Helsingfors: L. Heimbürger, 1868.
Hobsbawm, E. J. "The Crisis of the Seventeenth Century." In *Crisis in Europe*, ed. Aston, 5–58.
Hoefer, Jean Chrétien Ferdinand, ed. *Nouvelle biographie universelle* [also *Nouvelle biographie générale*]. 46 vols. Paris: Firmin Didot Frères, 1852–68.
Holmes, F. L. "Analysis by Fire and by Solvent Extractions: The Metamorphosis of a Tradition." *Isis* 62 (1971): 129–48.
———. "Tradition and Invention in the Art of Organic Analysis, 1500–1800." Manuscript kindly lent by the author.
Hoppen, K. T. *The Common Scientist in the Seventeenth Century: A Study of the Dublin Philosophical Society, 1683–1708*. London: Routledge & Kegan Paul, 1970.
———. "The Nature of the Early Royal Society." *British Journal for the History of Science* 9 (1976): 1–24, 243–73.
Howard, Rio. *La bibliothèque et le laboratoire de Guy de La Brosse au Jardin des plantes à Paris*. École Pratique des Hautes Études, Histoire et civilisation du livre, 13. Geneva: Droz, 1983.
———. "Guy de La Brosse: Botanique et chimie au début de la révolution scientifique." *RHS* 31 (1978): 301–26.
———. "Medical Politics and the Founding of the Jardin des Plantes in Paris." *Journal of the Society for the Bibliography of Natural History* 9 (1980): 395–402.
Howe, Herbert M. "A Root of van Helmont's Tree." *Isis* 56 (1965): 408–19.
Huguet, Edmond. *Petit glossaire des classiques français du dix-septième siècle, contenant les mots et locutions qui ont vieilli ou dont le sens s'est modifié*. Paris: Librairie Hachette, n.d. [1907].
Hunt, Rachel McMasters M. *Catalogue of Botanical Books in the Collection of Rachel McMasters Miller Hunt*. Compiled by Jane Quinby. 2 vols. 1: *Printed*

Books, 1477–1700, with Several Manuscripts. Pittsburgh: Hunt Botanical Library, 1958.

Huppert, George. *Les Bourgeois Gentilshommes: An Essay on the Definition of Elites in Renaissance France.* Chicago: University of Chicago Press, 1977.

Institut de France; Académie des Sciences. *Index biographique des membres et correspondants de l'Académie des Sciences du 22 décembre 1666 au 15 décembre 1967.* Paris: Gauthier-Villars, 1968.

———. *Troisième centenaire, 1666–1966.* 2 vols. Paris: Gauthier-Villars, 1967.

———. *Histoire et prestige de l'Académie des sciences, 1666–1966.* See Musée du Conservatoire National des Arts et Métiers, Paris.

Joravsky, David. *The Lysenko Affair.* Russian Research Center Studies, 61. Cambridge: Harvard University Press, 1970.

Juillard, Alain. "La Société Royale de Londres et ses correspondents français, 1660–1670." In *Le XVIIe siècle et la recherche,* 79–84. See Centre Méridional de Rencontres sur le XVIIe Siècle.

Kaufman, Paul. *Borrowings from the Bristol Library 1773–1784: A Unique Record of Reading Vogues.* Charlottesville: Bibliographical Society of the University of Virginia, 1960.

———. *The Community Library: A Chapter in English Social History.* Transactions of the American Philosophical Society, n. s. 57, pt. 7. Philadelphia, 1967.

———. *Libraries and Their Users: Collected Papers in Library History.* London: The Library Association, 1969.

Kearns, E. J. *Ideas in Seventeenth-Century France.* New York: St. Martin's, 1980.

Keynes, Geoffrey. *John Ray: A Bibliography.* London: Faber & Faber, 1951.

Kiernan, Colm. *Science and the Enlightenment in Eighteenth-Century France.* Studies on Voltaire and the Eighteenth Century, 59. Geneva: Librairie Droz for Institut et Musée Voltaire Les Délices, 1968.

King, James E. *Science and Rationalism in the Government of Louis XIV, 1661–1683.* The Johns Hopkins University Studies in Historical and Political Sciences, ser. 66, no. 2. Baltimore: The Johns Hopkins Press, 1949.

Klaits, Joseph. *Printed Propaganda under Louis XIV: Absolute Monarchy and Public Opinion.* Princeton: Princeton University Press, 1976.

Koenigsberger, H. G. "Republics and Courts in Italian and European Culture in the 16th and 17th Centuries." *Past and Present* 83 (1979): 32–56.

Laissus, Joseph, and Laissus, Yves. "Joseph Pitton de Tournefort (1656–1708) et ses portraits." *90e Congrès des sociétés savantes* 3 (Nice, 1965): 17–46.

Laissus, Yves, and Monseigny, Anne-Marie. "*Les Plantes du Roi:* note sur un grand ouvrage de botanique préparé au XVIIe siècle par l'Académie royale des sciences." *RHS* 22 (1969): 193–236.

Larousse. *Grand Larousse de la langue française.* See Guilbert et al., eds.

Lavisse, Ernest. *Louis XIV.* 2 vols. Paris: Librairie Jules Tallandier, 1978 [1911].

Lebrun, François. *Les hommes et la mort en Anjou aux XVIIe et XVIIIe siècles: Essai de démographie et de psychologie historiques.* Paris and The Hague: Mouton, 1971.

Lemmon, K. *The Golden Age of Plant Hunters.* South Brunswick and New York: A. S. Barnes, 1969.

Leopold, J. H. "Christiaan Huygens and His Instrument Makers." In *Studies on Christiaan Huygens*, ed. Bos et al., 221–33.
Leroy, Jean-F. "Tournefort (1656–1708)." *RHS* 9 (1956): 350–54.
Le Roy Ladurie, Emmanuel. *Carnival in Romans*. Trans. Mary Feeney. New York: George Braziller, Inc., 1979.
Lery, Edmond. "Le P. Sébastien Truchet membre honoraire de l'Académie des sciences (1657–1729). Ses travaux à Versailles et à Marly." *Revue de l'histoire de Versailles et de Seine-et-Oise* (Oct.–Dec. 1929) (Versailles: Librairie Léon Bernard, 1929).
Lindroth, Sten. "Urban Hiärne och Laboratorium Chymicum." *Lychnos* (1946–47): 51–116.
Littré, Émile. *Dictionnaire de la langue française*. 5 vols. Paris: Librairie Hachette, 1878–1879. 7 vols. N.p.: Jean Jacques Pauvert, 1956–58.
Livet, Charles-Louis. *Lexique de la langue de Molière comparée à celle des écrivains de son temps avec des commentaires de philologie historique et grammaticale*. 3 vols. Paris: Imprimerie nationale, 1895–97.
Lough, John. *France Observed in the Seventeenth Century by British Travellers*. Stocksfield: Oriel Press, 1985.
———, ed. *Locke's Travels*. See Locke, *Travels*.
Maddison, R. E. W. "Studies in the Life of Robert Boyle, F. R.S., pt. 1, Robert Boyle and Some of his Foreign Visitors." *Notes and Records of the Royal Society of London* 9 (1951): 1–35.
Mahoney, Michael S. "Christiaan Huygens: The Measurement of Time and of Longitude at Sea." In *Studies on Christiaan Huygens*, ed. Bos et al., 234–70.
Maindron, Ernest. *L'Académie des Sciences. Histoire de l'Académie. Fondation de l'Institut National. Bonaparte membre de l'Institut National*. Paris: Félix Alcan, 1888.
Marion, Michel. *Recherches sur les bibliothèques privées à Paris au milieu du XVIIIe siècle (1750–1759)*. Ministère des universités, Comité des travaux historiques et scientifiques, Mémoires de la section d'histoire moderne et contemporaine, 3. Paris: Bibliothèque nationale, 1978.
Mariotte, savant et philosophe (†1684): Analyse d'une renommée. Preface by Pierre Costabel. L'histoire des sciences: Textes et études. Paris: Librairie Philosophique J. Vrin, 1986.
Marsak, L. M. *Bernard de Fontenelle: The Idea of Science in the French Enlightenment*. Transactions of the American Philosophical Society, n. s. 49, 7. Philadelphia, 1959.
Martin, Henri-Jean. *Livre, pouvoirs et société à Paris au XVIIe siècle (1598–1701)*. 2 vols. Histoire et civilisation du livre, Centre de Recherches d'Histoire et de Philologie, 3. Geneva: Droz, 1969.
Marx, Jacques. "Alchimie et palingénésie." *Isis* 62 (1971): 275–89.
Maury, L.-F. Alfred. *Les académies d'autrefois. L'ancienne Académie des sciences*. 2d ed. Paris: Didier & Cie, 1864.
McClellan, James E., III. "The Académie Royale des Sciences, 1699–1793: A Statistical Portrait." *Isis* 72 (1981): 541–67.
McKeon, Robert M. "Une lettre de Melchisédech Thévenot sur les débuts de l'Académie Royale des Sciences." *RHS* 18 (1965): 1–6.

McKie, Douglas. "James, Duke of York, F.R.S." *Notes and Records of the Royal Society of London* 13 (1958): 6–18.
McLaughlin, Trevor, and Picolet, Guy. "Un exemple d'utilisation du Minutier central de Paris: La bibliothèque et les instruments scientifiques du physicien Jacques Rohault selon son inventaire après décès." *RHS* 29 (1976): 3–20.
Méjanès, Jean-François. "Le Cabinet du Roi et la Collection des planches gravées de Louis XIV." Chalcographie du Musée du Louvre. Offprint seen courtesy of Simone Balayé.
Mendelsohn, Everett I. *Heat and Life: The Development of the Theory of Animal Heat.* Cambridge: Harvard University Press, 1964.
———. "Philosophical Biology vs. Experimental Biology: Spontaneous Generation in the Seventeenth Century." *Actes du XIIe Congrès International d'Histoire des Sciences* (Paris, 1968), 1, B: 201–27; discussion by Jacques Roger, Lucien Plantefol: 227–29. Paris, 1971.
Metzger, Hélène. *Les doctrines chimiques en France du début du XVIIe à la fin du XVIIIe siècle.* Paris: A. Blanchard, 1969 [1923].
Michaud, Louis Gabriel, ed. *Biographie universelle ancienne et moderne.* 3d ed. 45 vols. Paris: A. Thoisnier Desplaces & Michaud, 1854–65.
Michel, Marie-José. "Clergé et pastorale jansénistes à Paris (1669–1730)." *Revue d'histoire moderne et contemporaine* 26 (1979): 177–97.
Middleton, W. E. Knowles. *The Experimenters: A Study of the Accademia del Cimento.* Baltimore: The Johns Hopkins Press, 1971.
———. *The History of the Barometer.* Baltimore: The Johns Hopkins Press, 1964.
Millburn, John R. *Benjamin Martin: Author, Instrument-Maker and "Country Showman."* Science in History, 2. Leiden: Noordhoff, 1976.
Millington, E. C. "Studies in Capillarity and Cohesion in the Eighteenth Century." *Annals of Science* 5 (1947): 352–69.
———. "Theories of Cohesion in the Seventeenth Century." *Annals of Science* 5 (1945): 253–69.
Ministère de la Culture. *Colbert, 1619–1683.* Alençon: Imprimerie Alençonnaise, 1983.
Monseigny, Anne-Marie. See Yves Laissus.
Moréri, Louis. *Le grand dictionnaire historique.* 18th ed. 8 vols. Amsterdam: P. Brunel, R. Wetstein, et al., 1740 [1674]. Also 10 vols. Paris: Libraires Associés, 1759.
Mousnier, Roland. *The Institutions of France under the Absolute Monarchy, 1598–1789.* Trans. Brian Pearce. 2 vols. Chicago: University of Chicago Press, 1979, 1984.
———. *Paris au XVIIe siècle.* Les Cours de Sorbonne. Histoire Certificat L. Paris: Centre de Documentation Universitaire, n. d. [1961].
———. *Peasant Uprisings in Seventeenth-Century France, Russia, and China.* Trans. Brian Pearce. New York: Harper & Row, 1970 [1969].
Multhauf, Robert. "J. B. Van Helmont's Reformation of the Galenic Doctrine of Digestion." *Bulletin of the History of Medicine* 29 (1955): 154–63.
———. "Medical Chemistry and 'the Paracelsians.'" *Bulletin of the History of Medicine* 28 (1954): 101–26.
———. *The Origins of Chemistry.* London: Oldbourne, 1967.

———. "The Significance of Distillation in Renaissance Medical Chemistry." *Bulletin for the History of Medicine* 30 (1956): 329–46.
Mungello, D. E. *Curious Land: Jesuit Accommodation and the Origins of Sinology.* Honolulu: University of Hawaii Press, 1989 [1985].
Murray, James; Bradley, Henry; Craigie, W. A.; and Onions, C. T., eds. *The Oxford English Dictionary.* 13 vols. Oxford: Clarendon Press, 1961 [1933].
Musée du Conservatoire National des Arts et Métiers, Paris. *Histoire et prestige de l'Académie des sciences 1666–1966.* Paris, 1966.
Muséum National d'Histoire Naturelle, Paris. *Exposition du troisième centenaire.* Paris: Société des Amis du Muséum, [1935].
Nef, John U. *Industry and Government in France and England, 1540–1640.* Memoirs of the American Philosophical Society, 15. Philadelphia, 1940.
Neveu, Bruno. "La vie érudite à Paris à la fin du XVIIe siècle, d'après les papiers du P. Léonard de Sainte-Catherine (1695–1706)." *Bibliothèque de l'École des Chartes* 124 (1966): 432–511.
Neville, Roy G. "Christophe Glaser and the *Traité de la Chymie*, 1663." *Chymia* 10 (1965): 25–52.
———. "The 'Pratique de chymie' of Sébastien Matte La Faveur." *Ambix* 10 (1962): 14–27.
Nicéron, Jean Pierre. *Mémoires pour servir à l'histoire des hommes illustres dans la république des lettres.* 44 vols. Paris: Briasson, 1727–45.
Nicolas, Michel. *Histoire littéraire de Nîmes et des localités voisines.* Nîmes: Chez Ballivet & Fabre, 1854.
Nierenstein, M. "The Early History of the First Chemical Reagent." *Isis* 16 (1931): 438–46.
Nissen, Claus. *Die botanische Buchillustration. Ihre Geschichte und Bibliographie.* 2d ed. 3 vols. in 1. Stuttgart: A. Hiersemann, 1966.
North, J. D., and Roche, J. J., eds. *The Light of Nature: Essays in the History and Philosophy of Science presented to A. C. Crombie.* International Archives in the History of Ideas, 110. The Hague: Martinus Nijhoff, 1985.
Nouvelle biographie générale. See Hoefer, ed.
Nyrop, Kristoffer. *Grammaire historique de la langue française.* 6 vols. 4th ed. Copenhagen: Gyldendalske Bokhandel, Nordisk Forlag, n.d. [1930–1960].
Oliver, F. W., ed. *Makers of British Botany: A Collection of Biographies by Living Botanists.* Cambridge: Cambridge University Press, 1913.
Olmsted, John W. "The Scientific Expedition of Jean Richer to Cayenne (1672–1673)." *Isis* 34 (1942): 117–28.
———. "The Voyage of Jean Richer to Acadia in 1670: A Study in the Relations of Science and Navigation under Colbert." *Proceedings of the American Philosophical Society* 104 (1960): 612–34.
Ornstein, Martha. *The Rôle of Scientific Societies in the Seventeenth Century.* Chicago: University of Chicago Press, 1928 [1913].
The Oxford English Dictionary. See Murray et al., eds.
Pagel, Walter. *Paracelsus: An Introduction to Philosophical Medicine in the Era of the Renaissance.* Basel: Karger, 1958.
———. *William Harvey's Biological Ideas. Selected Aspects and Historical Background.* Basel and New York: S. Karger, 1967.

Partington, J. R. *A History of Chemistry.* 4 vols. London: Macmillan; New York: St. Martin's Press, 1961–70.
Paul, Charles B. *Science and Immortality: The Éloges of the Paris Academy of Sciences (1699–1791).* Berkeley, Los Angeles, London: University of California Press, 1980.
Pedley, Mary Sponberg. "The Map Trade in Paris, 1650–1825." *Imago Mundi* 33 (1981): 33–45.
Pelcher, Richard B. "Boyle's Laboratory." *Ambix* 2 (1938): 17–20.
Pelletier, Monique. "Les globes de Louis XIV: Les sources françaises de l'oeuvre de Coronelli." *Imago Mundi* 34 (1982): 72–89.
Pelseneer, Jean. "Petite contribution à la connaissance de Mariotte." *Isis* 42 (1951): 299–301.
Peter, J.-P. "Disease and the Sick at the End of the Eighteenth Century." Trans. E. Forster. In *Biology of Man in History,* ed. Forster and Ranum, 95–100; from *Annales* (1967): 711–51.
Peumery, Jean-Jacques. "Conversations médico-scientifiques de l'Académie de l'abbé Bourdelot (1610–1685)." *Histoire des sciences médicales* 12 (1978): 127–35.
Plantefol, Lucien. "L'Académie des Sciences durant les trois premiers siècles de son existence." In *Troisième centenaire,* 56–139. See Institut de France; Académie des Sciences.
Porcher, Jean. "La création du Cabinet des Planches gravées à la Bibliothèque du Roi." *Papyrus* 10 (31 juillet 1929): 475–82.
Postan, M. M., and Habbakkuk, H. J., eds. *The Cambridge Economic History of Europe.* 2d ed. 7 vols. Cambridge: Cambridge University Press, 1966.
Poynter, F. N. L., ed. *Chemistry in the Service of Medicine.* London: Pitman, 1963.
Prévost, Anne-Marie. "Sémantique de 'herbe' et 'plante': Pierre Belon (1555) et Tournefort (1708). Référence aux Éléments de Botanique (1694)." *CRA* 258 (1964): 5697–5700.
———. "Sur la sémantique des mots herba et herbe, vue à travers les traductions de l'Histoire naturelle de Pline l'Ancien." *CRA* 257 (1963): 1211–14.
Puech-Milhau, M. L. "An Interview on Canada with La Salle in 1678." *Canadian Historical Review* 18 (1937): 163–77.
Purver, Margery. *The Royal Society: Concept and Creation.* Introduction by H. R. Trevor-Roper. London: Routledge & Kegan Paul, 1967.
Pyenson, Lewis. "'Who the Guys Were': Prosopography in the History of Science." *History of Science* 15 (1977): 155–88.
Quemada, B. ed. *Matériaux pour l'histoire du vocabulaire français.* 3 vols. Annales Littéraires de l'Université de Besançon. Besançon: Centre d'Étude du Vocabulaire Français, 1959, 1960, 1965.
Quinby, Jane, comp. See Hunt.
Ranum, Orest. *Artisans of Glory: Writers and Historical Thought in Seventeenth-Century France.* Chapel Hill, N. C.: University of North Carolina Press, 1980.
———. "Islands and the South in a Ludovician Fête." In *Sun King: The Ascendancy of French Culture during the Reign of Louis XIV,* ed. Rubin.
———. *Paris in the Age of Absolutism.* Bloomington and London: Indiana University Press, 1979.

Raven, C. E. *John Ray, Naturalist. His Life and Works*. Cambridge: The University Press, 1942.
Reeds, Karen Meier. "Renaissance Humanism and Botany." *Annals of Science* 33 (1976): 519–42.
Riley, Philip F. "Louis XIV: Watchdog of Parisian Morality." *The Historian* 36 (1973): 19–33.
———. "Police and the Search for *Bon Ordre* in Louis XIV's Paris." *Proceedings of the Annual Meeting of the Western Society for French History* 7 (1979): 11–20.
Roberts, Hazel van Dyke. *Boisguilbert. Economist of the Reign of Louis XIV.* New York: Columbia University Press, 1935.
Roche, Daniel. "Milieux académiques provinciaux et société des lumières." In *Livre et société dans la France du XVIIIe siècle,* Geneviève Bollème et al. 2 vols. Paris and The Hague: Mouton, 1965.
Rochot, B. "Roberval, Mariotte et la logique." *Archives internationales de l'histoire des sciences* 6 (1953): 38–43.
Roger, Jacques. *Les sciences de la vie dans la pensée française du XVIIIe siècle: La génération des animaux de Descartes à l'Encyclopédie.* 2d ed. Paris: Armand Colin, 1971 [1963].
Roland, F. "Alexis-Hubert Jaillot, géographe du roi Louis XIV (1632–1712)." *Mémoires de l'Académie des Sciences, Belles Lettres et Arts de Besançon* (1918): 45–76.
Rooseboom, Maria. "Christiaan Huygens et la microscopie." *Archives néerlandaises de zoologie* 13 (1958), suppl.
———. "Concerning the Optical Qualities of Some Microscopes made by Leeuwenhoek." *Journal of the Royal Microscopical Society* 59 (1939): 117–83.
———. "The History of the Microscope." *Proceedings of the Royal Microscopical Society* 2 (1967).
Ross, Sydney. "Scientist: The Story of a Word." *Annals of Science* 18 (1962): 65–85.
Rubin, David Lee, ed. *Sun King: The Ascendancy of French Culture during the Reign of Louis XIV.* Cranbury, N.J. and London: Associated University Presses, 1991 (tentative).
Rudwick, Martin J. S. "Charles Darwin in London: The Integration of Public and Private Science." *Isis* 73 (1982): 186–206.
Ruestow, Edward G. *Physics at Seventeenth and Eighteenth-Century Leiden: Philosophy and the New Science in the University.* International Archives of the History of Ideas, 11. The Hague: Martinus Nijhoff, 1973.
Sachs, J. von. *History of Botany, 1530–1860.* Trans. Henry E. F. Garnsey, rev. I. B. Balfour. Oxford: Clarendon Press, 1906 [1890].
Sainte-Beuve, C.-A. *Port-Royal.* 7 vols. Paris: Hachette, 1867–71.
Saisselin, Rémy G. *The Literary Enterprise in Eighteenth-Century France.* Detroit: Wayne State University, 1979.
Salomon-Bayet, Claire. *L'institution de la science et l'expérience du vivant: Méthode et expérience à l'Académie royale des sciences, 1666–1793.* Paris: Flammarion, 1978.
———. "Opiologia, imposture et célébration de l'opium." *RHS* 25 (1972): 125–50.

———. "Un préambule théorique à une Académie des Arts. Académie royale des Sciences, 1693–1696. Présentation et textes." *RHS* 23 (1970): 229–50.
Sarton, George. *Six Wings: Men of Science in the Renaissance.* Bloomington: Indiana University Press, 1957.
Saunders, Elmo Stewart. *The Decline and Reform of the Académie des Sciences à Paris, 1676–1699.* Ph.D. diss., The Ohio State University, 1980.
Saveney, Edgard. "Histoire des sciences. L'ancienne Académie et les académiciens." *Revue des deux mondes* 84 (1869): 199–226. Review of J. Bertrand, *L'Académie des Sciences* (Paris, 1869).
Schaeper, Thomas J. *The Economy of France in the Second Half of the Reign of Louis XIV.* Montreal: Interuniversity Centre for European Studies, 1980.
Schiller, Joseph. "Les laboratoires d'anatomie et de botanique à l'Académie des sciences au XVIIe siècle." *RHS* 17 (1964): 97–114.
Sealy, Robert J., S. J. *The Palace Academy of Henri III.* Geneva: Librairie Droz, 1981.
Sedgwick, Alexander. *Jansenism in Seventeenth-Century France: Voices from the Wilderness.* Charlottesville: University Press of Virginia, 1977.
Sée, Henri. *Esquisse d'une histoire économique et sociale de la France.* Paris: F. Alcan, 1929.
Sgard, Jean; Gilot, Michel; and Weil, Françoise, eds. *Dictionnaire des journalistes (1600–1789).* Grenoble: Presses Universitaires de Grenoble, 1976.
Shapin, Steven. "The House of Experiment in Seventeenth-Century England." *Isis* 79 (1988): 373–404.
Solovine, M. "À propos d'un tricentenaire oublié: Edme Mariotte (1620–1920)." *Revue scientifique* (24 Dec. 1921): 708–9.
Sommer, E. *Lexique de la langue de Madame de Sévigné.* 2 vols. Vols. 13 and 14 of *Lettres de Madame de Sévigné, de sa famille et de ses amis,* ed. M. C. R. Richard de Cendrecourt, dame de Saint-Surin, then Mme de Mommerqué. Paris: Hachette, 1866.
Stevenson, Ian P. "John Ray and His Contributions to Plant and Animal Classification." *Journal of the History of Medicine* 2 (1947): 250–61.
Stimson, Dorothy. *Scientists and Amateurs: A History of the Royal Society.* New York: Henry Schuman, 1948.
Stone, Lawrence. "Literacy and Education in England 1640–1900." *Past and Present* 42 (1969): 69–139.
Stroup, Alice. "Christiaan Huygens and the Development of the Air-Pump." *Janus* 68 (1981): 129–58. The editor of *Janus* published an incomplete and uncorrected version of this article over the author's objections; a corrected offprint may be obtained from the author.
———. "Louis XIV as Patron of the Académie Royale des Sciences." In *Sun King: The Ascendancy of French Culture during the Reign of Louis XIV,* ed. Rubin.
———. *Royal Funding of the Parisian Académie Royale des Sciences During the 1690s.* Transactions of the American Philosophical Society, 77, pt. 4. Philadelphia, 1987.
———. "Some Assumptions Behind Medicine for the Poor during the Reign of Louis XIV." In *The Light of Nature,* ed. North and Roche, 35–55.

———. "Wilhelm Homberg and the Search for the Constituents of Plants at the 17th-Century Académie Royale des Sciences." *Ambix* 26 (1979): 184–201.

Stubbs, J. G. "Chemistry at l'Académie Royale des Sciences from Its Foundation in 1666 to the Middle of the Eighteenth Century." Ph.D. thesis, University College, London, 1939.

Taton, René, ed. *Histoire générale des sciences.* 2 vols. 2: *La science moderne, 1450 à 1800.* Paris: Presses Universitaires de France, 1958–66.

———. *Les origines de l'Académie royale des sciences.* Paris: Palais de la Découverte, 1966.

Thirsk, Joan. *Economic Policy and Projects. The Development of a Consumer Society in Early Modern England.* Oxford: Clarendon Press, 1978.

Thorndike, Lynn, Jr. *A History of Magic and Experimental Science.* 8 vols. New York and London: Columbia University Press, 1923–1958.

Tilly, Louise A. "La révolte frumentaire, forme de conflit politique en France." *Annales* 27 (1972): 731–57.

Todériciu, Doru; Delorme, Suzanne; and Costabel, Pierre. "Notes sur trois hommes de science du XVIIe siècle: Samuel Duclos, Henri-Louis Habert de Montmor, et Flormond de Beaune." *RHS* 27 (1974): 63–75. Todériciu, "Sur la vraie biographie de Samuel Duclos (Du Clos) Cotreau," 64–67.

Tournefort. Paris: Muséum National d'Histoire Naturelle, 1957.

Turner, G. L'E. *Essays on the History of the Microscope.* Oxford: Senecio, 1980.

Ultee, Maarten. *The Abbey of St. Germain des Prés in the Seventeenth Century.* New Haven: Yale University Press, 1981.

Ulyatt, K. W. "Further Studies in the History of Mineral Waters." Ph.D. thesis, University College, London, 1954.

Viguerie, Jean de, and Saive-Lever, Evelyne. "Essai pour une géographie socio-professionelle de Paris dans la première moitié du XVIIe siècle." *Revue d'histoire moderne et contemporaine* 20 (1973): 424–29.

Viner, Jacob. "Power versus Plenty as Objectives of Foreign Policy in the Seventeenth and Eighteenth Centuries." In *Revisions in Mercantilism,* ed. Coleman, 61–91.

Vines, S. H., and Druce, G. Claridge. *An Account of the Morisonian Herbarium in the Possession of the University of Oxford together with Biographical and Critical Sketches of Morison and the two Bobarts and the Works and the Early History of the Physic Garden 1619–1720.* Oxford: Clarendon Press, 1914.

Watson, E. C. "The Early Days of the Académie des Sciences as Portrayed in the Engravings of Sébastien Le Clerc." *Osiris* 7 (1939): 556–87.

Webster, Charles. *The Great Instauration: Science, Medicine and Reform, 1626–1660.* London: Duckworth, 1975.

———. "The Recognition of Plant Sensitivity by English Botanists in the Seventeenth Century." *Isis* 57 (1966): 1–23.

———. "Water as the Ultimate Principle of Nature: The Background to Boyle's Sceptical Chymist." *Ambix* 13 (1966): 96–107.

Whitmore, P. J. S. *The Order of Minims in Seventeenth-Century France.* International Archives of the History of Ideas, 20. The Hague: Martinus Nijhoff, 1967.

Wilson, C. H. "Trade, Society and the State." In *Cambridge Economic History,* ed. Postan and Habbakkuk, 4: 486–575.
Wolf, Charles Joseph Étienne. *Histoire de l'Observatoire de Paris de sa fondation à 1793.* Paris: Gauthier-Villars, 1902.
Yates, Frances. *The French Academies of the Sixteenth Century.* Studies of the Warburg Institute, vol. 15. London: The Warburg Institute, University of London, 1947.
Ziggelaar, August, S. J. *Le physicien Ignace Gaston Pardies S. J. (1636–1673).* Acta Historica Scientiarum Naturalium et Medicinalium, 26. Copenhagen: Odense University Press, 1971.
Zilsel, Edgar. "The Sociological Roots of Science." *American Journal of Sociology* 47 (1941–42): 544–62.

Index

Academicians: books by for sale in Paris, 191–192; as collectors, 195; in conflict with Academy, 199, 204–209; compared with assistants, 211; as Fellows of Royal Society, 202; housing of, 20, 32, 39, 41, 43, 52, 55, 294 n. 41; as inventors, 156; naturalization of, 23; origins and careers of, 15–25, 26, 32, 36, 171–174, 186; and religion, 19, 23, 178, 190, 200, 201; as teachers of mathematics and science, 16, 38, 114, 194, 197. *See also names of individuals;* Académie royale des sciences; Community; Pensions
Académie française, 22, 23, 25, 74, 183, 191
Académie physique (Caen), 205
Académie royale d'architecture, 21, 197
Académie royale des inscriptions, 23, 24
Académie royale des sciences. *See also* Academicians; Bibliothèque du roi; Colbert; Jardin royal; Louvois; Louvre; Observatoire; Patronage; Pensions paid to academicians; Pontchartrain
—accomplishments of, 3, 10, 221–226, and passim
—assistants to, 14, 34, 40, 41, 48, 55, 209, 210–211, 295 n. 2
—audience for work of, 5, 26, 44, 73–74, 101
—books dedicated to, 210
—dynasties in, 17 (*see also* Bourdelin; Cassini; Couplet; La Hire; Marchant)
—exchange, scholarly, and, 101, 103–104, 169, 170, 180–198, 199–217; effect of, 212–213, 215–217; restrictions on, 199, 204–217; unequal character of, 210–212
—expectations about, 4, 8, 9, 14, 27, 28–33, 60–61, 222, 225
—failures of: in natural history of plants, 65, 69, 88, 102, 103–116; response to, 65, 89–90, 91–93, 94–95, 96, 97–98, 98–100, 100–101, 137–144, 225 (*see also* Analogical reasoning; Analogies)
—finances of, 24, 34–45, 46–61, 88, 111, 208, 224; *jetons*, 291 n. 3; for mathematical sciences, 47–48, 51, 52, 53–54, 56, 57, 58, 60; for natural philosophy, 47, 48, 51, 52, 54–55, 56, 57, 58–59, 60, 80, 104, 114; for Observatory, 43, 47, 52, 57, 60; overview, 47, 51–53, 56–58, 60; for pensions, 34, 35–38, 46, 47, 52, 57, 60; for practical projects, 47, 48–50, 51, 52, 55–56, 57, 59, 60; for repairs to equipment and physical plant, 51, 52; for shared expenses, 47, 52, 57, 60; for small expenses, 47, 48, 51, 52, 53, 57
—foundation of, prototypes for, 28–29, 30–33
—functions and goals of, 10, 27, 37, 50, 79, 110, 169, 221; apologist for use of vernacular in science, 223; arm of

371

government, 178; clearing-house of information, 212, 217; disseminator of knowledge, 197–198, 204, 210–211, 217, 222–223; educator, 44, 197–198, 204, 211, 217, 223; instrument of royal propaganda, 8, 31, 223; "moral person," 222; referee of ideas, inventions, and prospective publications, 59, 60, 115–116, 169, 175, 205, 210, 216, 222, 290–291 n. 27; reform science, medicine, and technology, 9, 29–30, 60–61, 98, 99, 108, 109, 110, 117, 119, 154, 169–179, 209, 217, 221–222; research institution, 45; technical consultant to government, 50, 51, 169, 178, 223
— hierarchy of, 13–14, 16, 23–24, 83, 182, 211, 224; associate memberships, 14; celebrities and regulars, 37; corresponding memberships, 14; honorary memberships, 14; ministerial spokesmen, 23–24; and pensions, 35–36; president, 14, 24; secretary, 13, 14, 334 n. 81; students, 14, 35, 36, 182, 211; treasurer, 14; usher, 211
— laboratory of, 18–19, 20, 34, 39, 48, 58, 71, 82, 101, 109, 210, 304 n. 36; crisis in, 54–55; portrait of, 40
— Louis XIV, and: attitude toward science and Academy, 22, 25–26, 31, 32, 48, 49, 223; flattered by Academy, 110, 113; founder and patron of Academy, 3, 13; visits Academy and Observatory, 5–8, 26, 32, 80, 107, 208, 294 n. 40
— medical interests of, 19, 98, 99, 108, 109, 110, 169–179, 204, 212
— meetings of, 14–15, 82, 96, 110, 132, 209, 285–286 n. 5, 333–334 n. 80; attendance at, 103, 223, 291 nn. 3, 12; function of changes, 59–60; minutes of, 15, 83, 107, 208
— membership in: aspirants to, 210, 217, 224; compared with being assistant to, 211; composition of, 15–25, 48, 53, 54, 58; criteria for, 15, 17, 32, 224; effect of lifetime appointment, 224; kinds of, 13–14, 22–23, 24, 36, 291 n. 12; prestige and privileges of, 200, 211
— ministerial protectors, 23–24, 24–25, 26, 28, 216, 223, 224, 225, 226; control Academy's research, 46–61, 103–116; influence Academy, 8–9, 23–25, 32, 37–38, 39, 107–111, 113, 174, 208, 222, 224, 225; live near Academy, 38–39; policies of compared, 13, 36–37, 46–47, 51, 107
— morale of, 15, 19–20, 22, 24, 25, 34, 37–38, 52–53, 55, 56, 59, 83, 107, 111, 174, 204
— names of, 28, 30
— patronage of, 5–8, 8–9, 30–31, 44, 170; advantages of, 65, 71, 75, 80, 223; effects of, 9, 10, 25, 34, 43, 45, 61, 69, 107–111, 115–116, 169, 207–208, 221–226; control of Academy through, 23, 32, 37–38; extent of, 34–64; motives of, 3, 29–30, 30–31, 109, 110
— physical quarters of, 7, 38–45, 59
— portraits of, 5–8, 39–42
— publication program of, 21–22, 33, 34, 100–101, 103–116, 205–209, 212–213, 216, 217, 222–223; Colbert, Louvois, and Pontchartrain compared, 25, 46, 49, 52, 54, 56, 57, 58, 59; restrictions on, 56, 87, 97, 104, 107, 205, 206, 208, 223
— regulations of, 9, 13, 28, 58, 170, 199, 200, 204–209, 211–212, 216, 223
— relations of with others, 18, 44, 170, 174–175, 178, 179, 180–198, 199–217; Bibliothèque du roi, 38–43; Compagnie des arts et métiers, 59, 197; Dutch scientists, 203; Jardin royal, 38–43, 107, 114; Malpighi, 201; medical faculties, 173; other royal institutions, 217; other scientific societies, 165, 172, 196–197, 199–200, 202, 204, 205, 206, 223; public, 45, 79, 101, 103–104, 179, 212–213, 222; scientific community, 9–10, 101, 169, 170, 178, 186, 198, 205, 210, 216–217, 222
— religion and, 17, 19, 21, 23, 24, 43, 44, 55, 170, 172, 178, 190, 200, 201
— reputation of protected, 48, 87, 97, 101, 170, 205–207, 208, 216
— research of, 5–7, 60, 65, 222, 223, 224, 225; in botany, 65, 67–88, 89–102, 103–116, 203–204; in chemistry, 89–102, 108–109; control of, 111, 223; how planned, 59–60, 60–61, 83, 110–111; individual and collective, 29, 59, 60–61, 80, 82, 87–88, 111, 115, 157, 207, 217, 224, 225; long-term, 59, 83
— rivalry: within Academy, 9, 69, 83, 87, 102, 113–115, 174, 206, 225; be-

Index 373

tween Academy and others, 202, 204, 205–207, 307–308 n. 29
— size of, 15, 37, 52, 57; as "working Academy," 13–14, 15, 37
— traits of, 4–5, 9–10, 13, 35, 37, 45, 50, 59, 60–61, 65, 67, 69, 71, 74, 115–116, 154, 165, 169–170, 171–174, 180, 199, 204–209, 209–217, 221–226; agnosticism of, 48, 211, 222; Baconianism of, 147, 165; as collector, 38, 44, 49, 55, 56, 80, 156, 195, 210; as company, 28–29, 31–33; corporate esprit of, 217; experimentalism of, 100, 222; medical empiricism of, 99, 173; mistrust of Platonism and Paracelsianism, 87; Parisian, 186; pragmatism of, 99, 102; proprietor of equipment, 44–45; as society in microcosm, 38, 45
— utility of, 3, 25–26, 32, 46, 47, 48–50, 51, 52, 55–56, 57, 59, 61, 101, 108, 109, 110, 222, 223
— war, political hostilities and, 50, 51, 53, 54, 56, 60, 80, 103, 104, 107, 108, 201
Academy of Inscriptions. *See* Académie royale des inscriptions
Academy of Sciences. *See* Académie royale des sciences
Accademia del Cimento (Florence), 22, 29, 34, 156, 206, 208, 291 n. 12, 317 n. 3, 331 n. 56, 335 n. 101
Acids and alkalis: Mariotte translates Boyle's study of, 214; theories of, 173
Agriculture, interest of Academy in, 49, 169, 194
Air: nature and functions of, 154, 160, 161, 165, 336 n. 102; role of in vegetable physiology, 141, 150, 160–165, 166
Airlessness, effects of on plants, 203–204
Air pressure, 18, 44, 138, 139, 140, 141, 142, 312–313 n. 34
Air pump, 118, 141, 155, 156, 193, 203, 211, 212, 320 n. 27; illustration of Huygens's, 6; use of in Academy, 154, 159–166
Alchemy. *See* Duclos
Aldrovandi, Ulisse, 66
Alkahest, Duclos's recipe for, 97
Alpin, Prosper, 70
Amielle, seller of cylindrical mirrors, 194
Amsterdam, publishing in, 97, 206
Analogical reasoning, 117–130, 131–144; biological and technological models in, 119, 128, 129, 144; in biology, 119, 120, 122–123; in Har-

veian model, 128–130; merits and disadvantages of, 119, 120, 121–124, 144; problem of crucial dissimilarities, 123, 132–133, 136–139, 140–142, 144; problem of material similarities, 121, 123, 132, 134–137; problem of pretheoretic similarities, 121, 123, 132, 133–134
Analogies: in botany, 51, 61, 146, 147, 164, 166, 225; in chemistry, 90–91; failure of, 124; types and functions of, 120, 121, 123, 131, 133, 134, 135, 137, 142, 143, 144; in zoology, 148
Anatomical research of Academy. *See under* Natural philosophical research of Academy
Anatomy. *See also Histoire des animaux;* Sap
— animal and human, 21, 48, 54, 66, 142, 225, 311 n. 9; influence of on botany, 65, 126–127, 133–134
— vegetable, 67, 118, 136–137, 140, 141, 142, 146–149, 157–159, 215; putative air vessels and valves in, 136–137, 142, 144, 214–215
Ancients, 71,120,122,123,147; Aristotle, Aristotelianism, 66, 67, 117, 131, 133; Euclidean model, 222
Animalcules, 148, 159, 158, 204
Animals: in air pump, 159–160; compared to plants, 135, 148, 165; distinguished from plants, 214; illustrations of, 194–195, 213. *See also* Anatomy; Chemical analysis; *Histoire des animaux;* Natural history; Physiology; Plant-animals
Antidotes. *See* Poisons and antidotes
Antipathy and sympathy, theory of. *See* Sympathy and antipathy, theory of
Aqueducts: of Arcueil, 44; for water supply of Versailles, 50, 55. *See also* Frontinus
Architecture, 18, 21, 22, 43, 44, 49, 185. *See also* Blondel; Perrault, Claude; Vitruvius
Arnauld, Antoine, 19, 201
Arts and crafts, 59
Astronomy, 20, 53, 56, 58, 225; assistants for, 210; Copernican system, 44, 48, 211; fictionalism of Academy about, 48; information about supplied by outsiders, 210, 212–213; maps of moon and stars, 44, 211; observations of satellites of Jupiter,

53, 58; Ptolemaic system, 48; Tychonic system, 48, 211
Atomism. *See* Corpuscularian and atomist theories
Attraction, concept of, 140, 141
Audience for science. *See* Science
Auzout, Adrien, 41, 156, 203, 295 n. 3, 327 n. 57, 330 n. 36; Fellow of Royal Society, 202; helps Furetière, 184

Bacon, Sir Francis, Baconianism, 9, 28, 29, 30, 66, 70, 79, 95, 155, 182, 185, 199, 298 n. 12. *See also* Solomon, House of
Baluze, Étienne, scholar, 192
Baudelot de Dairval, Charles César, as collector, 195
Bauhin, Caspar and Jean, 66, 67, 69, 70, 151, 300 n. 47
Beeswax, theory of formation of, 158
Behn, Aphra, playwright, 205, 206
Bellay, N., physician, 174, [204]
Bessé de La Chapelle, Henri de, seigneur de La Chapelle-Milon, 14, 36, 55, 109, 110, 207; career of, 24; harangue by in 1686, 108–110, 172, 225; as Louvois's spokesman, 108–110; urges study of roots, 110
Betbeder, Pierre, 127
Beze, de, 54
Bibliothèque du roi, 7, 17, 28, 43, 44, 45, 51, 107, 160, 190, 192, 196, 203, 295 n. 7, 297 n. 7, 329 n. 22; as headquarters for Academy, 14, 15, 38–43; as living quarters for savants, 20, 41–42, 334 nn. 80–81; refurbishment of houses of, 39; visit of Louis XIV to, 80
Bignon, abbé Jean Paul, 14, 20, 36, 59, 99, 111, 113; family of, 196; role of in Academy, 24
Biological sciences. *See under* Natural philosophical research of Academy
Black, Max, on metaphor, 122
Bled-cornu. *See* Ergot
Blegny, Nicolas, 185, 191, 192, 193, 194; as collector, 196; director of Société royale de la médecine, 196
Blondel, Nicolas François, sieur des Croisettes et de Gallardon, 44, 49, 202, 206, 208, 331 n. 61; books of in Parisian libraries, 185; as collector, 195; helps Furetière, 184; teaches architecture, 197
Blood, theories about circulation of, 21, 59, 119, 120, 123, 125–128, 130, 132; Cartesian model, 127, 312 n. 33; Harveian model, 125–126, 127–128; Riolan's model, 126–127; theory established in France, 127–128, 130
Boccone, Paolo, 70, 207, 307 n. 29, 332 n. 66
Book trade: in Paris, 184–185, 187, 191–192, 325 nn. 24, 27; problems of under Louis XIV, 191, 201. *See also* Censorship
Borelli, Giovanni Alfonso, 139, 141, 142, 206, 317 n. 4, 331 n. 56, 335 n. 96
Borelly, Jacques, 18, 95, 96, 97, 102, 110, 150, 151, 293 n. 24, 304 n. 35, 327 n. 57, 332 n. 65, 333 n. 71, 336 n. 102; and air pump, 160, 319 n. 22; career of, 20; helps Furetière, 184; lodged in Royal Library, 41, 55; role of in Academy, 44, 54, 55, 95, 96, 97
Bosse, Abraham, 76, 80, 81, 84; academicians criticize his illustrations of plants, 82
Botany: audience for in Paris, 185; "botanique," 18, 70; a changing discipline, 61, 65–67, 70, 117, 118, 119, 131, 153, 165, 224–225; develops independently in England and France, 214; eclecticism of, 153–154; study of by Academy, 9, 17–23, 65–88, 89–102, 103–116, 117–130, 131–144, 145–154, 155–165, 203–204, 212–213, 215, 217, 225. *See also under* Natural philosophical research of Academy
Boucot, Nicolas, collector, 195
Boulduc, Simon, 111, 172; teaches at Jardin royal, 38
Bourdelin, Claude, 18, 41, 132, 285 n. 2; career of, 16, 19, 171, 201, 298–299 n. 18, 334 n. 86; medical interests of, 173, 175, 204; method of, 91–92, 93–94, 95, 99, 100; notebooks of, 19, 50, 96, 99, 292–293 nn. 24, 25; possible portrait of, 40; as purchasing agent for crown, 50, 285 n. 4; responsible for chemical analysis, 74, 79, 80, 82, 89, 99, 100, 101, 102, 104, 107, 109, 110, 146, 150–151, 301 n. 53, 305 n. 57; role of in Academy, 19, 96, 151, 182; work of assessed by Dodart, Homberg, and Tournefort, 20, 82, 87, 90, 99, 111, 113, 114, 152; works at home, 55
Bourdelot, Pierre Michon, 20, 21, 22, 195, 196, 205

Index 375

Bournonville, Alexandre Hippolyte Balthazar and Jeanne Françoise d'Aremberg, prince et princesse de, 44
Boyle, Robert, 152, 204, 215; and air pump, 160, 164, 312–313 n. 34, 320 n. 27; book by for sale in Paris, 191; and capillary tubes, 139; and Huygens, 203; and Mariotte, 18, 91; sees chemistry as central, 102, 124; work of reviewed by Academy, 83, 213–214, 217, 325 n. 17
Brice, Germain, 44
Butterfield, Michael, maker of mathematical instruments, 45, 48, 192, 193, 210–211

Cabinet du roi, 39
Calendars: for sale in Paris, 194; study of, 58
Campani, Giuseppe, lens maker, 44
Canguilhem, Georges, on analogy, 120, 121, 122, 144
Capillarity, 138, 139–140, 141, 142, 157
Carcavi, Pierre de, 14, 60, 294 n. 44, 327 n. 57; associated with both Academy and Bibliothèque du roi, 39, 51; handles finances of Academy, 50, 285 n. 4; lodged at Bibliothèque du roi, 41, 334 n. 80
Cartesianism. See Descartes
Cartography: influence of Academy on, 192, 198, 211, 223; map of France by Academy, 21, 22, 32, 49–50; map of généralité de Paris by Academy, 32; map of world by Academy, 22, 44, 49–50, 53; popular among Parisian readers, 185; study of by Academy, 42, 48, 49–50, 55, 59, 169, 222
Cassini, Jean Dominique, 18, 20, 23, 32, 39, 49, 203, 206, 211, 291 n. 13, 294 n. 44, 331 n. 61; access of to king, 22, 26; attends meetings sponsored by Geoffroy, 196; and biological sciences, 22; celebrity in Academy, 13, 35, 36; designs planisphere, 45, 48; family of, 43, 58; as Fellow of Royal Society, 202; has network of informants, 212, 333 n. 78, 334 n. 90; lodging of, 43, 190; and Observatoire, 41, 43; portrait of, 6; role of in Academy, 182; work of, 43, 44, 48, 53, 58
Causation. See Explanation
Censorship, 191, 223
Chaillou, Jacques, 127, 128
Chapelain, Jean, adviser to Colbert, 31

Chapotot, maker of mathematical instruments, 192
Charas, Moyse, 18, 23, 172, 173, 202, 203, 209, 214, 331 n. 61; appointment of helps revive Academy, 111; career of, 20–21; teaches at Jardin royal, 38
Charles I, king of England, as collector, 195
Charles II, king of England, 34, 329 n. 29, 330 n. 32
Chassebras du Breau, M. le Chevalier, sponsor of scientific meetings, 196
Chastillon, Louis Claude, engraver, 54, 58, 68, 72, 80, 82, 106, 107, 112, 299 n. 36, 300 n. 38, 301 n. 49
Chatton, surgeon, 175, [204]
Chazelles, Jean Mathieu de, 36, 211
Chemical analysis. See also Bourdelin; Chemical constituents; Distillants; Distillation; Dodart; Duclos; Homberg; Medicine, and pharmacology
— of animals, 18, 96
— methods of, 19, 20, 73, 93, 94, 95–97, 100, 151, 214, 333 n. 71
— of plants, 18, 19, 20, 70, 73, 74, 75, 79, 80, 83, 89–102, 152; goals of, 82, 87, 91, 94, 98–100, 102, 225; important to Academy, 82, 89–90; medical applications of, 173; outsiders asked to send data, 212; problematic, 69, 82, 87, 88–89, 92, 95, 100, 104, 108, 110; selection of plants for, 100
— of saps, 136
— of seeds, 152
— of soils, 96, 109, 113, 151
— of testes-mortes, 151
— of waters, 18, 48, 50, 82, 83, 89, 108, 109, 172, 305 n. 46, 315 n. 27, 317 n. 3
— of wines, 108
Chemical constituents: in organic matter, 102, 156; in plants, 73, 90, 91, 94, 98–99, 100, 101, 113, 134, 152, 204, 225; in waters, 169. See also Chemical analysis
Chemistry, 58, 83, 145, 319 n. 22; apparatus for on sale in Paris, 193; and botany, 65; central to scientific explanation, 67, 73, 102, 118, 124; French and English different, 214, 215. See also Chemical analysis; Chemical constituents; Distillants; Distillation; Literature; Medicine
Cipolla, Carlo M., 184

Circulation, general concept of, 123, 128, 129, 143. *See also* Blood; Sap
Clandestine activities by savants, 201
Clark, Timothy, 132
Clave, Estienne de, 171
Claviceps purpurea, 174
Clément, Nicolas, 43, 292 n. 16; lodged at Bibliothèque du roi, 41
Clérambault, Nicolas, as purchasing agent for Academy and Royal Library, 51
Clocks, 21, 156; for Jesuits, 53; pendulum, 44, 48, 58; and watches, 193
Cochineal, seed or insect, 214
Colbert, Jean Baptiste, 25, 71, 104, 107, 160, 196, 205, 221, 223, 302 n. 62, 319 n. 20; advisers of, 18, 31; appoints academicians, 18–19, 20–23, 48, 223; and Bibliothèque du roi, 190; designs Academy, 3, 13, 14, 17, 24–25, 29–33, 34, 36–37, 51, 55, 59, 70, 221, 223; encourages Louis XIV to visit Academy, 7, 26; family of, 25, 39; funds Academy, 7, 46, 47–51, 52, 56, 57, 58, 60, 80; plans to reform kingdom, 30; portraits of, 6, 42. *See also* Patronage
Collaboration. *See* Ideals
Collecting and collections, 190, 194, 195–196, 199, 215; by academicians, 20, 21, 38, 195. *See also* Académie royale des sciences, traits of
Collège royal, 17, 20, 21, 22, 23, 35; offers instruction in mathematics, 197
Colson, taxidermist, 194, 293 n. 26
Columna, Fabius (Colonna, Fabio), 70
Community, scholarly and scientific, 28, 79, 155–156, 159, 169, 170, 180–182, 200, 210; eclipsed by Academy, 182, 186, 205, 209, 223, 224; ideals of, 27–29, 297 n. 40; intellectual sectarianism in, 201; of mathematicians, 205; in Paris, 180–198; rivalry in, 200, 202, 206–207; role of monasteries in, 22, 24, 194, 195, 196, 327 n. 52. *See also* Académie royale des sciences, relations of with others; *Gloire*
"Compagnie," 28
Compagnie des arts et métiers, 59, 197
Compton, Henry, bishop of London, 203, 213
Condé, Louis II de Bourbon, prince de ("le grand Condé"), portrait of, 6
Cornut, Jacques Philippe, 70
Corpuscularian and atomist theories of matter, 98, 102, 214, 317 n. 4; and theories of generation, 147, 148, 149
Correspondence, 18, 199, 210–202, 203, 209, 210; to Academy, 212; embargoes on, 201; to Nicolas and Jean Marchant, 204
Couplet, Claude Antoine, 14, 211, 294 n. 35; as purchasing agent for Academy, 51
Couplet, Pierre, 58
Curiosities, 26, 80, 110, 184, 194, 210, 212. *See also* Collecting and collections; Louis XIV; Plants, rare and foreign
"Curious research," deprecated, 108
Cusset, 44

Dalancé, maker of air pumps, 320 n. 27
Dalesma, André, 55, 193, 211
Davison, William, 89
Deaubonne, Daniel, monk and instrument maker, 194
Deglos, received advance for expenses, 50, 295 n. 4, 333 n. 72
Deiconti, 97
Delamare, Nicolas, 192
Denis, Jean Baptiste, sponsor of scientific meetings, 21, 196
Descartes, René, 9, 29, 119, 125, 126, 127, 130, 182, 191; influence of, 23, 28, 29, 30, 141, 196, 199, 222. *See also* Blood
Des Hayes, 295 n. 4, 333 n. 72
Dew, 152–153, 164
Dictionaries, "battle of," 183. *See also* Académie française; Furetière
Digby, Sir Kenelm, 160, 164, 203, 331 n. 45
Dippy, interpreter, 298 n. 14
Dissections, 39, 41, 42; of animals, 38, 334 n. 81; assistants for, 210; autopsies of academicians and others, 38, 54, 292 n. 14; of fish, 32; risks of, 95; supply of animals for, 39, 51, 52, 54, 59
Distillants, 90, 91, 100, 101, 151, 157, 305 n. 53; debates about, 91, 92; identification of, 93, 94, 95, 109; oils in plants, 90, 92, 94, 99–100, 151, 152; salts compared in animals, earths, plants, and waters, 150, 152–153; salts in plants, 90, 93, 94, 99–100, 153; spirits in plants, 90, 94; *testes-mortes* of plants, 93, 95, 151
Distillation, 73, 82, 89, 104, 151; choice of plants for, 95, 103; controversial, 90–93, 94, 96; Dodart's influence on,

Index 377

95; effects of, 90, 92; goals of, 94, 101–102; limits of, 94, 98, 99; method of, 90, 91–92, 93–95; results of, 101. *See also* Medicine, and pharmacology
Divini, Eustachio, lens maker, 44
Do, enameler and instrument maker, 193
Dodart, Denis, 18, 19, 23, 104, 178, 185, 190, 209, 212, 214, 327 n. 57, 328 n. 9, 330 n. 34, 333 n. 70, 334–335 n. 91; career and education of, 19, 26, 103, 172; and chemical analyses of plants, 82, 89, 90, 91–92, 93–94, 96, 98, 99, 100, 101, 104, 109, 303–304 n. 34, 304–305 n. 46, 331 n. 61; and Duclos, 83, 87, 97, 103, 104, 110, 111, 225; family of, 20; a Jansenist, 19, 23, 178, 201; and natural history of plants, 48, 55, 75, 79, 80, 103, 107, 110, 111, 114, 115, 300 n. 40, 302 n. 64, 317 n. 5; and neologisms, 184; and poor, 171, 176, 178–179; possible portrait of, 40; protégé of Perrault, 83; role of in Academy, 83, 87, 94–95, 96, 114, 182; sends *Mémoires des plantes* to Grew, Locke, and Morison, 202, 214; studies ergotism, 171, 174–176, 178–179, 204; studies nutrition and medicine, 96, 173; studies plant anatomy and physiology, 146, 147, 149, 150, 157; and Tournefort, 113, 114–115
Dubé, Paul, physician, 174, 178, [204]
Duclos, Samuel Cottereau, 17, 18–19, 20, 23, 48, 201, 293 n. 25, 299 n. 32, 301 n. 50, 305 n. 46, 311 n. 13, 312–313 n. 34, 316 n. 37, 331 n. 61; career of, 18–19, 172; and chemical analysis, 18, 48, 50, 73, 82, 83, 89, 90, 92, 93, 96–97, 98, 102, 108, 109, 169, 172; and circulation of sap, 119, 132, 164; controversial views of, 18–19, 43, 83, 87, 96–97, 173, 205, 206; and Dodart, 55, 83, 87, 97, 103, 225; lodged at Bibliothèque du roi, 41, 55; and natural history of plants, 73–75, 79, 80, 82–83, 87, 90, 97, 98, 103; and plant physiology, 150, 152–153; Protestant, 19, 43, 55; possible portrait of, 40; reviews work of Boyle, 214, 335 n. 101; role of in Academy, 83, 87, 182; and Tournefort, 114
Du Hamel, Jean Baptiste, 14, 18, 19, 25, 36, 104, 107, 200, 206, 288–289 n. 42, 292 n. 24, 331 n. 61, 336 n. 101; career of, 22–23; as historian of Academy, 23, 184, 298 n. 15; in London, 51, 202; possible portrait of, 40; as purchasing agent for Academy, 51, 285 n. 4; role of in Academy, 13, 14, 18, 22, 110, 182, 318 n. 14
Du Verney, Joseph Guichard, 18, 26, 59, 172, 214, 293 n. 29, 295 n. 15, 331 n. 61; attends meetings sponsored by Geoffroy, 196; career of, 21; and Jardin royal, 38, 41
Du Vivier, David, royal geographer, 211, 333 n. 76; as collector, 195

Eau de vie, used during dissections and as remedy, 41, 95
Education, scientific, 197–198, 200–201. *See also* Académie royale des sciences, functions and goals of
Elephant, scientific and public interest in, 26, 194, 335 n. 100
Emboîtement, 148, 149, 157
England, 101, 119, 209; ties of academicians to, 21, 22, 23, 58, 99, 172
Engravers and engravings, 79, 157; of animals, 59; of arts and crafts, 59; expenditure on, 48, 51, 54, 58; of plants, 22, 51, 54, 58, 207. *See also* Bosse; Chastillon; Illustrations of plants; Robert
"Engyscope," 318 n. 6
Ergot, 174, 175, 204, 212; as cause of ergotism, 174, 176, 178; description of, 174, 176; illustration of, 177; obstetric uses of, 174
Ergotism, 171, 174–179, 204
"Ethereal matter," existence of posited by Homberg, 162, 163
Exchange, scholarly, 180, 198, 201–203, 205; Anglo-French, 202–203; of ideas and of data, 217; limits on, 198, 200–201, 215–217; methods of, 196, 198, 199, 211, 215, 217; between scientific societies, 205; as struggle between contestants, 215–216. *See also* Académie royale des sciences; Correspondence
Expeditions, scientific, 32, 34, 50, 51, 53, 58, 107, 209, 293 n. 25, 333 n. 75. *See also* Jesuits
Experiments, 119, 210, 211. *See also* Ideals
—botanical, 142, 149, 153, 160–166; in air pump, 159–166; models for, 119; relating to circulation of sap, 119, 134–135, 136–137, 141–142; role of, 146, 225

—medical, 99, 173, 179
Explanation, 109, 121, 139–142, 144; chemical, 67, 118, 127, 129–130, 138–139, 141, 142, 145–154, 225; of exceptions, 148; by "faculties" or souls, 131, 150, 164, 312–313 n. 34; mechanistic, 67, 95, 118, 125–127, 129–130, 134, 131, 137, 138, 139–141, 145–154, 225; by multiple causation, 141, 142, 145; by reduction, 120, 121; teleological, 90, 125–126; and theology, 146, 149; vitalist, 125–126. *See also* Analogical reasoning; Analogies; Corpuscularian and atomist theories

Fagon, Guy Crescent, 20, 38, 107, 201, 207, 296 n. 33, 330 n. 32
Falling bodies, studies of, 44
Fantet de Lagny. *See* Lagny
Far East, 58; coordinates of cities in, 49; scientific data and specimens from, 53–54, 111, 112, 195, 212–213
Fer, Nicolas de, cartographer and royal geographer, 192
Ferry, Paul, 201
Flora, geographical origins of, 71, 74, 82, 111, 213, 334 n. 81. *See also* Far East
Florence, scholarly center, 31, 208
Foley, Samuel, 149
Fontaney, de, sponsor of scientific meetings, 196
Fontenay, Henri François or Claude de, painter of flowers and animals, 195
Fontenelle, Bernard Le Bovier de, 18, 19, 90, 92, 201, 207, 296 n. 21, 315 n. 16, 319 n. 25; career of, 22–23; collects reimbursements for other academicians, 50; historian of Academy, 18, 22, 23, 184, 202, 212, 298 n. 15, 301 n. 57, 309 n. 49; role of in Academy, 14, 182
Food chain, study of, 99, 153
Fouquet, Nicolas, 172
Frenicle de Bessy, Bernard, 17
Frontinus, treatise on aqueducts, 22, 55
Furetière, Antoine, 20, 74, 183

Gaignières, François Roger de, collector, 195
Galen, 126. *See also* Medicine
Galileo, 119; books by read in Paris, 185; telescope of, 155, 195
Gallois, abbé Jean, 14, 18, 23, 187, 327 n. 57, 334 n. 81; career of, 22; edits *Journal des sçavans*, 191; possible portrait of, 40; teaches at Collège royal, 197; urges study of roots, 110
Galls, cause of, 110
Garden, Royal. *See* Jardin royal; *Petit jardin*
Gardens and gardening, 25, 191, 194, 195, 197; 196, 203, 213, 289 n. 45; horticulture not an interest of Academy, 194
Gassendi, Pierre, books by read in Paris, 185
Gaston, duke of Orléans, 19, 71, 73, 127, 194, 202, 223; influence of on Academy, 71, 75
Gayant, Louis, 171, 172, 334 n. 81; possible portrait of, 42
Généralité de Paris, map of, 49, 51
Generation of plants and animals, 70, 145; spontaneous, 145–149. *See also Emboîtement;* Physiology; Preformation
Gennes, Jean Baptiste de, 332 n. 67, 334 n. 81
Geoffroy, Étienne François, 293 n. 25, 325 n. 19, 328 n. 16, 336 n. 106
Geoffroy, Mathieu François, sponsor of scientific meetings, 196, 328 n. 16
Geography: books on popular among Parisian readers, 185, 191, 192; royal geographer, 192, 211
Gérard, André and André Guillaume, cutlers, 39
Germination, 146, 209, 214; mechanical explanation of, 149; roles of external and internal factors in, 147, 149–151, 160–165, 166. *See also* Physiology
Gesner, Conrad, 298 n. 13
Ginseng, 111, 213
Glaser, Christophe, chemist, 89, 94
Gloire, 200; desire for, 206, 216
Gobelins, 32, 297 n. 38
Gosselin, Georges, mathematical instrument maker, 44, 192
Gouye, Thomas, S.J., 54, 208, 332 n. 65
Goyton, Jean, engraver, 6, 292 n. 16
Grande écurie, 16
Gresham College, 139, 203
Grew, Nehemiah, 102, 117, 147, 149, 154, 158, 202, 203, 214, 215, 317 n. 4; on anatomy of plants, 67, 132, 307 n. 29, 311 n. 5, 336 n. 106; on circulation of sap, 119, 137, 312 n. 33
Guglielmini, Domenico, 36, 172
Guiffart, Pierre, 127, 128

"Guinea pigs," human and animal, 173, 179
Guise, duchesse de, 201

Hales, Stephen, 141
Haller, Albrecht von, 123
Hartsoeker, Nicolas, 58, 203, 204, 210; makes lenses, 44; and spherical microscope, 158, 319 n. 20; supervises production of glass for scientific instruments, 53
Harvey, William, 9, 119, 123, 132, 133, 138, 144, 311 n. 9, 315 n. 19; defenders and detractors of in France, 127–128; influence of, 120, 124, 125–126, 127, 128–130, 131, 134–137, 138, 142. *See also* Analogical reasoning; Blood
Hautefeuille, abbé Jean de, 332 n. 66
Heart, 138; functions of, 125, 126–127; motions of, 125, 126–127
Helmont, Joan Baptista van, 91, 97, 124, 152
Herbelot de Molainville, Barthélemy d', sponsor of learned society, 327 n. 51
Hesse, Mary, on analogy, 120, 121, 132–133, 137, 143
Hevelius, Johannes, 31
Hierarchy, of forms of life, 132, 136, 137. *See also* Académie royale des sciences; Community
Histoire des animaux, 18, 59, 207; installments of planned, 54. *See also* Anatomy; Natural history
Holland, regarded as rival of France, 31, 32. *See also* Hartsoeker; Huygens; Leeuwenhoek
Homberg, Guillaume, 18, 23, 98; and air pump, 160, 161–165, 297 n. 37, 308 n. 36, 336 n. 101; appointment of helps revive Academy, 19–20, 24, 111, 225; attends meetings sponsored by Geoffroy, 196; career of, 20, 172, 202, 211–212; collects reimbursements for other academicians, 50; and distillation of plants, 90, 92, 95, 99, 287 n. 21, 305 n. 53; and medicine, 99; and natural history of plants, 103, 111, 114; and plant physiology, 146, 152, 153, 160, 161–165; role of in Academy, 59; travels of, 202
Hooke, Robert, 67, 136, 139, 140, 156, 203, 317 n. 4, 320 n. 27, 335 n. 100; reviews *Mémoires des plantes*, 303 n. 28; tests seeds in air pump, 160

Hôtel des invalides, 41, 187; supplies corpses for dissection, 59
Hubin, enameler and instrument maker, 44, 193, 317 n. 3, 333 n. 68
Huilliot, Claude, painter of plants and animals, 195
Huygens, Christiaan, 14, 18, 20, 36, 43, 49, 60, 156, 202, 208, 213, 289 n. 42, 326 n. 38, 327 n. 57, 331 n. 61; and air pump, 160–162, 162, 163–164, 165, 193, 212; and capillary tubes, 139; celebrity in Academy, 13, 35; and Charas, 172; as collector, 195; correspondents of, 203, 331 n. 45; enemies of, 159, 206; lodged in Bibliothèque du roi, 41, 43, 334 n. 81; and natural history, 70, 71, 74, 79; portrait of, 42; receives information about Academy, 22, 95, 109, 110, 207, 288 n. 38; on rise of sap, 140, 142; role of in Academy, 14, 182; and spherical microscope, 158, 159, 204; and study of plants, 21, 132, 158, 160–161, 312–313 n. 34
Huygens, Constantyn, 158
Hydraulics, 49, 50, 59, 222; hydraulic machines, 56
Hydrography, teaching of, 16
Hydrostatics, study of, 18

Ideals, scholarly, 27–29, 209; conflict between ideal and practice, 205–215, 224-225; cooperation, 29, 169, 180, 199–217; experiment, 5, 7, 115–116, 155–156, 181, 215, 222, 225; reform, 29–30, 113, 222; utility, 182. *See also* Académie royale des sciences, functions and goals of; Utility
Illustrations of plants, 213. *See also* Engravers and engravings; Gaston; Plants
— commissioned by Academy: 22, 69, 73, 75, 80, 88, 110, 111, 113, 223; published in eighteenth century, 80, 114, 115; scientific standards for, 71, 75, 80, 82, 107, 157, 298 n. 18, 300 n. 38, 317 n. 5; shortcomings of, 68, 73, 80–83, 86, 104, 112; size of, 82
Imprimerie royale, 191, 192, 208, 209; typeface for, 59
Imprimeurs du roi, 191, 208
Industry, interest of Academy in, 49, 110, 169. *See also* Machines; Practical projects of Academy
Instrument makers, scientific, 39, 44, 45, 187, 194, 198, 320 n. 27; and Academy, 192–193; essential to ex-

perimental science, 193, 200; shops of, 190, 192–194. See also names of individuals
Instruments and apparatus, scientific, 7, 18, 20, 22, 55, 118, 156, 192–194. See also names of instrument makers; Air pump; Chemistry; Clocks; Inventions; Lenses; Microscopy
— and botanical research, 61, 146, 154, 155–166, 225
— demonstrational, 44–45, 48, 210–211, 329 n. 25; burning mirrors, 26, 48, 53, 54; globes, 45; lodestone, 194; planisphere, 45, 48, 210–211; "talking" ephemerides, 44
— observational, 23, 44–45, 329 n. 25; aerometer, 156, 165, 193; astronomical and mathematical, 44, 155, 156, 159, 193, 194; compass, 207; magnetic needle, 44, 58; meteorological, 155, 156, 165, 193; microscope, 23, 51, 157, 159; supplied to Jesuit missionaries, 53, 210; surveying, 18, 156, 193
— portrayed, 6, 40, 42
— purchase and maintenance of for Academy, 34, 43, 44–45, 48, 50, 51, 53, 58, 165, 210–211
— role of in new science, 154, 155–156, 159, 160, 166
Interdisciplinary links amoung the sciences, 118, 144, 153–154, 156, 165, 166, 225. See also Analogies; Sap
Inventions, 156, 193, 222; assessed by Academy, 49, 55, 56, 59. See also Académie royale des sciences, functions of; Dalesme; Machines
James II, king of England, visits Academy, 7, 294 nn. 40, 43, 296 n. 21, 335 n. 96
Jansenism, Jansenists, 19, 23, 178, 187, 190, 200, 201, 223; pedagogues, 197
Jardin royal, 7, 16, 17, 20, 21, 23, 41, 45, 147, 171–172, 187, 193, 213, 217; scientific collections of, 80, 195; ties of with Academy, 15, 38–39, 79, 82, 107, 114, 195, 296 n. 33; teaching at, 38, 89, 93, 114, 127, 197–198, 332 n. 63. See also Petit jardin
Jesuits, 17, 23, 201; colleges of, 197; missionary-scientists in Far East, 22, 53–54, 58, 208, 210, 212–213
Joblot, attends meetings sponsored by Geoffroy, 196–197
Joubert, Jean, engraver for Tournefort's Élémens, 300 n. 38

Journal des sçavans, 22, 24, 172, 191, 208, 209, 213
Jussieu, Antoine de, 303 n. 11
Justel, Henri, 203, 205, 213, 298 n. 9, 313 n. 37

Kepler, Johannes, 128
Kermes, seed or insect, 214
King's Library. See Bibliothèque du roi
Kircher, Athanasius, 206
Kunckel, Johann, 202

La Beurthe, surgeon, 293 n. 26
Laboratories, private, 27, 55. See also Académie royale des sciences
La Brosse, Guy de, 71
La Chambre, Marin Cureau de, 17, 171; books by read in Paris, 185; opposes neologisms, 183
La Chapelle. See Bessé de la Chapelle
La Faye, 294 n. 40
Lagny, mathematical instrument maker, 44, 192
Lagny, Thomas Fantet de, 36; Fellow of Royal Society, 202
La Hire, Gabriel Philippe de, 18, 22, 23, 111
La Hire, Philippe de, 18, 19, 44, 55, 157, 158, 165, 182, 197, 206, 207, 208, 214, 222, 294 n. 35, 296 n. 21, 302 n. 62, 324 n. 5, 326 n. 33, 331 n. 61, 337 n. 112; career of, 21–22; on circulation of sap, 119, 136–137, 140, 141–142, 144; dissects fish, 32; keeps Huygens informed, 95, 109, 110; revives study of plant physiology, 110; role of in Academy, 182; studies fig and orange trees, 111; teaches at Collège royal, 197
La Hyre, Laurent, 21
La Londe, engineer, his arithmetic for engineers for sale, 191
Langlade, de, 172
Language, scientific, 22, 27, 223, 318 n. 6; in botany, 73, 214, 298 n. 18; shift to vernacular, 73–74, 75, 88, 127–128, 153, 182, 183–184, 223; in technology, 183
Languages: knowledge of, 200, 298 n. 15, 325 n. 17, 335 n. 100; study or teaching of, 20, 21, 328 n. 65
Lannion, abbé Pierre de, 14, 307 n. 16
Lanterns, magic, 194
Lantin, Jean Baptiste, 201, 204
La Quintinie, Jean de, 194, 203, 329 n. 28

Index   381

La Roque, abbé, sponsor of scientific meetings, 196
Le Bas, mathematical instrument maker, 44, 192
Le Brun, Charles, autopsy of by Academy, 54
Le Clerc, Sébastien, 5–8, 15, 41; portraits of Academy by, 6, 40, 42
Lecture-demonstrations, scientific and medical, 182, 196. See also Jardin royal
Leeuwenhoek, Antony van, 203, 204
Le Febvre, Jean, 36
Le Febvre, Nicaise, chemist, 19, 90, 93, 132, 201
Le Gallois, Pierre, 181
Le Guern, mathematical instrument maker, 44
Leibniz, Gottfried, Wilhelm, 36, 332 nn. 65, 66
Lémery, Nicolas, 20, 21, 90, 98, 209; book for sale in Paris, 191; offers instruction in chemistry, 197
Lenses, 20, 44, 194; for telescopes, 58, 210; tested, 53. See also Borelly; Campani; Divini; Hartsoeker
Léonard (de Sainte Cathérine), père, 207
L'Hospital, Guillaume François Antoine, marquis de Sainte-Mesme, 16, 36
Lhwyd, Edward, 202
Librarian, royal, 24
Libraries, private, 182, 185, 215. See also Bibliothèque du roi
Life, ideas about nature of, 133
Linnaeus, Carl von, 20
Lister, Martin, 41, 44, 186, 191, 194, 196, 200, 201, 202, 211, 214, 292 n. 18, 294 n. 40, 321 n. 7, 325 n. 19, 328 n. 16, 337 n. 113
Literacy in France, 184
Literature, scientific and technical: by academicians for sale in Paris, 191, 192; Academy's use of, 213–214, 215, 309 n. 1; herbals, 117; manuals for practitioners, 191; periodicals, 199; specialized bookshops for in Paris, 191–192; tastes in, 184–185. See also Journal des sçavans; Language; Philosophical Transactions
Locke, John, 44, 159, 187, 193, 194, 200, 202–203, 305 n. 58, 312 n. 26, 327 n. 46, 336 n. 110
London, 181, 187; as scholarly and cultural center, 16, 51, 203
Longitude, problem of determining at sea, 32, 49, 52
Louis XIII, 171

Louis XIV, 20, 23, 34, 169, 170, 172, 178, 193, 201, 221, 226, 293 n. 32, 298 n. 15; and individual academicians, 26, 103, 172; illness of, 103, 108, 109, 172, 225; nurseries and orangerie of, 74, 110; patronage of sought by English savants, 203; portrait of, 6; visits Gobelins, 32. See also Académie royale des sciences: and Louis XIV; and war
Louvois, Michel François Le Tellier, marquis de, 13, 14, 18, 19, 20, 24, 32, 82, 208, 225; appoints academicians, 52, 53, 54, 56, 223; compared with Colbert and Pontchartrain, 46, 47, 52, 57, 58, 59, 111; garden of, 194; hôtel of, 38, 190; and natural history of plants, 25, 52, 56, 80, 107–111; policies of regarding Academy, 25, 36–38, 46, 51–56, 59, 110, 223; utilitarian demands of, 52, 55, 56, 107–110, 172
Louvre, headquarters of Academy from 1699, 39
LSD poisoning. See Ergotism
Luxembourg, Madeleine Charlotte de Clermont Tonnerre, duchesse de, 44
Lysergic acid diethylamide. See Ergot

Mabillon, Dom Jean, 201
Macard, instrument maker, 193
Machines, 20, 55–56; hydraulic, 50, 55, 56; models of, 44, 49, 156, 210; salle des machines, 44, 195
Magnetic variation, 207
Magnol, Pierre, 213
Major, Johann Daniel, 67, 119, 132
Malpighi, Marcello, 21, 154, 201, 203, 214, 215, 315 n. 19, 317 n. 4; books by read in Paris, 185, 311 n. 5; study of anatomy and physiology of plants by, 67, 117, 119, 132, 307 n. 29, 312 n. 30
Mandrake root, 82, 84, 85
Maps. See Cartography
Marchant, Nicolas and Jean, 146, 147, 204, 213, 309 n. 1, 329 n. 28; coin neologisms, 184; library of, 185, 214, 308 n. 29, 337 n. 115; revise Bauhin's Pinax, 70, 111; and Tournefort, 113, 114, 115; work on natural history of plants, 74, 75, 79, 80, 82, 104, 107, 110, 111, 114, 302 n. 64. See also Plants, rare
—Jean, son of Nicolas, 18, 19, 23, 82, 104, 111, 148, 203, 214, 300 n. 43, 301 n. 53, 301–302 n. 60; career of,

19; and Jardin royal, 38, 80, 296 n. 33; role of in Academy, 96
—Nicolas, 18, 19, 104, 202, 213; career of, 19, 71, 73, 172; cultivates plants, 175, 314 n. 7; and circulation of sap, 119; establishes *petit jardin*, 79; and Jardin royal, 38; possible portrait of, 40; and Robert Morison, 214; role of in Academy, 96; writes *Description de quelques plantes nouvelles*, 75, 80, 87, 329 n. 26
Margrave [Marcgrave or Markgrav or Marggraf], Georg, 70
Mariotte, abbé Edme, 18, 19, 23, 43, 44, 50, 90, 165, 200, 203, 206, 209, 222, 325 n. 17, 327 n. 57, 329 n. 30, 331 n. 61, 331–332 n. 63; autopsy of, 54; on chemical analyses of plants, 90, 91, 92, 98, 113, 287 n. 21; on circulation of sap, 119, 131, 132–138, 140–141, 142; on plant anatomy, 157–158; on plant physiology, 147–150, 151, 152, 153, 164, 201, 204, 214; possible portrait of, 42; role of in Academy, 132, 182
Marly, royal palace at, 194
Marly tower, 43, 44, 56, 57
Martet, Jean, 127, 128
Martin, Henri-Jean, 185
Mathematical sciences studied at Academy, 21, 58, 225. *See also* Académie royale des sciences, finances of; Astronomy; Instruments and apparatus; Literature
Matte La Faveur, Sébastien, 89
Mechanics, study of principles of, 49, 222
Medici, Cosimo, Giovanni Carlo, and Leopoldo de', 156, 206, 334 nn. 80, 81
Medicine, 25, 48, 99, 102, 135, 178, 215, 221, 222, 225; autopsies, 38, 54; and botany, 22, 66, 70, 87, 89, 91, 94, 96, 98, 100, 101–102, 108, 109, 113, 114, 170, 171, 172; chemical, 171, 172; faculties of, 20, 127–128, 171–172, 174, 202; Galenic, 172, 173; and pharmacology, 20, 171, 173, 210; for poor, 19, 170, 174–176, 178–179, 184, 204, 212, 295 n. 13; and public health, 171, 176, 178; writings on by academicians, 172. *See also* Académie royale des sciences; Ergotism; "Guinea pigs"; Literature; Société royale de la médecine

Mémoires des animaux. See *Histoire des animaux*
Mémoires des plantes, 75, 79, 80, 83, 87, 90, 93, 95, 100, 101, 103, 202, 207, 213, 303 n. 28; installments of planned, 54, 103–104, 107, 111, 113, 114; presented to Grew and Morison, 214; as smokescreen, 101
Ménage, Gilles, 286 n. 9
Menageries, 194. *See also* Versailles
Mentel, Jacques, 127, 128
Meridian, extension of in France, 21, 22, 49–50; canceled, 43, 55; revived, 58, 59
Méry, Jean, 21, 41, 59, 172, 293 n. 29; taught at Jardin royal, 38
Meteorological observations, 44
Meurisse, received advance for expenses, 50
Microcosm-macrocosm, 119, 120, 129, 131
Micrographia, 67, 203
Microscopy, 149; and anatomy of plants, 137, 146, 157–159; ancillary to other ways of studying plants, 159; and illustration of plants, 75, 82; instruments used (compound and spherical microscope, hand lens, loupe), 82, 136, 137, 155, 156, 157, 158–159, 194, 204; and physiology of plants, 149, 157–159, 163; value of in botanical research, 65, 70, 118, 154, 158–159, 165–166. *See also* "Engyscope"; Instruments and apparatus
Migon, maker of globes and mathematical instruments, 44, 45
Military: fortification a popular subject among Parisian readers, 185; interest of academicians in inventions and techniques for, 49, 169, 208, 222. *See also* Literature
Mimosa. *See* Plants, sensitive
Mineralogy, 21, 59
Mineral waters, analysis of. *See* Chemical analysis of waters
Mining, 20; interest in of academicians, 202
Mirrors, cylindrical, for sale in Paris, 194
Model. *See* Analogies
Montmor, Henri Louis Habert de, 20, 202, 205; Academy of, 206
Moray, Sir Robert, 203, 331 n. 45
Morin (de Toulon), 18, 111, 172; career of, 21
Morin, Louis, 200; as collector, 195; garden of, 203

Index

Morison, Robert, 70, 80, 147, 202, 214, 302 n. 60, 311 n. 9; and classification of plants, 67, 75

Natural history, 32, 46, 48, 51, 54–55, 56, 212–213. *See also* Chemical analysis; Illustration of plants; Natural philosophy
— of animals, 70, 95, 222
— of arts and crafts, 59
— audience for, 185, 194–195
— data about solicited, 212–213
— of plants, 19, 25, 48, 54–55, 65–88, 89–102, 103–116, 215, 222, 225; conflicting views about, 70–79, 83, 87, 109; choice of plants for, 69, 73, 74, 88; Colbert and, 65–88, 103–107; collaboration on, 67, 69, 74, 79–80, 82–83; cultivating and describing plants for, 69, 73, 75, 79–80, 88, 104, 110, 111, 299 n. 33; directors of, 73, 75, 83, 87; failure of, 65, 69, 88, 102, 103–116, 225; geographical scope of, 73, 74, 114, 115; goals of, 108; installments of planned, 54, 103–104, 107, 111, 113, 114; Louvois and, 107–111; Pontchartrain and, 111, 113–115; review of literature for, 80; patronage and, 69; theft of manuscripts for, 103, 104, 225; traditional and innovative elements in, 67, 69, 74; and Tournefort, 113, 114
— related to natural philosophy and medicine, 65–67, 69, 70, 79, 101, 108, 109, 173–174
Natural philosophical research of Academy. *See* Académie royale des sciences, finances of; Air pump; Illustrations; Microscopy; *names of individual academicians*; Natural history; Natural philosophy; Physiology; Plants
Natural philosophy: distinguished from natural history, 65–66, 70, 79, 101, 118, 145, 146, 154. *See also* Anatomy; Natural history; Physiology
Navigation, 191; interest of Academy in, 48, 169
Newton, Sir Isaac, 119, 335 n. 96
Niquet, Antoine de, 182, 211, 291 n. 6
Nolin, Jean Baptiste, engraver, 294 n. 40
Nomenclature, anatomical, 122–123. *See also* Plants, names of
Nutrition of plants. *See* Physiology, vegetable
Nutrition of animals and humans: diet and social hierarchy, 176, 178; Dodart's study of, 19, 96, 99, 100, 103–104; Dodart's theory of digestion, 173; study of food chain, 99, 153; value of plants in, 91, 98, 102, 108, 110, 171. *See also* Chemical analysis

Observatoire, 7, 18, 22, 26, 31, 34, 38, 41, 43–45, 56; cost of, 43, 47, 51, 52, 53, 57, 187, 190, 195, 197, 198, 203, 210; grounds of, 43–44; porter of, 57; purpose and functions of, 15, 43–45, 197; as symbol of Academy's corporate identity, 223; visitors to, 44. *See also* Cartography, map of world by Academy
Observatories, private, 27
Oldenburg, Henry, 202, 203, 205, 215, 330 n. 36
Orangerie, 74, 110; study of trees in, 110, 111
Orléans, dukes of. *See* Gaston; Philippe
Oury, watchmaker, 193
Ozanam, Jacques, mathematician, 197

Papin, Denis, 193, 212, 317 n. 1, 332 n. 66
Paracelsus, Theophrastus Philippus Aureolus Bombastus von Hohenheim; Paracelsianism, 17, 83, 87, 90, 96–97, 120, 124, 173
Pardies, Ignace Gaston, S.J., 22, 337 n. 116
Paris, 16, 21, 22, 82, 171, 175; as magnet for ambitious, 16
— addresses serving scientific community, 327 n. 55; Arsenal, 195; faubourg Saint Antoine opposite rue de Charonne, 194; faubourg Saint Honoré, 194; faubourg Saint Jacques, 187, 193; Galeries du Louvre, 190, 192, 193; Palais de justice, grande salle of, 191; place de l'Hôtel de Cluny, 193; porte Montmartre, 294 n. 35; quai des Augustins, 192; quai de l'Horloge, 190, 192; quai de Morfundus, 193; quai de Nesle, 191; quai Peletier (now quai de Gesvres), 190, 193; rue Bourlabé, 195; rue Dauphine, 194; rue du faubourg Saint Jacques, 187; rue du Harlay, 190, 193; rue de La Harpe, 190; rue Mazarine, 187, 193; rue neuve des Fossés, 187, 326 n. 28; rue des Postes, 294 n. 35; rue Saint Denis across from rue aux Ours, 193; rue

Saint Denis near Queen's Fountain, 193; rue Saint Honoré opposite church of Saint Roch, 193; rue Saint Jacques, 190, 191, 192; rue Saint Pierre, 190; rue Vivienne, 38, 190; Saint Hilaire, 194; Saint Severin fountain, 191; Sainte Chapelle, steps of, 191
—description, map, and topography of, 186–190; administrative and regal center of, 187, 188, 190; Cité of, 187, 188, 190; faubourg Saint Germain of, 203; Hôtel de Ville, 190; Jansenist landmarks in, 187, 190; Left Bank of, 187, 189, 190; Marais of, 190; Palais royal district of, 38, 190, 194; Right Bank of, 187, 188, 190; university quarter of, 187, 191. See also Bibliothéque du roi; Hôtel des invalides; Jardin royal; Observatoire
—hospitals in, 41, 99, 173, 187, 203
—setting for scholarly community, 159, 164, 170, 180, 181, 186, 190–198; Academy's immediate theater of operations, 43, 186; center of French book trade, 191–192; center of scientific and mathematical education, 198; cultural center of France, 186; intellectual center of shifting, 190; scientific goods and services available in, 185–198; scholarly landmarks in, 187, 190; topography of scientific community in, 187, 190–195; university quarter of, 187, 190
—University of, 16, 187; medical faculty of, 127, 128, 172
Parkinson, John, 70, 308 n. 34
Patigny, Jean, engraver, 43
Patronage. See also Académie royale des sciences, patronage of; Societies, learned and scientific
—private, 196; by upper robe, 190
—royal: character and effects of, 30–33, 80, 113, 172, 192, 203, 216, 217, 221–226; on Paris, 190; on publishing, 208; on science and scientific community, 217, 223
Pecquet, Jean, 41, 127, 128, 172, 311 n. 9, 327 n. 57, 334 n. 81; portrait of, 42
Peiresc, Nicolas Claude Fabri de, 126
Pensions
—paid to academicians, 4, 23, 24, 103, 223, 224; changes in, 36–37; converted into annuities, 37, 57, 60; cost of, 47, 52, 56, 57; nature and effects of, 35–38, 45, 50
—paid to individual scholars: *pensions et gages*, 35–36; *pensions et gratifications*, 30–31, 35
Pépinerie, 194, 217
Perrault, Claude, 16, 18, 19, 23, 43, 48, 49, 90, 175, 184, 203, 204, 206, 209, 211, 214, 305 n. 46, 311 n. 5, 331 n. 61, 331–332 n. 63, 333 n. 70, 334 n. 81, 334–335 n. 91; autopsy of by Academy, 54; career of, 18; on chemical analysis of plants, 95, 96, 98; on circulation of sap, 119, 132–138, 141, 142, 164, 312 n. 30; death of, 95; on describing plants, 75; and Dodart, 83; family of, 18, 19, 41, 140, 286 n. 9; on illustrating plants, 71; on natural history of plants, 70–71, 73, 74, 79, 286 n. 18; on physiology of plants, 146, 149, 150, 152–153, 157; portraits of, 6, 42; as purchasing agent for Academy, 51; role of in Academy, 182; traits of as scientist, 132, 135
Perrault, Pierre, 140
Peter the Great, 195
Petit jardin, 38, 54, 58, 79, 308 n. 40
Pharmacology. See Medicine
Philippe, duke of Orléans, portrait of, 6
Philosophical Transactions, 202, 209, 214–215; translated for but neglected by Academy, 213, 215. See also Journal d'Angleterre
Phosphorus, 202, 211, 336 n. 101
Physiology. See also Analogical reasoning; Analogies; Anatomy; Animals; Blood; Botany; Chemical Analysis; Explanation; Generation; Germination; Natural philosophy; Nutrition; Plants; Sap
—animal, 54, 59, 66, 142, 145, 225; influence on botany, 65
—vegetable, 70, 79, 91, 110, 117–130, 131–144, 145–154, 155–166, 214; chemical processes inside plant, 139, 141, 142, 152; germination and growth, 149–151, 160–165, 166; growth and nutrition, 151–153; influences on, 117, 118, 150; leaves, functions of, 135; life cycle, 145
Picard, abbé Jean, 44, 203, 294 n. 35, 317 n. 1, 335 n. 101; and spherical microscope, 158–159; studies pollen, 158
Pinax. See Bauhin, Caspar and Jean; Marchant, Jean and Nicolas
Pivert, 182

Index 385

Plant-animals, 137–138
Plantes du roi, Les, 115. *See also* Engravers and engravings
Plants: active or passive, 152–153, 165; behavior of in air pump, 160–166; choice of for study, 104; classifying, 20, 59, 70, 71, 73, 75, 113, 209, 214, 222; compared with animals, earths, insects, sponges, 66, 67, 79, 131–132, 135, 136, 137, 141, 144, 149, 150, 157, 164–165, 166, 312 n. 30; cultivating, 19, 69, 74, 79, 80, 104, 110; describing, 19, 73, 74; distinguishing among, 98; distinguishing from animals, 214; explaining effects of, 70, 91; guttation of in a vacuum, 163–164; illustrated, 67, 194–195; information about sent to Academy, 210; names of, 69, 71, 73, 80, 82, 113, 308 n. 34; nature of sought, 70, 98, 100; nutrition of, 20, 204; pollen of, 158, 204; propagation of, 70, 75, 146–147; rare and foreign, 19, 69, 71, 74, 79, 82, 111, 112; research on exemplifies traits of Academy, 61; role of external factors in development of, 149–151; roots of, 110; seeds of, 146–147; sensitive, 67, 68, 82, 297 n. 1; sexuality of, 145; species of known, 70, 117; uses of, 19, 22, 69, 70, 71, 110. *See also under* Natural philosophical research of Academy; Physiology
Plato, Platonism, 28, 87, 206
Plumier, Charles, 71, 194, 201, 214
Poisons and antidotes, 20, 91, 99, 102, 173
Pontchartrain, Louis Phélypeaux de, 13, 14, 17, 24, 47, 52, [225]; appoints academicians, 19, 20–21, 22, 56, 58, 111, 223; compared with Colbert and Louvois, 46, 47, 52, 54, 57, 58, 59, 111; effect of on Academy, 46, 56–60, 80, 110–111, 113, 223; *hôtel* of, 38, 190; policies of, 24, 25, 36–38, 46, 47, 56–60, 330 n. 39
Poor. *See* Medicine, for
Porcelain, 21, 195
Pouilly, mathematical instrument maker, 194
"Pousser," 134
Practical projects of Academy, 44, 48, 49, 51, 56, 59. *See also* Académie royale des sciences, finances of; Agriculture; Cartography; Industry; Machines; Military; Navigation; Porcelain; Versailles
Practitioners, scientific and technical, 182, 183, 210–211; trained by Academy, 223. *See also* Literature
Preformation, 145–149, 157
Propaganda, 30, 31, 65, 223

Racine, Jean, friend of Dodart and Bourdelin, 19
Rainwater. *See* Water
Ray, John, 147, 200, 204, 299 n. 23, 307–308 n. 29, 312 n. 30, 314 n. 8, 328–329 n. 16, 336 n. 110; classification of plants by, 67, 75; on distillation of plants, 90; on need for illustrated natural history of plants, 70, 80
Réaumur, René Antoine Ferchault de, 193
Referee. *See* Académie royale des sciences, functions and goals of
Reform of science and technology. *See* Académie royale des sciences, functions of; Ideals
Régis, Pierre Sylvain, 154, 311 n. 5, 314 n. 9
Reméde des pauvres. *See* Medicine, for poor
Renaudot, Théophraste, 172
Reneaume, Michel Louys, 306 n. 8; tries to revive natural history of plants in 1709, 114, 302 n. 65
Revolution, scientific, 3–4, 8, 67, 155; scientific community during, 145, 181, 221, 225
Richelieu, Armand Jean du Plessis, cardinal, 17
Richer, Jean, 50, 213, 293 n. 25, 295 nn. 4, 6, 333 n. 75, 335 n. 94
Riolan, Jean, 132; on circulation of blood, 125, 126–127, 130, 136
Roannez, duke of, 19
Robert, Nicolas, painter and engraver, 80, 194; drawings and engravings of plants, 72, 77, 78, 86; watercolors of plants, 71, 300 nn. 41, 43
Roberval, Gilles Personne de, 17, 156, 327, n. 5
Robin, Vespasian, 213
Roemer, Ole, 44, 45, 202, 203, 294 n. 35; and spherical microscope, 159; trip of to England, 99
Rohault, Jacques, 196, 205; books by read in Paris, 185; and capillary action, 139; offers instruction in natural philosophy, 197
Rolle, Michel, offers instruction in algebra, 197

Royal Academy of Architecture. *See* Académie royale d'architecture
Royal Garden. *See* Jardin royal
Royal Society (London), 22–23, 29, 34, 156, 182, 203, 205, 207, 215, 291 n. 12, 330 n. 37, 331 n. 45, 332 n. 65; academicians as Fellows of, 202; and air pump, 160; contacts with Academy, 202; medical practitioners in, 172
Rudwick, Martin, 181, 216
Rye, infected with ergot, 174, 175, 176, 177, 178, 179

Sablé, Madeleine de Souvré, marquise de, 201
Saint Anthony's fire. *See* Ergotism
Saint Hilaire, 332 n. 67
Saint Pierre, abbé Charles Irénée Castel de, 190
Saint-Simon, Louis de Rouvroy, duc de, 19
Salomon-Bayet, Claire, on analogy, 120, 121
Salt, crystals of resemble corpuscles, 98. *See also* Distillants
Sanson, family of cartographers, 192
Sap: compared with blood, 132, 133–134, 135–136; flow of, 21, 214; functions of, 135–136, 141, 146, 151–152; theory that it circulates, 67, 70, 79, 118, 119, 124, 131–144, 158, 164, 317 n. 4. *See also* Chemical analysis of sap
Sauveur, Joseph, 327 n. 57
Scarron, Paul, poet, 186
Science: attitudes toward, 20, 28, 45; audience for, 170, 181, 182–186, 190, 194–198; and culture, 226; mocked, 205–206; and popular culture, 135, 311 n. 6; as a profession, 35, 45, 181; 223–224; as recreation, 182, 185, 193–194; reform of, 30; and religion, 201, 203; teaching of, 44, 190, 197–198, 204, 217, 223; and war, 201
"Scientist," word not coined yet, 4, 181
Sébastien, père. *See* Truchet
Secrecy, 223; impossibility of, 207; risks of, 207
Sédileau, 18, 44, 55, 294 n. 40, 333 n. 70; role at Academy, 22; studies disease of orange trees, 110, 111
Seeds, 70, 82; active or passive, 152–153; behavior of in evacuated receiver, 160, 162–163; collected from around world for cultivation, 79; dispersal of,

147; functions of, 146, 148; maturation and release of, 151–152; physiology of, 164–165; study of, 146–148, 158, 160
Séguier, Pierre, chancellor of France, 17
Sevin, Pierre, mathematical instrument maker, 44
Shadwell, Thomas, playwright, 205
Signatures, theory of, ridiculed by Dodart, 97
Sloane, Sir Hans, and Tournefort, 328–329 n. 16
Société royale de la médecine, 196
Societies, learned and scientific, 27, 28–29, 34, 156, 199, 205; communication between, 205; contacts between, 202; correspondence with, 203; private, 20, 21, 22, 155, 165, 181, 190, 195, 196, 199, 202, 303 n. 5; publications of in vernacular, 184. *See also names of societies or sponsors*
Soils, 147, 150, 151
Solomon, House of, 28, 43, 65, 292 n. 23
Solvents. *See* Alkahest; Chemical analysis, methods of
Sorbière, Samuel, 215
South, Robert, 205
Specimens of unusual plants and curiosities, 80, 194–195, 204, 207; collected by Academy, 213; exchanged, 203
Spontaneous generation. *See* Generation
Sprat, Thomas, 215
Spurred rye. *See* Rye
Stubbe, Henry, 205
Superstition, 79, 117, 184
Surveying. *See* Instruments and apparatus; Versailles, water supply of
Sweden, and academicians, 202, 209
Sympathy and antipathy, theories of, 119, 120

Tardy, Claude, 127–128
Tauvry, Daniel, 18, 21, 172, 315 n. 16
Taxidermy, 210. *See also* Colson
Technology. *See* Practical projects of Academy
Temperaments, theory of, ridiculed by Dodart, 97
Terrasson, abbé Jean Baptiste, tries to revive natural history of plants in 1709, 114
Teste-morte. *See* distillants
Theophrastus, 70, 73, 146; theories of challenged, 75
Thévenot, Melchisédech, 36, 201, 202,

Index 387

203; associated with both Academy and Bibliothèque du roi, 39, 51, 285 n. 4; role of in Academy, 182; as sponsor of scientific meetings, 196

Thuret, Isaac and Jacques, clockmakers, 43, 44, 193, 317 n. 1

Torricelli, Evangelista, 155; Torricellian void, 155, 156

Tournefort, Joseph Pitton de, 18, 23, 25, 58, 184, 194, 201, 202, 209, 214, 215, 222, 225, 300 n. 40, 309 n. 1, 328–329 n. 16, 331 n. 61, 331–332 n. 63, 337 n. 115; on anatomy and physiology of plants, 141, 142, 146, 147, 148, 150, 151, 153–154, 158, 165; and Bourdelin, 90, 94, 111; career of, 20, 107, 172; and classification of plants, 59, 75; as collector, 195; death of, 187; describes plants, 111, 113; and Dodart, 103, 111, 113–115; *Élémens de botanique* of, 58, 80, 113, 222; expeditions of, 107, 209; *Histoire des plantes qui naissent aux environs de Paris* of, 113, 114–115; and Jardin royal, 38, 197–198; and natural history of plants, 111, 113, 114, 115; role of in Academy, 24, 111, 113–115

Transpiration: in humans, 99; in plants, 141, 164

Travel, 23, 199, 201–202, 202–203; books about for sale in Paris, 191; grand tour of continent, 198, 200; inspires scientific inquiry, 20, 22–23, 58, 59, 212–213. *See also* Expeditions

Truchet, Jean, called père Sébastien, 195, 196, 293 n. 25

Tschirnhaus, Ehrenfried Walther, count of, 36

Tuillier, father and son, physicians, 174–175, 176, [204]

Utility, 46, 55, 61, 101, 110, 111, 169, 170, 171, 179, 226; practice related to theory, 60, 99, 108, 109, 111, 222, 224. *See also* Ideals; Practical projects of Academy

Vacuum: abhorence of in nature, 140–141; debate over, 154, 159–160, 166, 335 n. 101; experiments with puzzling, 161, 162, 163, 164. *See also* Torricelli

Val, Mlle de, seller of maps and geographies, 192

Vallant, physician, 201

Varignon, Pierre, 36, 55, 190; Fellow of Royal Society, 202; teaches mathematics, 197

Varin, 295 n. 4, 333 n. 72

Vauban, Sébastien Le Prestre, seigneur de, 325 n. 19; treatise by on fortifications for sale, 191

Vernacular. *See* Language

Vernon, Francis, 32, 211, 291 n. 13, 334 n. 81

Versailles, gardens of, 194; menagerie of, 194; menagerie of supplies animals for dissection, 39, 54, 59, 194; water supply of, 21, 32, 50, 55, 56, 156, 169, 210, 337 n. 112

Villette, instrument maker, 44

Vines, how they grip walls, 111

Vitruvius, 16, 49

Void. *See* Torricelli; Vacuum

Wallis, John, 164

Water: desalination of, 108, 109, 332 n. 67, 336 n. 102; role of in growth of plants, 150, 152; as ultimate source of matter, 152. *See also* Chemical analysis; Versailles, water supply of

Watercolors. *See* Illustrations; Gaston

Willughby, Francis, 200, 214

Winds, studies of, 203

Wines. *See* Chemical analysis

Zoology. *See under* Natural philosophical research of Academy

Designer:	U.C. Press Staff
Compositor:	Harrison Typesetting, Inc.
Text:	10/13 Sabon
Display:	Sabon
Printer:	Bookcrafters
Binder:	Bookcrafters